AF766916

Proceedings
of the
Steklov Institute of Mathematics

edited by
S. M. Nikol'skiĭ

Number 131 (1974)

Theory and applications of differentiable functions of several variables. V

edited by

S. M. Nikol'skiĭ

Providence, Rhode Island
AMERICAN MATHEMATICAL SOCIETY
1975

Академия Наук
Союза Советских Социалистических Республик

ТРУДЫ
ордена Ленина
МАТЕМАТИЧЕСКОГО ИНСТИТУТА
имени В. А. СТЕКЛОВА
CXXXI

Ответственный редактор
академик С. М. НИКОЛЬСКИЙ

ИССЛЕДОВАНИЯ ПО ТЕОРИИ ДИФФЕРЕНЦИРУЕМЫХ ФУНКЦИЙ МНОГИХ ПЕРЕМЕННЫХ И ЕЕ ПРИЛОЖЕНИЯМ. V

Сборник работ
под редакцией
С. М. НИКОЛЬСКОГО

Издательство "Наука"
Москва 1974

AMS (MOS) subject classifications (1970). Primary 26A15, 26A54, 26A57, 26A84, 26A86, 31B10, 35A35, 35B45, 35D05, 35D10, 35J25, 35J60, 41A10, 41A25, 41A30, 41A50, 42A08, 42A68, 42A92, 46E15, 46E30, 46E35, 47F05, 65H10; Secondary 26A15, 26A24, 26A69, 26A93, 35A40, 35J20, 41A50, 42A08, 42A18, 46E30, 46E35, 65D99.

ISBN 0-8218-3031-7

PROCEEDINGS OF THE STEKLOV INSTITUTE OF MATHEMATICS IN THE ACADEMY OF SCIENCES OF THE USSR

(Труды математического института им. В. А. Стеклова, т. CXXXI, 1974)

TABLE OF CONTENTS

Trudy Mat. Inst. Steklov.
131 (1974)

Proc. Steklov Inst. Math.
131 (1974)

MULTIPLICATIVE ESTIMATES FOR INTEGRAL NORMS OF DIFFERENTIABLE FUNCTIONS OF SEVERAL VARIABLES

UDC 517.518.22

O. V. BESOV

ABSTRACT. Multiplicative estimates are established connecting the L_p-norms of the derivatives of a function defined on n-dimensional Euclidean space.

Bibliography: 12 items.

We will consider the functions $f(x)$ defined on a domain $G \subset E^n$ and belonging to the space $L_p(G)$, $p = (p_1, \ldots, p_n)$, with norm

$$\|f\|_{p,G} = \left(\int \cdots \left(\int \left(\int |f_G(x)|^{p_1} dx_1\right)^{\frac{p_2}{p_1}} dx_2\right)^{\frac{1}{p_2}} \cdots dx_n\right)^{\frac{1}{p_n}} < \infty,$$
$$1 \leqslant p_j < \infty \qquad (j = 1, \ldots, n),$$

where $f_G(x) = f(x)$ on G and $f_G(x) = 0$ on $E^n \backslash G$, and having various derivatives with finite norms. When $G = E^n$, we will simply write $\|f\|_p$ instead of $\|f\|_{p,G}$. A vector p with equal components will be called a scalar and the L_p-norm will be respectively called a vector or scalar norm.

A number of works, connected with the names of Jacques Hadamard, G. E. Šilov, A. N. Kolmogorov, A. M. Rodov and others and corresponding to the case of one and the same metric in all factors (a scalar $p = \infty$), have been devoted to multiplicative estimates for functions of one variable. Multiplicative estimates have been obtained in the papers of Emilio Gagliardo [1] and Louis Nirenberg [2] for the case of several variables. They have the form

$$\sum_{|\alpha|=m} \|D^\alpha f\|_p \leqslant C \|f\|_{p_1}^{\theta} \left(\sum_{|\alpha|=l} \|D^\alpha f\|_{p_2}\right)^{1-\theta} \tag{1}$$

for scalar p, p_1, p_2 and bear the name Gagliardo-Nirenberg estimates.(1)

We obtain below estimates sharpening the estimates of type (1) in the sense that they contain, instead of sums of norms of derivatives, norms of individual derivatives, which permits one, for example, to generalize estimate (1) to the anisotropic case (the case of assigning different derivatives with respect to different coordinate directions).(2) Moreover,

---

AMS (MOS) subject classifications (1970). Primary 46E30, 26A84, 26A86.

(1) We note that, under certain assumptions concerning p_1, p_2 amd p, for example, when $p_1 = p_2$ or when $p_1 \leqslant p$, $p_2 \leqslant p$, inequality (1) and its analog were established for the special case of a mixed L_p-norm by V. P. Il'in (1951), Ehrling (1954, $p = 2$) and Nirenberg (1955).

(2) After the manuscript of the present paper had been sent to the printer there appeared the paper of V. A. Solonnikov [3] containing, in particular, a (different) proof of our Theorem 3 for the case of scalar norms and $\mu_0 > 0$.

the right side of a multiplicative estimate obtained below contains an arbitrary number $N \geqslant 2$ of factors. But a shortcoming is the presence of a restriction on the index of the spaces L_p: $1 < p < \infty$, whereas the results of papers [1]–[3] are free of this restriction in many cases.

We also note that the results given below pertain to vector L_p-norms.

We will denote the vector L_p-norm with respect to a period for $1 \leqslant p_j \leqslant \infty$ ($j = 1, \ldots, n$) of a periodic function $f(x)$, $x = (x_1, \ldots, x_n)$, of period 2π in all of its arguments by

$$\| f \|_p^* = \left(\int_0^{2\pi} \cdots \left(\int_0^{2\pi} \left(\int_0^{2\pi} | f(x) |^{p_1} dx_1 \right)^{\frac{p_2}{p_1}} dx_2 \right)^{\frac{1}{p_2}} \cdots dx_n \right)^{\frac{1}{p_n}}.$$

If $p_j = \infty$ for some j, we take in place of the corresponding "partial" norm with respect to x_j the essential supremum of the modulus of the function. The space so normed is denoted by L_p^*.

For a vector $k = (k_1, \ldots, k_n)$ with integral coordinates we denote by $\square_k$ the rectangular parallelepiped with edges parallel to the coordinate axes whose projection onto the jth coordinate axis is the interval with end points at

$$2^{|k_j|-1} \operatorname{sign} k_j, \qquad (2^{|k_j|} - 1) \operatorname{sign} k_j \qquad (j = 1, \ldots, n).$$

A parallelepiped $\square_k$, which is generally n-dimensional, degenerates into a parallelepiped of lesser dimension if some of the k_j are equal to 0 or ± 1.

For vectors $m = (m_1, \ldots, m_n)$ and $k = (k_1, \ldots, k_n)$ with integral coordinates and a multiple sequence c_m we put

$$c_{m,k} = \begin{cases} c_m & \text{for} \quad m \in \square_k; \\ 0 & \text{for} \quad m \notin \square_k. \end{cases} \tag{2}$$

For

$$f(x) = \sum c_m e^{imx} \in L_p^* \tag{3}$$

and a vector $r = (r_1, \ldots, r_n)$ with nonnegative components we put

$$f^{(r)}(x) = \sum (im_1)^{r_1} \cdots (im_n)^{r_n} c_m e^{imx},$$

where $(im_j)^{r_j} = |m_j|^{r_j} \exp\{\tfrac{1}{2}\pi i r_j \operatorname{sign} m_j\}$, and we put

$$\delta_k(f) = \sum_m c_{m,k} e^{imx},$$

$$\delta_k^{(r)}(f) = \begin{cases} 2^{\bar{k}r} \delta_k(f) & \text{for } \operatorname{supp} k \supset \operatorname{supp} r; \\ 0 & \text{for } \operatorname{supp} k \not\supset \operatorname{supp} r, \end{cases}$$

where $\bar{k} = (|k_1|, \ldots, |k_n|)$ and the support (supp) of a vector is the set of indices of its nonzero components.

For $f(x) = \Sigma c_m e^{imx}$ we put, in addition,

$$\| f^{(r)} \|_p^{**} = \left\| \left(\sum_k | \delta_k^{(r)}(f) |^2 \right)^{1/2} \right\|_p^*, \qquad 1 \leqslant p \leqslant \infty.$$

LEMMA 1.[3] *For* $1 < p_j < \infty$ $(j = 1, \ldots, n)$

$$\| f^{(r)} \|_p^{**} \sim \| f^{(r)} \|_p^{*}. \tag{4}$$

PROOF. Let us first explain that the equivalence (4) is understood in the sense that the finiteness of one side of (4) implies the finiteness of the other side and the existence of two positive constants (depending on p and r but not on f) bracketing the ratio of the two sides of (4). Suppose to begin with that $r = 0$. When p is a scalar, this case constitutes the content of a Littlewood-Paley theorem the multidimensional version of which is given in [4], §1.5.2.1. A generalization of the Littlewood-Paley theorem to the case of a mixed L_p^*-norm can be carried out under the plan of proof of the scalar case presented in [4], §1.5.2, (6), with the following additions. Inequality 1.5.2(8) is obtained for a vector p by a successive n-multiple application of its one-dimensional case. In place of the chain of inequalities 1.5.2(10) we have with the use in addition of Minkowski's inequality the following chain of inequalities for $p_{\min} = \min_{1 \leqslant j \leqslant n} p_j$, $p_{\max} = \max_{1 \leqslant j \leqslant n} p_j$, $\Omega = \{\theta = (\theta_1, \ldots, \theta_n): 0 \leqslant \theta_j \leqslant 1 \ (j = 1, \ldots, n)\}$:

$$\| f \|_p^* = \left\{ \int_\Omega (\| f \|_p^*)^{p_{\max}} d\theta \right\}^{\frac{1}{p_{\max}}} \leqslant C_1 \left\{ \int_\Omega \left(\left\| \sum \omega_k(\theta)\, \delta_k(f) \right\|_p^* \right)^{p_{\max}} d\theta \right\}^{\frac{1}{p_{\max}}}$$

$$\leqslant C_2 \left\| \left(\int_\Omega \left| \sum \omega_k(\theta)\, \delta_k(f) \right|^{p_{\max}} d\theta \right)^{\frac{1}{p_{\max}}} \right\|_p^*$$

$$\leqslant C_3 \left\| \left(\sum \delta_k^2(f) \right)^{\frac{1}{2}} \right\|_p^* \leqslant C_4 \left\| \left(\int_\Omega \left| \sum \omega_k(\theta)\, \delta_k(f) \right|^{p_{\min}} d\theta \right)^{\frac{1}{p_{\min}}} \right\|_p^*$$

$$\leqslant C_5 \left\{ \int_\Omega (\| f \|_p^*)^{p_{\min}} d\theta \right\}^{\frac{1}{p_{\min}}} \leqslant C_6 \| f \|_p^*,$$

where the constants $C_1, \ldots, C_6$ depend only on p.

The proof of inequality 1.5.2(13) for a vector p differs from the proof given in [4] for a scalar p only in that the argument connected with induction should be replaced by an n-multiple application of the corresponding one-dimensional inequality.

Suppose now $r \neq 0$. Relation (4) is established with the use of the same Littlewood-Paley inequality for a vector p and the following theorem of Marcinkiewicz on multipliers of Fourier series.

THEOREM (Marcinkiewicz). *Suppose for a sequence* $\{\lambda_m\}$ *of multipliers there exists a number* M *such that*

$$\sum_m | \Delta_1 \ldots \Delta_n \lambda_{m,k} | \leqslant M \quad \forall k, \tag{5}$$

where the $\lambda_{m,k}$ *are constructed with respect to the* λ_m *by means of formula* (2), *the difference* $\Delta_j \lambda_{m,k} = \lambda_{m+e_j,k} - \lambda_{m,k}$, e_j *is the jth basis vector and* $\Delta_1 \cdots \Delta_n \lambda_{m,k}$ *is the*

---

[3] Established by N. S. Nikol'skiĭ for $r \neq 0$ and a scalar p.

correspondingly defined multiple difference.

Then for $f(x) = \Sigma c_m e^{imx} \in L_p^*$ $(1 < p_j < \infty;\ j = 1, \ldots, n)$ *the function* $F(x) = \Sigma \lambda_m c_m e^{imx} \in L_p^*$ *and* $\|F\|_p^* \leqslant c_p M \|f\|_p^*$, *where* c_p *does not depend on* f.

The proof for a scalar p can be found, for example, in [4], §1.5.3. Let us show how to generalize it to the case of a vector p. Following the proof given in [5] (Chapter XV, Theorem (4.14)), we first generalize Lemma (2.10) of [5], Chapter XV, to the case of several variables and a vector p, $1 \leqslant p_j < \infty$. To this end we give a derivation of (a weakened) inequality (2.11) in [5], Chapter XV, that carries over in an obvious way to the case of a vector norm for functions of several variables. It is necessary in this connection to replace M in (2.11) by $A_p M$, where $1 \leqslant p < \infty$ and A_p does not depend on the number n. For

$$\varphi(x) = \sum_1^n \varphi_k(x) z_k, \qquad \psi(x) = \sum_1^n \psi_k(x) z_k, \qquad z \in E^n,$$

$$\Phi(x) = \sqrt{\sum_1^n \varphi_k^2(x)}, \qquad \Psi(x) = \sqrt{\sum_1^n \psi_k(x)}$$

we have, using in addition to the notation and arguments of [5], p. 225, the inequalities of Hölder and Minkowski,

$$\int_{|z|<1} |z_1|\, dz \left\{\int_a^b |\Psi(x)|^p dx\right\}^{\frac{1}{p}} = \left\{\int_a^b \left[\int_{|z|<1} |\psi(x)|\, dz\right]^p dx\right\}^{\frac{1}{p}}$$

$$\leqslant \int_{|z|<1} \left\{\int_a^b |\psi(x)|^p dx\right\}^{\frac{1}{p}} dz \leqslant M \int_{|z|<1} \left\{\int_a^b |\varphi(x)|^p dx\right\}^{\frac{1}{p}} dz$$

$$\leqslant \left\{\int_{|z|<1} dz\right\}^{\frac{1}{q'}} \left\{\int_{|z|<1} \left[\int_a^b |\varphi(x)|^p dx\right]^{\frac{q}{p}} dz\right\}^{\frac{1}{q}}$$

$$\leqslant \left\{\int_{|z|<1} dz\right\}^{\frac{1}{q'}} \left\{\int_a^b \left[\int_{|z|<1} |\varphi(x)|^q dz\right]^{\frac{p}{q}} dx\right\}^{\frac{1}{p}} = J_1(n)\, J_2(n) \left\{\int_a^b |\varphi(x)|^p dx\right\}^{\frac{1}{p}},$$

where $1 \leqslant p < q = 2k + 1 < \infty$, k is a natural number, $1/q + 1/q' = 1$, and

$$J_1^{q'}(n) = \int_{|z|<1} dz, \qquad J_2^q(n) = \int_{|z|<1} |z_1|^q dz.$$

To complete the proof, it remains to show that

$$J_1(n)\, J_2(n) \leqslant A_p \int_{|z|<1} |z_1|\, dz, \qquad n = 2, 4, 6, \ldots.$$

We will consider three integrals. Let

$$J(n) = \int_{|z|<1} |z_1|\, dz = 2\int_0^1 z_1 c_{n-1} (1 - z_1^2)^{\frac{n-1}{2}} dz_1 = c_{n-1} \frac{2}{n+1},$$

where c_{n-1} is the volume of a unit ball in E^{n-1}, and let

$$J_1^{q'}(n) = 2c_{n-1}\int_0^1 (1-z_1^2)^{\frac{n-1}{2}}\,dz_1 = 2c_{n-1}\int_0^{\frac{\pi}{2}} \sin^n t\,dt = \pi c_{n-1}\frac{(n-1)!!}{n!!} = \frac{\pi c_n n!}{2^n\,[(n/2)!]^2}.$$

From Stirling's formula ($n! = \sqrt{2\pi n}(n/e)^n \epsilon_n$, $\epsilon_n \to 1$ as $n \to +\infty$) we have

$$J_1^{q'}(n) = \frac{c_{n-1}\sqrt{2\pi}}{\sqrt{n}}\tilde{\varepsilon}_n, \qquad \tilde{\varepsilon}_n^{q'} \to 1 \quad \text{as} \quad n \to \infty.$$

Let

$$I_{n,q} = \int_0^{\frac{\pi}{2}} \sin^n t \cos^q t\,dt, \qquad q = 2k+1.$$

From formulas for lowering the index we have

$$I_{n,q} = \frac{(q-1)(q-3)\dots 2}{(n+q)(n+q-2)\dots(n+3)} I_{n,1}, \qquad I_{n,1} = \frac{1}{n+1}.$$

Hence

$$J_2^q(n) = 2c_{n-1}\int_0^1 |z_1|^q (1-z_1^2)^{\frac{n-1}{2}}\,dz_1 = 2c_{n-1}I_{n,q} \leqslant c_{n-1}2\,(q-1)!!\,\frac{1}{n^{k+1}}$$

$$= c_{n-1}2\,(q-1)!!.$$

Consequently

$$J_1(n)\,J_2(n) \leqslant c_{n-1}\,(2\pi)^{\frac{1}{2q'}}\,[2\,(q-1)!!]^{\frac{1}{iq}}\left(\frac{1}{n}\right)^{\frac{1}{2q'}}\left(\frac{1}{n}\right)^{\frac{1}{2}+\frac{1}{2q}}\tilde{\varepsilon}_n$$

$$= c_{n-1}\,(2\pi)^{\frac{1}{2q'}}\,[2\,(q-1)!!]^{\frac{1}{q}}\,\frac{\tilde{\varepsilon}_n}{n} \leqslant (2\pi)^{\frac{1}{2q'}}\,[2\,(q-1)!!]^{\frac{1}{q}}\,\tilde{\varepsilon}_n J(n).$$

Q.E.D.

It remains now to generalize Lemma 2.15 of [5], Chapter XV, to the case of a vector L_p-norm, $1 < p_j < \infty$ ($j = 1, \dots, n$). We note that this generalization is actually needed only for trigonometric polynomials, i.e. for the functions $f(x) = \Sigma c_m e^{imx}$ with a finite number of nonzero terms. Following the proof of Lemma (2.15) and taking into account the fact that the convolution with respect to x_j of $\cos k_j x_j$ and a function belonging to L_p^* is a (uniformly in k_j) bounded operator in L_p^*, we reduce the proof of Lemma (2.15) with a vector p to the estimate (representing in the one-dimensional case a theorem of Riesz)

$$\|\tilde{f}_j\|_p^* \leqslant A_p \|f\|_p^*, \qquad 1 < p_\nu < \infty \ (\nu = 1, \dots, n),$$

where $\tilde{f}_j(x)$ is constructed with respect to $f(x)$ by going over to the trigonometric conjugate function with respect to the variable x_j:

$$\tilde{f}_j(x) = -\frac{1}{\pi}\int_0^{\pi} \frac{f(x+te_j) - f(x-te_j)}{2\tan\frac{t}{2}}\,dt.$$

It clearly suffices to assume that $n = j$. In this case the function (trigonometric polynomial) $f(x) = f(x', x_n)$, $x' = (x_1, \ldots, x_{n-1})$, can be regarded as an (abstract) function of the variable x_n with values in the space $L^*_{p'}$, $p' = (p_1, \ldots, p_{n-1})$. Since the difference

$$\varepsilon(t) = \frac{1}{2\tan\frac{t}{2}} - \frac{1}{t}$$

is continuous on $(0, \pi)$ and continuously extendable to $[0, \pi]$, by taking into account the estimate in L^*_p of a convolution, we can reduce our problem to the estimate

$$\| f_{(n)} \|_p \leqslant B_p \| f \|, \qquad 1 < p_\nu < \infty \qquad (\nu = 1, \ldots, n),$$

where

$$f_{(n)}(x) = -\frac{1}{\pi}\int_0^{\pi} \frac{f(x + te_n) - f(x - te_n)}{t}\, dt = -\frac{1}{\pi}\lim_{\varepsilon\to 0}\int_{\varepsilon<|t|<\pi} \frac{f(x+te_n)}{t}\, dt.$$

This estimate for $n = 1$ is contained, for example, in [6], §5. It can be obtained for an arbitrary $n > 1$ by induction with respect to the dimension of the space analogously to the way it was done for a similar case in Theorem 1 of [7].

In our case the theorem of Marcinkiewicz is applied to the multipliers $\{\lambda_m\}$, which for all $m \in \square_k$, supp $k \supset$ supp r, are equal to

$$\lambda_{m,k} = \frac{|m_1|^{r_1} \ldots |m_n|^{r_n}}{2^{kr}}, \qquad \lambda_{m,k} = \frac{2^{kr}}{|m_1|^{r_1} \ldots |m_n|^{r_n}}.$$

Condition (5) is easily verified by taking into account the monotonicity.

LEMMA 2. *Suppose for $j = 1, \ldots, N$ that r and $r^{(j)}$ are n-dimensional vectors with nonnegative components, $1 \leqslant p_\nu^{(j)} \leqslant \infty$,*

$$0 < \mu_j < 1, \ \sum_1^N \mu_j = 1, \quad r = \sum_1^N \mu_j r^{(j)}, \quad \frac{1}{p} = \sum_1^N \frac{\mu_j}{p^{(j)}}.$$

Then

$$\| f^{(r)} \|^{**}_p \leqslant \prod_1^N \| f^{(r^{(j)})} \|^{**\,\mu_j}_{p^{(j)}}. \tag{6}$$

PROOF. Inasmuch as supp $r \supset$ supp $r^{(j)}$ for all $j = 1, \ldots, N$, we have

$$|\delta_k^{(r)}(f)|^2 \leqslant \prod_{j=1}^N |\delta_k^{(r^{(j)})}(f)|^{2\mu_j},$$

which implies by virtue of Hölder's inequality ([8], Theorems 11 and 12) that

$$\sum_k |\delta_k^{(r)}(f)|^2 \leqslant \prod_{j=1}^N \left(\sum_k |\delta_k^{(r^{(j)})}(f)|^2\right)^{\mu_j}.$$

It follows by virtue of Hölder's inequality for an integral of a product of N functions that in the case of finite $p^{(j)}$ $(j = 1, \ldots, N)$

$$\|f^{(r)}\|^{**}_{p} = \Big\|\Big(\sum_k |\delta_k^{(r)}(f)|^2\Big)^{\frac{1}{2}}\Big\|^{*}_{p} \leqslant \Big\|\prod_{j=1}^{N}\Big(\sum_k |\delta_k^{(r^{(j)})}(f)|^2\Big)^{\frac{p_1^{(j)}}{2}\frac{\mu_j}{p_1^{(j)}}}\Big\|^{*}_{p} \leqslant \ldots$$

$$\ldots \leqslant \prod_{j=1}^{N}\Big\|\Big(\sum_k |\delta_k^{(r^{(j)})}(f)|^2\Big)^{\frac{1}{2}}\Big\|^{*\,\mu_j}_{p^{(j)}},$$

which is equivalent to (6). If, on the other hand, some or all of the $p^{(j)}$ are infinite, it is easily seen that the latter chain of inequalities remains valid under an appropriate writing of the penultimate term in the chain, and we again arrive at inequality (6).

LEMMA 3. *Suppose* $1 < p_j \leqslant q_j < \infty$, $s_j \geqslant 0$, *and* $s_j - 1/q_j = r_j - 1/p_j$ $(j = 1, \ldots, n)$. *Then*

$$\| f^{(s)} \|^{*}_{q} \leqslant c_{p,q} \| f^{(r)} \|^{*}_{p} \tag{7}$$

for all functions $f(x)$ of the form

$$f(x) = \sum_{\operatorname{supp} m \supset \operatorname{supp} r} c_m e^{imx}, \tag{8}$$

i.e. the functions $f(x)$ with zero mean values with respect to those x_j for which $r_j > 0$ (for almost every collection of the remaining variables).

PROOF. It suffices to confine oneself to the case when p and q differ only by one (the jth) component and $r_\nu = s_\nu = 0$ for $\nu \neq j$, $r_j = 1/p_j - 1/q_j > 0$.

Consider the function

$$K^{(j)}(x_j) = K^{(j)}_{\frac{1}{p_j} - \frac{1}{q_j}}(x_j) = \sum_{m_j \neq 0} \frac{1}{(im)^{r_j}} e^{im_j x_j}.$$

We have $f^{(s)} = K^{(j)} * f^{(r)}$, where the convolution is taken only with respect to the variable x_j. Thus it suffices to show that

$$\| |K^{(j)}| * \varphi \|^{*}_{q} \leqslant \tilde{c}_{p,q} \| \varphi \|^{*}_{p}. \tag{9}$$

Since $p_{j+1} = q_{j+1}, \ldots, p_n = q_n$, estimate (9) follows from its special case $n = j = 1$, which we will analyze further.

An estimate of $K^{(j)}(t)$ is easily obtained with the use of Abel's identity. However, it is well known ([9], p. 70) and has the form

$$|K^{(j)}(t)| \leqslant \frac{\varkappa}{|t|^{1-r_j}},$$

so that estimate (9) can be obtained for $n = 1$ by making use of a Hardy-Littlewood inequality ([8], §10.17). This completes the proof of Lemma 3.

For a vector $p = (p_1, \ldots, p_n)$ we put $1/p = (1/p_1, \ldots, 1/p_n)$.

THEOREM 1. *Suppose for $j = 1, \dots, N$ that l and $r^{(j)}$ are n-dimensional vectors with nonnegative components, $1 < p_\nu^{(j)}, q_\nu < \infty$ ($\nu = 1, \dots, n$), and*

$$0 < \mu_j < 1, \qquad \sum_1^N \mu_j = 1,$$

$$\frac{1}{q} \leqslant \sum_1^N \frac{\mu_j}{p^{(j)}}, \qquad l - \frac{1}{q} = \sum_1^N \mu_j \left(r^{(j)} - \frac{1}{p^{(j)}} \right).$$

Then the estimate

$$\| f^{(l)} \|_q^* \leqslant C \prod_{j=1}^N \| f^{(r^{(j)})} \|_{p^{(j)}}^{*\,\mu_j} \tag{10}$$

is valid for a periodic function $f(x)$ of period 2π in all of its arguments and with zero mean values with respect to those x_ν for which $l_\nu > 0$.

The proof follows by virtue of Lemma 1 from estimates (6) and (7).

Let $D^\alpha f$, where $\alpha = (\alpha_1, \dots, \alpha_n)$ is a multi-index with nonnegative integral components, denote a generalized mixed derivative of a locally summable function f.

THEOREM 2. *Suppose for $j = 1, \dots, N$ that β and $\alpha^{(j)}$ are n-dimensional multi-indices with nonnegative integral components, $1 < p_\nu^{(j)}, q_\nu < \infty$ ($\nu = 1, \dots, n$), $0 < \mu_j < 1$, and* $(\nu = 1, \dots, n)$, $0 < \mu_j < 1$,

$$\sum_1^N \mu_j = 1, \qquad \frac{1}{q} \leqslant \sum_1^N \frac{\mu_j}{p^{(j)}}, \qquad \beta - \frac{1}{q} = \sum_1^N \mu_j \left(\alpha^{(j)} - \frac{1}{p^{(j)}} \right).$$

Then for $f(x) \in C_0^\infty(E^n)$

$$\| D^\beta f \|_q \leqslant C \prod_{j=1}^N \| D^{\alpha^{(j)}} f \|_{p^{(j)}}^{\mu_j}. \tag{11}$$

PROOF. Suppose to begin with that $f(x)$ is concentrated in the cube $\square_1 = \{0 < x_\nu < \pi\ (\nu = 1, \dots, n)\}$. We extend it from $\square_1$ to the cube $\square = \{|x_\nu| < \pi\ (\nu = 1, \dots, n)\}$ so that it is odd with respect to each variable x_ν, and then from $\square$ to all of E^n so that it is periodic with period 2π in each variable. Clearly, for the function $f_*(x)$ obtained in this way we have

$$\| D^\alpha f_* \|_p^* = 2^{\frac{1}{p_1} + \dots + \frac{1}{p_n}} \| D^\alpha f \|_p, \qquad p = (p_1, \dots, p_n).$$

Estimate (11) now follows from estimate (10) for f_*.

Suppose now $f(x) = f(x_1, \dots, x_n)$ is an arbitrary function from $C_0^\infty(E^n)$. Then for any sufficiently small $\epsilon > 0$ the function

$$f_\varepsilon(x) = f\left(\frac{x_1 - \pi/2}{\varepsilon}, \dots, \frac{x_n - \pi/2}{\varepsilon} \right)$$

is concentrated in $\square_1$. Substituting f_ϵ in (11) and going over to the norms of the derivatives of f, we obtain the assertion of the theorem in the general case after dividing out a power of the parameter ϵ.

REMARK 1. By virtue of the fact that the generalized differentiation operator is closed, estimate (11) carries over to the locally summable functions $f(x)$ that can be approximated with any accuracy by a sequence of functions from $C_0(E^n)$ converging simultaneously in the sense of $L^{\mathrm{loc}}(E^n)$ and in the sense of the seminorms the product of the powers of which is in the right side of (11).

As an example, we establish the following theorem.

THEOREM 3. *Suppose the components of the vector* $l = (l_1, \ldots, l_n)$ *are natural numbers,* $1 < p_\nu^{(j)}, q_\nu < \infty$ $(\nu = 1, \ldots, n; j = 0, 1, \ldots, n)$, $0 \leqslant \mu_0 < 1$, $0 < \mu_j < 1$ $(j = 1, \ldots, n)$, $\Sigma_0^n \mu_j = 1$, $\alpha = (\alpha_1, \ldots, \alpha_n)$, *the* α_j *are nonnegative integers,*

$$\frac{1}{q_\nu} \leqslant \sum_{j=0}^{n} \frac{\mu_j}{p_\nu^{(j)}}, \qquad \alpha_\nu - \frac{1}{q_\nu} = \mu_\nu l_\nu - \sum_{j=0}^{n} \frac{\mu_j}{p_\nu^{(j)}} \qquad (\nu = 1, \ldots, n).$$

Then for $f \in L_{p^{(0)}}$

$$\| D^\alpha f \|_q \leqslant C \| f \|_{p^{(0)}}^{\mu_0} \prod_{j=1}^{n} \| D_j^{l_j} f \|_{p^{(j)}}^{\mu_j}. \tag{12}$$

PROOF.(4) By the conditions of the theorem, inequality (12) is satisfied for all functions $f \in C_0^\infty(E^n)$. In view of the fact that the generalized differentiation operator D^α is closed, it suffices to show that an arbitrary function with finite norm

$$\| f \|_{p^{(0)}} + \sum_{j=1}^{n} \| D_j^{l_j} f \|_{p^{(j)}} = \| f \|_{W^l_{p^{(0)};\, p^{(1)}, \ldots, p^{(n)}}} \tag{13}$$

can be approximated with any accuracy in this norm by functions from $C_0^\infty(E^n)$. It can be assumed in addition without loss of generality that the function $f \in C^\infty(E^n)$ and $\Sigma_{j=0}^n \|D^l f\|_{p^{(j)}} < \infty$, since otherwise it can be replaced with small error in $W^l_{p^{(0)};p^{(1)},\ldots,p^{(n)}}$ by a weighted moving average of it with sufficiently small steps in each variable.

Let $B_j^{(\delta,H)}$ $(j = 1, \ldots, n)$ denote the operator associating with a function $f(x)$, $x = (x_1, \ldots, x_n)$, the following difference of weighted moving averages of it with step δ and with step H $(0 < \delta < H < \infty)$ in the variable x_j:

(4) The case of scalar $p^{(j)}$, q and the strict inequality $1/q < \Sigma_0^n \mu_j / p^{(j)}$ is treated by A. D. Džabrailov ([10], p. 137); the case of scalar $p^{(j)}$, q, $p^{(1)} = \cdots = p^{(n)} = p$, $\mu_0 > 0$ is treated by V. A. Solonnikov [3], the possibility of any of the values of p, q, or $p^{(0)}$ being equal to unity not being excluded under the conditions

$$| \alpha : l | + (1/p - 1/q) - \frac{1}{l} < 1, \quad p \leqslant q.$$

$$B_j^{(\delta,H)}f(x)=\int\omega_j(t)\,f(x+\delta te_j)\,dt-\int\omega_j(t)\,f(x+Hte_j)\,dt$$
$$=\int_\delta^H\int h^{l_j-2}\Phi_j\left(\frac{t}{h}\right)D_j^{l_j}f(x+te_j)\,dtdh, \tag{14}$$

where $\omega_j(t)\in C_0^\infty(E^1)$ is a special averaging kernel concentrated in a preassigned segment of the interval (0, 1) and such that $\int\omega_j(t)dt=1$. The second equality in (14) is a consequence of the special choice

$$\omega_j(t)=-\frac{d^{s+1}}{dt^{s+1}}\left(\frac{t^s}{s!}\mu_j(t)\right)$$

($s\geqslant l_j$, $\mu_i(t)\in C^\infty(E^1)$, $\mu_j(t)=1$ for $t\leqslant 0$, $\mu_j(t)=0$ for $t\geqslant 1$), so that $\Phi_j(t)\in C_0^\infty(E^1)$ is concentrated in a preassigned segment of the interval (0, 1) (concerning this, see, for example, [11], page 11 = 8]).

We will show that

$$\|f-B_j^{(\delta,H)}f\|_{W^l_{p^{(0)};\,p^{(1)},\dots,p^{(n)}}}\to 0\quad\text{as}\quad\delta\to 0, H\to\infty. \tag{15}$$

As $\delta\longrightarrow 0$, the minuend of the middle part of (14) tends to $f(x)$ in $W^l_{p^{(0)};p^{(1)},\dots,p^{(n)}}$, which follows from the permutability of the operations of differentiating and averaging and from the continuity with respect to a translation in $L_{p^{(j)}}(E^n)$. Thus it suffices to show that

$$\left\|\int\omega_j(t)\,\varphi(\cdot+Hte_j)\,dt\right\|_p\to 0\quad\text{as}\quad H\to\infty,$$

for $\varphi\in L_p(E^n)$, $1<p_j<\infty$, applying this assertion for $\varphi=f$, $p=p^{(0)}$ as well as for $\varphi=D_i^{l_i}f$, $p=p^{(i)}$. In view of the fact that $C_0^\infty(E^n)$ is dense in $L_p(E^n)$ it can be assumed that $|\varphi(x)|\leqslant N$ on E^n and $\varphi(x)=0$ outside $\square=\{x\colon |x_i|<R\ (i=1,\dots,n)\}$.

But in this case

$$\left|\int\omega_j(t)\,\varphi(x+Hte_j)\,dt\right|\leqslant\frac{C}{H}\int|\varphi(x+te_j)|\,dt\leqslant\frac{C_1R}{H},$$
$$\int\omega_j(t)\,\varphi(x+Hte_j)\,dt=0$$

outside the parallelepiped $\square_1=\{x\colon |x_i|<R\ (i\neq j),\ |x_j|<R+H\}$. It follows by a direct calculation that

$$\left\|\int\omega_j(t)\,\varphi(\cdot+Hte_j)\,dt\right\|_p\leqslant\frac{C(R)}{H}(R+H)^{\frac{1}{p_j}}\to 0\quad(H\to\infty).$$

Q.E.D.

Successively (for $j=1,\dots,n$) applying (15), we see that a function $f\in W^l_{p^{(0)};p^{(1)},\dots,p^{(n)}}$ can be approximated with any accuracy by functions of the form

$$B^{(\delta,H)}D^lf(x)=\left(\prod_{j=1}^n B_j^{(\delta_j,H_j)}\right)f(x)$$
$$=\int_{\delta_1}^{H_1}\dots\int_{\delta_n}^{H_n}\int_{E^n}\prod_1^n h_j^{l_j-2}\Phi_j\left(\frac{t_j-x_j}{h_j}\right)D^lf(t)\,dtdh_1\dots dh_n, \tag{16}$$

where $\delta = (\delta_1, \ldots, \delta_n)$ and $H = (H_1, \ldots, H_n)$.

Consider now the function $B^{(\delta,H)}\chi_R D^l f$, where $\chi_R(x)$ is the characteristic function of the ball $|x| < R$. Clearly, $B^{(\delta,H)}\chi_R D^l f \in C_0^\infty(E^n)$. On the other hand, $\chi_R D^l f \longrightarrow D^l f$ in $L_{p(j)}$ as $R \longrightarrow \infty$ $(j = 0, 1, \ldots, n)$. Hence, clearly,

$$B^{(\delta,H)}\chi_R D^l f \to B^{(\delta,H)} D^l f \text{ in } W^l_{p^{(0)};\, p^{(1)},\ldots,p^{(n)}} \quad \text{as} \quad R \to \infty.$$

The theorem is proved.

Our aim now is to show that the condition $f \in L_{p(0)}$ in Theorem 3 can be removed for $\mu_0 = 0$ if one takes instead of the function $f(x)$ the difference $f(x) - P(f; x)$, where where $P(f; x)$ is a polynomial. A corresponding result for scalar $p^{(1)} = \cdots = p^{(n)}$ is given in [12].

THEOREM 4. *Suppose $l = (l_1, \ldots, l_n)$ is a vector whose components are natural numbers, $1 < p_\nu^{(j)}$, $q_\nu < \infty$, $0 < \mu_j < 1$ $(\nu = 1, \ldots, n; j = 1, \ldots n)$, $\Sigma_1^n \mu_j = 1$, $\alpha = (\alpha_1, \ldots, \alpha_n)$, the α_j are integers,*

$$\frac{1}{q_\nu} \leqslant \sum_{j=1}^n \frac{\mu_j}{p_\nu^{(j)}}, \qquad \alpha_\nu - \frac{1}{q_\nu} = \mu_\nu l_\nu - \sum_{j=1}^n \frac{\mu_j}{p_\nu^{(j)}} \quad (\nu = 1, \ldots, n)$$

and there exists a vector $\lambda = (\lambda_1, \ldots, \lambda_n)$ with positive components for which

$$(\alpha, \lambda) > \lambda_j l_j - \left(\lambda, \frac{1}{p^{(j)}}\right) \quad (j = 1, \ldots, n). \tag{17}$$

Then there exists a polynomial $P_{l-1}(x; f)$ of degree not greater than $l_j - 1$ in x_j, linearly depending on f and such that

$$\| D^\alpha (f - P_{l-1}(\cdot\,; f)) \|_q \leqslant C \prod_{j=1}^n \| D_j^{l_j} f \|_{p^{(j)}}^{\mu_j}. \tag{18}$$

PROOF. We first note that in a number of cases condition (17) follows from the preceding conditions and hence can be omitted in formulating the theorem. For example, in the case of scalar $p^{(j)} = (p_j, \ldots, p_j)$, $j = 1, \ldots, n$, one can take as λ the vector with components (under the condition that they be positive)

$$\lambda_j = \frac{1}{l_j}\left(1 - \sum_{i=1}^n \frac{1}{p_i l_i} + \frac{1}{p_j}\sum_{i=1}^n \frac{1}{l_i}\right) > 0 \quad (j = 1, \ldots, n).$$

These values of the λ_j are easily found from the condition that the right sides of (17) be independent of j. Here λ is the normal vector to the plane passing through the points $a^{(j)} = l_j e_j - 1/p^{(j)}$ $(j = 1, \ldots, n)$. In the present case the choice of λ is realized independently of α.

In another interesting case, when $\Sigma_1^n \alpha_j / l_j = 1$, one can take as λ the vector $\lambda = (1/l_1, \ldots, 1/l_n)$.

Let

$$\|f\|_{L^l_{p(1),\ldots,p(n)}} = \sum_{j=1}^{n} \|D_j^{l_j} f\|_{p(j)}.$$

The following relations are valid for $f \in L^l_{p(1),\ldots,p(n)}$ when $0 < \epsilon < \rho < \infty$ (see [11] and [12] for $\lambda_j = 1/l_j$; the derivation in the general case is analogous):

$$D^\alpha f_\rho(x) = (-1)^{|\alpha|} \int \rho^{-|\lambda|-(\alpha,\lambda)} \Phi_0^{(\alpha)}(y:\rho^\lambda) f(x+y)\,dy, \tag{19}$$

$$\begin{aligned} -D^\alpha f_{(\epsilon,\rho)}(x) &= D^\alpha f_\epsilon(x) - D^\alpha f_\rho(x) \\ &= (-1)^{|\alpha|} \sum_1^n \int_\epsilon^\rho \int h^{-1-|\lambda|+l_j\lambda_j-(\alpha,\lambda)} \Phi_j^{(\alpha)}(y:h^\lambda) D_j^{l_j} f(x+y)\,dy\,dh, \end{aligned} \tag{20}$$

where the $\Phi_j(x) \in C_0^\infty(E^n)$ $(j = 0, 1, \ldots, n)$ are concentrated in a preassigned cube in the positive orthant and satisfy the conditions

$$\int \Phi_0(x)\,dx = 1, \qquad \int \Phi_j(x)\,dx = 0 \ (j = 1, \ldots, n), \qquad h^\lambda = (h^{\lambda_1}, \ldots, h^{\lambda_n}).$$

The above relations are a modification of a corresponding representation by V. P. Il'in. For $f \in L^l_{p(1),\ldots,p(n)}$ we get from (20) with the use of Hölder's inequality that

$$\sup_x |D^\alpha f_\epsilon(x) - D^\alpha f_\rho(x)| \leqslant C \sum_1^n \int_\epsilon^\rho h^{-1-(\alpha,\lambda)-|\lambda:p(j)|+l_j\lambda_j} \|D_j^{l_j} f\|_{p(j)}\,dh,$$

which implies that $D^\alpha f_\rho(x)$ tends to a limit uniformly in $x \in E^n$ as $\rho \to \infty$ for at least all those α for which condition (17) is satisfied.

From (19) we easily get that $D^\alpha f_\rho \to 0$ (as $\rho \to \infty$) uniformly in $x \in E^n$ if $\alpha_j \geqslant l_j$ for at least one j.

We put

$$c_\alpha = \lim_{\rho\to\infty} \frac{1}{\alpha!} D^\alpha f_\rho(0), \qquad (\alpha,\lambda) > \varkappa, \qquad \varkappa = \max_j \left[\lambda_j l_j - \left(\lambda, \frac{1}{p^{(j)}}\right)\right],$$

so that $c_\alpha = 0$ if $\alpha_j \geqslant l_j$ for some j.

Consider now the polynomial

$$P_{l-1}(x;f) = \sum_{\substack{(\alpha,\lambda)>\varkappa \\ \min(l_j-\alpha_j)\geqslant 1}} c_\alpha x^\alpha \tag{21}$$

of degree not greater than $l_j - 1$ in the variable x_j. It follows from Taylor's formula that

$$D^\alpha f_\rho(x) \to D^\alpha P_{l-1}(x;f) \quad \text{as} \quad \rho \to \infty,\ (a,\lambda) > \varkappa \tag{22}$$

uniformly in $x \in K$ for every compactum $K \subset E^n$.

We will show that (18) is satisfied with the polynomial $P_{l-1}(x;f)$ defined by (21). We first note that it readily follows from Theorem 3 that $D^\alpha f$ locally belongs to L_q under the conditions of the present theorem. Thus the proof reduces to a derivation of the estimate

$$\|D^\alpha(f - P_{l-1}(\cdot\,;f))\|_{q,K} \leqslant C\|f\|_{L^l_{p(1),\ldots,p(n)}},$$

in which C does not depend on the compactum $K \subset E^n$.

By almost repeating the arguments used in proving Theorem 3 (cf. also [11], p. 12 = 9), we get

$$\| f_\rho \|_{L^l_{p^{(1)},\ldots,p^{(n)}}} \to 0 \qquad (\rho \to \infty).$$

By virtue of this fact and (22) it suffices for the proof of Theorem 4 to show that

$$\| D^\alpha (f_\varepsilon - f_\rho) \|_{q,K} \leqslant C \| f_\varepsilon - f_\rho \|_{L^l_{p^{(1)},\ldots,p^{(n)}}}$$

with the constant C not depending on f, K, ϵ or ρ $(0 < \epsilon < \rho < \infty)$.

By going over, if necessary, from the function f to a weighted moving average of it with small steps in all of the variables, it can be assumed without loss of generality that for a function $f \in C^\infty(E^n)$ satisfying the last inequality

$$F_j = D_j^{l_j} f \in W^l_{p^{(j)};\, p^{(1)},\ldots,p^{(n)}}, \qquad \| D^l F_j \|_{p^{(j)}} < \infty$$

$$(j = 1, \ldots, n;\ i = 1, \ldots, n).$$

We replace the function f in the latter inequality by the function $B^{(\delta,H)} D^l f$ of (16). Since the difference $f_\epsilon - f_\rho$ is by virtue of (20) a sum of convolutions of the F_j and kernels from C_0^∞ $(j = 1, \ldots, n)$, our change is equivalent to the replacement of each function F_j by the function $B^{(\delta,H)} D^l F_j$. As follows from the proof of Theorem 3, the quantity

$$\max_{1 \leqslant j \leqslant n} \| F_j - B^{(\delta,H)} D^l F_j \|_{W^l_{p^{(j)};\, p^{(1)},\ldots,p^{(n)}}}$$

is sufficiently small under a suitable choice of δ and H, which implies that the difference

$$[f_\varepsilon - f_\rho] - [(B^{(\delta,H)} D^l f)_\varepsilon - (B^{(\delta,H)} D^l f)_\rho]$$

and a derivative D^α of it are sufficiently small in $L^l_{p^{(1)},\ldots,p^{(n)}}$ and $C(E^n)$ respectively. Thus it suffices for the proof of the theorem to establish the inequality

$$\| D^\alpha [(B^{(\delta,H)} D^l f)_\varepsilon - (B^{(\delta,H)} D^l f)_\rho] \|_{q,K} \leqslant C \| (B^{(\delta,H)} D^l f)_\varepsilon - (B^{(\delta,H)} D^l f)_\rho \|_{L^l_{p^{(1)},\ldots,p^{(n)}}}$$

for an arbitrary function $f \in C^\infty(E^n)$, $\Sigma_{j=0}^n \| D^l f \|_{p^{(j)}} < \infty$, with the constant C not depending on f, K, ϵ, ρ, δ or H.

Since $D^l f$ can be approximated with any accuracy in $L_{p^{(j)}}$ $(j = 1, \ldots, n)$ by a function $\chi_R D^l f$ for sufficiently large R (where χ_R is the characteristic function of the ball $|x| < R$), it suffices to establish the inequality differing from the preceding one only in that $D^l f$ is replaced by $\chi_R D^l f$, with the constant C not depending on f, K, ϵ, ρ, δ, H or R. But such an inequality is an inequality for infinitely differentiable functions with compact support and therefore holds by virtue of Theorem 2. The proof of Theorem 4 is complete.

BIBLIOGRAPHY

1. E. Gagliardo, *Ulteriori proprietà di alcune classi di funzioni in più variabili*, Ricerche Mat. **8** (1959), 24–51. MR 22 #181.

2. L. Nirenberg, *On elliptic partial differential equations*, Ann. Scuola Norm. Sup. Pisa (3) **13** (1959), 115–162. MR 22 #823.

3. V. A. Solonnikov, *On some inequalities for functions of the classes* $W_p(R^n)$, Zap. Naučn. Sem. Leningrad. Otdel. Mat. Inst. Steklov. (LOMI) **27** (1972), 194–210 = J. Soviet Math. (to appear).

4. S. M. Nikol'skiĭ, *Approximation of functions of several variables and embedding theorems*, "Nauka", Moscow, 1969; English transl., Die Grundlehren der math. Wissenschaften, Band 205, Springer-Verlag, Berlin and New York, 1974. MR **46** #9714.

5. A. Zygmund, *Trigonometric series*, 2nd rev. ed., Vol. II, Cambridge Univ. Press, New York, 1959. MR **21** #6498.

6. O. V. Besov and P. I. Lizorkin, *Singular integral operators and sequences of convolutions in* L_p*-spaces*, Mat. Sb. **73 (115)** (1967), 65–88 = Math. USSR Sb. **2** (1967), 57–76. MR **36** #1936.

7. P. I. Lizorkin, *Multipliers of Fourier integrals and bounds of convolutions in spaces with mixed norms. Applications*, Izv. Akad. Nauk SSSR Ser. Mat. **34** (1970), 218–247 = Math. USSR Izv. **4** (1970), 225–254. MR **41** #7420.

8. G. H. Hardy, J. E. Littlewood and G. Pólya, *Inequalities*, Cambridge Univ. Press, New York, 1934.

9. A. Zygmund, *Trigonometric series*, 2nd rev. ed., Vol. I, Cambridge Univ. Press, New York, 1959. MR **21** #6498.

10. A. D. Džabrailov, *Study of differential-difference properties of functions defined on n-dimensional domains*, Doctoral Dissertation, Baku, 1971. (Russian)

11. O. V. Besov, *The continuation of functions in* L_p^l *and* W_p^l, Trudy Mat. Inst. Steklov. **89** (1967), 5–17 = Proc. Steklov Inst. Math. **89** (1967), 1–15. MR **35** #5920.

12. ———, *The behavior of differentiable functions at infinity and the density of functions with compact support*, Trudy Mat. Inst. Steklov. **105** (1969), 3–14 = Proc. Steklov Inst. Math. **105** (1969), 1–15. MR **42** #6603.

Translated by S. SMITH

Trudy Mat. Inst. Steklov.
131 (1974)

Proc. Steklov Inst. Math.
131 (1974)

ESTIMATES FOR THE MODULI OF CONTINUITY OF ABSTRACT FUNCTIONS GIVEN IN A REGION

UDC 517.518.22

O. V. BESOV

ABSTRACT. Certain inequalities for moduli of continuity of various orders which are known in the case of scalar functions are carried over to the case of abstract functions. Bibliography: 9 items.

Our goal in this report is to carry over to the case of functions with values in a Banach space certain inequalities for moduli of continuity of various orders which are known in the case of scalar functions (see [1]–[3]) and for analogs of moduli of continuity of elements of Banach spaces (see [4] and [5]).(1) We use the method of integral representations of functions in terms of differences, due to V. P. Il'in (regarding this, see for example, [6]), as well as a generalization of certain representations of functions of our own [7], [8]. As a result it will become apparent that consideration of abstract functions does not require significant modifications, either in method or in its application. We have found it practicable, however, to present the results below for two reasons: first, they contain as a special case the corresponding estimates of Brudnyĭ and Šalašov [5] for analogs of moduli of continuity of elements of rigged Banach spaces (furthermore their approach is less simple and natural);(2) second, we deduce here a generalization as to the nature of the domain of definition of the functions, and we consider a more general (than exponential) relation of increments of moduli of continuity with respect to different coordinate directions.

We shall consider continuous functions $f(x)$ defined on a set E (containing no isolated points) of n-dimensional Euclidean space E^n with values in a Banach space B, i.e. such that $\|f(x) - f(x^0)\| \longrightarrow 0$ as $x \longrightarrow x^0$, $x^0 \in E$, $x \in E$.

By $C(B; E)$ we shall denote a Banach space of (not necessarily all) functions uniformly continuous on each bounded subset of E with norm

$$\|f\|_{C(B;\,E)} = \sup_{x\in E}\|f(x)\|, \qquad \|f\|_{C(B)} = \|f\|_{C(B;E^n)}.$$

By the partial derivative $D_i f = \partial f/\partial x_i$ at an interior point of E we shall mean, as usual, the limit in B (if it exists) of the respective difference quotient. Similarly to the case of scalar functions, one defines the mixed derivatives

$$D^\alpha f = D_1^{\alpha_1}\ldots D_n^{\alpha_n} f, \text{ where } \alpha = (\alpha_1, \ldots, \alpha_n), \qquad |\alpha| = \alpha_1 + \ldots + \alpha_n.$$

AMS (MOS) subject classifications (1970). Primary 26A15.

(1) We had access to the results in [5] in manuscript form before its publication. The main results are formulated in [4].

(2) A detailed comparison with the results of Brudnyĭ and Šalašov will be given below.

For a boundary point x^0 of E, $D^\alpha f(x^0)$ is taken to be the limit of the $D^\alpha f(x)$ with respect to interior points of E.

In integrating abstract functions we shall consider only the abstract Riemann integral, i.e. the limit in B of the respective integral sums, or the improper abstract Riemann integral. For $x = (x_1, \ldots, x_n)$ and $y = (y_1, \ldots, y_n)$ set

$$\frac{x}{y} = x : y = \left(\frac{x_1}{y_1}, \ldots, \frac{x_n}{y_n}\right),$$

$$xy = (x_1 y_1, \ldots, x_n y_n),\ x^y = x_1^{y_1}, \ldots, x_n^{y_n}.$$

Let $\Delta^m(z)f(x)$ denote the difference of order m with increment $z \in E^n$, $\Delta_i^m(t)f(x) = \Delta^m(te_i)f(x)$, and for $\alpha = (\alpha_1, \ldots, \alpha_n)$ let $\Delta^\alpha(z)f(x)$ denote the mixed difference with increments z_i in the direction of the basis vector e_i. Let $\Delta^m(z; E)f(x)$, $\Delta_i^m(t, E)f(x)$, $\Delta^\alpha(z; E)f(x)$ denote the corresponding differences formed at points lying in E together with segments joining them, and let them denote zero otherwise.

Let us further set

$$\omega_y^m(\delta; E; f) = \sup_{|t| \leqslant \delta} \sup_{x \in E} \left\| \Delta^m \left(t \frac{y}{|y|}; E\right) f(x) \right\|,$$

$$\omega_i^m(\delta; E; f) = \omega_{e_i}^m(\delta; E; f),$$

$$\omega^\alpha(z; E; f) = \sup_{\substack{0 < y_i \leqslant z_i \\ 1 \leqslant i \leqslant n}} \sup_{x \in E} \| \Delta^\alpha(z; E) f(x) \|.$$

Let $\sigma = \sigma(h) = (\sigma_1(h), \ldots, \sigma_n(h))$ be a continuous vector-valued function on $[0, \infty)$ with $\sigma(0) = 0$, $\sigma_i'(h) > 0$ $(h \in (0, \infty), i = 1, \ldots, n)$. An example of such a function is $\sigma(h) = (h^{\lambda_1}, \ldots, h^{\lambda_n})$.

By a σ-horn we mean a region

$$V(\sigma) = \bigcup_{0 < h < h_0} \left\{ x : \frac{x_i}{a_i} > 0, \sigma_i(h) < \frac{x_i}{a_i} < (1 + \varepsilon)\sigma_i(h) \quad (i = 1, \ldots, n) \right\}. \tag{1}$$

Consider a function

$$\Psi(x) \in C_0^\infty(E^n), \qquad \operatorname{supp} \Psi \subset \{x : 0 < x_i < 1\ (i = 1, \ldots, n)\}. \tag{2}$$

Then the support of the function $\Psi(x/\sigma(h))$ for all $h \in (0, H)$ is contained in some σ-horn (1), which we denote by $V(\sigma; \Psi)$.

In the sequel we shall assume that the following inequality holds for all Banach spaces B_1 and B_2 under consideration:

$$\left\| \int \Psi(y) f(\sigma(h) y)\, dy \right\|_{B_2} \leqslant C_\Psi \Lambda(h) \sup_{z \in V(\sigma; \Psi)} \| f(z) \|_{B_1} \tag{3}$$

for any function $f \in C(B_1; V(\sigma; \Psi))$ and any $h \in (0, H)$. Here $\Psi(x)$ is any function of the form (2) and the σ-horn $V(\sigma; \Psi)$ is also arbitrary. In fact, in the discussions there will appear only a finite number of averaging kernels $\Psi(x)$ and a finite number of horns $V(\sigma; \Psi)$, so that in this sense condition (3) can be replaced by a less restrictive one.

Note further that (3) is automatically fulfilled for the special case $B_1 = B_2$ with $\Lambda \equiv 1$. For the sake of simplicity we shall assume below that $\underline{\lim}_{h \to +\infty} \Lambda(h) > 0$, which however does hold in familiar concrete cases. Without loss of generality one can assume that $\Lambda(h) \geqslant 1$ on some interval $(0, h_0)$.

Let $f(x)$ be a continuous function defined on Euclidean space E^n with values in a Banach space B_1. Consider the averaging $f_{\sigma(h)}(x)$ with radius $\sigma_i(h)$ in x_i

$$f_{\sigma(h)}(x) = \iint \prod_{i=1}^{n} \chi_i(y_i)\chi_i(z_i) f(x + \sigma(h)y + \sigma(h)z)\,dydz, \tag{4}$$

where

$$\chi_i(t) = \frac{1}{A_i}\sum_{k=0}^{m_i} \frac{(-1)^{m_i-k}}{(1+k\delta_i)^2}\binom{m_i}{k}\eta_i\left(\frac{t}{1+k_i\delta}\right), \quad 0<\delta_i<1,$$

and the infinitely differentiable functions $\eta_i(t)$ are concentrated in a prescribed interval

$$[\gamma_i', \gamma_i''] \subset (0, 1), \qquad \zeta_i(t) = t\eta_i(t),$$

$$A_i = \int_0^1 (1 - t^{\delta_i})^{m_i}\,dt = \sum_{k=0}^{m_i} \frac{(-1)^k}{1+k\delta_i}\binom{m_i}{k} > 0.$$

Since $f(x)$ is continuous, $f_{\sigma(h)}(x) \longrightarrow f(x)$ as $h \longrightarrow 0$.

Just as in deriving the representation (8) in [6] (see also [1]), we obtain the following representation:

$$f(x) = f_{\sigma(H)}(x) + \int_0^H \sum_{i=1}^{n} \int_{E^n}\int_0^\infty \frac{\sigma_i'(h)}{\sigma_i^2(h)} \prod_{j=1}^{n} \frac{1}{\sigma_j(h)} \Psi_i\left(\frac{y}{\sigma(h)}\right)\zeta_i\left(\frac{t}{\sigma_i(h)}\right) \times \Delta_i^{m_i}(\delta_i t) f(x + y + te_i)\,dtdydh. \tag{5}$$

Here the $\Psi_i(x) \in C_0^\infty(E^n)$ are also concentrated on a prescribed cube $\square = \{x\colon 0 < x_i < 1;\ i = 1, \dots, n\}$.

From (4) and (5) we see that the right-hand side of (5) involves only values of f at points of the form $x + z$ for z in some horn $V(\sigma)$ (prescribed in advance and, in this case, with $a_i > 0$), which will be called the field of the representation (5). Thus (5) is valid for functions which are defined and continuous on the horn $x + V(\sigma)$ and at its vertex x.

We let U denote a degenerate or nondegenerate closed and bounded parallelepiped $\{x\colon a_i \leqslant x_i \leqslant b_i\ (i = 1, \dots, n)\}$ where $a_i \leqslant b_i$.

The principal objects for the estimates here will be continuous functions defined on the set $U + V(\sigma)$ (the symbol $U + V$ will denote throughout the arithmetic sum of the sets U and V). Of interest to us are estimates of their differences or differences of their derivatives on U.

THEOREM 1. *Let the Banach spaces B_1 and B_2 be continuously embedded in a Hausdorff linear topological space, and let condition* (3) *hold. Let $\tau = (\tau_1, \dots, \tau_n)$, $0 \leqslant \tau_i \leqslant H_0$, $U = \{x\colon 0 \leqslant x_i \leqslant \mu_i\tau_i\ (i = 1, \dots, n)\}$, $0 \leqslant T \leqslant H \leqslant H_0$, $0 \leqslant \nu \leqslant \mu$, $0 \leqslant \beta \leqslant \alpha$, and let $V = V(\sigma)$ be the σ-horn* (1). *Then*

$$\|\Delta^{\mu}(\tau)D^{\alpha}f(0)\|_{B_2}\leqslant C\left[\frac{\tau}{\sigma(H)}\right]^{\mu-\nu}\sigma(H)^{-\alpha+\beta}\Lambda(H)$$
$$\times\sup_{x\in U+V}\|\Delta^{\nu}(\tau;\,U+V)\,D^{\beta}f(x)\|_{B_1}+C\sum_{i=1}^{n}\left\{\int_0^T+\int_T^H\left[\frac{\tau}{\sigma(h)}\right]^{\mu}\right\}\sigma^{-\alpha}(h)\,\Lambda(h)\tag{6}$$
$$\times\int_{\delta_i\varepsilon}^{\delta_i}\sup_{x\in U+V}\|\Delta_i^{m_i}(t\sigma_i(h);\,U+V)\,f(x)\|_{B_1}dt\,\frac{\sigma_i'(h)}{\sigma_i(h)}\,dh,$$

where $\epsilon\in(0,1)$ *and the* δ_i *are any prescribed numbers in* $(0,1)$.

PROOF. Let $\tau=\tau_je_j$ and $\mu=Me_j$.

We apply to both sides of (5) the operations of taking the derivative D^{α} and the difference $\Delta_j^{\mu}(\tau_j)$, transfering them to the kernel and expressing for $h\in(T,H)$ the difference of the kernel in terms of the integral of the derivative:

$$\Delta_j^{M}(\tau_j)\,D^{\alpha}f(0)=\Delta_j^{M}(\tau_j)\,D^{\alpha}f_{\sigma(H)}(0)+\int_0^T\sum_{i=1}^{n}(-1)^{|\alpha|}\int_{E^n}\int_0^{\infty}\frac{\sigma_i'(h)}{\sigma_i^2(h)}\prod_{k=1}^{n}\sigma_k(h)^{-1-\alpha_k}$$
$$\times\Psi_i^{(\alpha)}\left(\frac{y}{\sigma(h)}\right)\zeta_i\left(\frac{t}{\sigma_i(h)}\right)\Delta_j^{M}(\tau_j)\,\Delta_i^{m_i}(\delta_it)\,f(y+te_i)\,dt\,dy\,dh$$
$$+\int_T^H\sum_{i=1}^{n}(-1)^{|\alpha|+M}\int_{E^n}\int_0^{\infty}\frac{\sigma_i'(h)}{\sigma_i^2(h)}\prod_{k=1}^{n}\sigma_k(h)^{-1-\alpha_k}\left[\frac{\tau}{\sigma_j(h)}\right]^{M}\tag{7}$$
$$\times\int_0^1\cdots\int_0^1\Psi_i^{(\alpha+Me_j)}\left(\frac{y-\tau_j(u_1+\ldots+u_M)e_j}{\sigma(h)}\right)du_1\ldots du_M$$
$$\times\zeta_i\left(\frac{t}{\sigma_i(h)}\right)\Delta_i^{m_i}(\delta_it)\,f(y+te_j)\,dt\,dy\,dh.$$

From this we have

$$\|\Delta_j^{M}(\tau_j)\,D^{\alpha}f(0)\|_{B_2}\leqslant\|\Delta_j^{M}(\tau_j)\,D^{\alpha}f_{\sigma(H)}(0)\|_{B_2}+C\sum_{i=1}^{n}\left\{\int_0^T+\int_T^H\left[\frac{\tau_j}{\sigma_j(h)}\right]^{M}\right\}$$
$$\times\frac{\sigma_i'(h)}{\sigma_i(h)\,\sigma(h)^{\alpha}}\Lambda(h)\int_0^{\infty}\frac{1}{\sigma_i(h)}\zeta_i\left(\frac{t}{\sigma_i(h)}\right)$$
$$\times\|\Delta_i^{m_i}(\delta_it;\,U+V)\,f\|_{C(B_1)}dt\,dh.$$

Estimating the first term on the right in the same way, we finally obtain that for $N\leqslant M$

$$\|\Delta_j^{M}(\tau_j)\,D^{\alpha}f(0)\|_{B_2}\leqslant C\left[\frac{\tau_j}{\sigma_j(H)}\right]^{M-N}\sigma(H)^{-\alpha+\beta}\Lambda(H)\,\|\Delta_j^{N}(\tau_j;\,U+V)\,D^{\beta}f\|_{C(B_1)}$$
$$+C\sum_{i=1}^{n}\left\{\int_0^T+\int_T^H\left[\frac{\tau}{\sigma_j(h)}\right]^{M}\right\}\sigma(h)^{-\alpha}\Lambda(h)\tag{8}$$
$$\times\int_{\delta_i(\varepsilon)}^{\delta_i}\|\Delta_i^{m_i}(t\sigma_i(h),\,U+V)\,f\|_{C(B_1)}dt\,\frac{\sigma_i'(h)}{\sigma_i(h)}\,dh.$$

For $M=0,1$ and $B_2=B_1$ it follows from (8) under the hypotheses of the theorem that $D^{\alpha}f\in C(B_1,U)$. Further, for $M=0$, (8) shows that the improper integral $\int_0^T\ldots dh$ on the right in (7) converges in B_2. Since B_1 and B_2 are continuously embedded in a Hausdorff space, then $D^\alpha f(x) \in B_2$ also, and from (8) for $M = 1$ it follows that $D^\alpha f \in C(B_2, U)$. Estimate (8) gives the assertion of the theorem when μ is a coordinate direction vector. The case of a general μ is treated similarly.

From Theorem 1 several corollaries may be derived, such as estimates of the moduli of continuity of functions, which due to their obvious nature we shall not formulate.

THEOREM 2. *Let* $0 < h < H$, $\tau = (\tau_1, \ldots, \tau_n)$, $0 \leqslant \tau_i \leqslant \sigma_i(h)$, $l = [M(n-1)+1] \cdot |\mu|\,|\sigma(H)|$,

$$s(t) = \max_i \frac{\sigma_i'(t)}{\sigma_i(t)}, \quad U = \{x : 0 \leqslant x_i \leqslant \mu_i \tau_i \quad (i = 1, \ldots, n)\},$$

Then for functions $f(x)$, $x \in U$, *with values in a Banach space* B *we have the estimate*

$$\| \Delta^\mu(\tau) f(0) \| \leqslant C \sum_{i=1}^n \omega_i^M(y_i; U + l\square; f) + C \int_h^H \frac{\tau^\mu}{\sigma(t)^\mu} \sum_{i=1}^n \omega_i^M(y_i; U + l\square; f) \times s(t)\,dt + C \frac{\tau^\mu}{H^\mu} \| f \|_{C(B;\, U + l\square)}. \tag{9}$$

In some cases the first term on the right in (9) can be estimated by the second one and may be omitted. For $H = \infty$ the last term on the right is zero.

PROOF. Let us first derive the necessary integral representation of functions. For a continuous B-valued function $f(x)$, $x \in U$, introduce the averaging

$$f_{\sigma(h)}(x) = \iint K(y) K(z) \Delta^N(\sigma(h) y) \Delta^N(\sigma(h) z) f(x)\,dy\,dz + f(x), \tag{10}$$

where the kernel $K(y) \in C_0^\infty(E^n)$ is concentrated in a prescribed cube in $\square = \{x: 0 < x_i < 1 \ (i = 1, \ldots, n)\}$ and $\int K(y)\,dy = 1$.

Writing $f_{\sigma(h)}(x)$ in the form

$$f_{\sigma(h)}(x) = \frac{1}{\prod_{i=1}^n \sigma_i(h)^2} \iint K(y : \sigma(h)) K(z : \sigma(h)) \Delta^N(y) \Delta^N(z) f(x)\,dy\,dz + f(x),$$

we find that

$$\frac{\partial}{\partial h} f_{\sigma(h)}(x) = \frac{2}{\prod_{i=1}^n \sigma_i(h)^2} \iint \left\{ - \sum_{i=1}^n \left(\frac{\sigma_i'(h)}{\sigma_i(h)} \right) K(z : \sigma(h)) - \sum_{i=1}^n D_i K(z : \sigma(h)) \frac{z_i \sigma_i'(h)}{\sigma_i(h)^2} \right\}.$$

The last expression remains unaltered if $\Delta^N(z) f(x)$ is replaced by $(-1)^N f(x) + \Delta^N(z) f(x)$, and hence we have

$$\frac{\partial}{\partial h} f_{\sigma(h)}(x) = \frac{1}{\prod_{i=1}^{n} \sigma_i(h)^2} \iint \sum_{i=1}^{n} \frac{\sigma_i'(h)}{\sigma_i(h)} K_i(y:\sigma(h))\, K(z:\sigma(h)) \times \Delta^N(y) \sum_{s=1}^{N} a_s f(x+sz)\, dy\, dz. \tag{11}$$

By the Leibniz-Newton formula

$$f(x) = f(x) - f_{\sigma(h)}(x) + \int_h^H \frac{\partial}{\partial t} f_{\sigma(t)}(x)\, dx - f_{\delta(H)}(x).$$

If in this equality we substitute $f_{\sigma(H)}(x)$ from (10) and $\partial f_{\sigma(t)}(x)/\partial t$ from (11) we obtain the desired representation of $f(x)$:

$$\begin{aligned} f(x) = &-\iint K(y)\, K(z)\, \Delta^N(\sigma(h)\, y)\, \Delta^N(\sigma(h)\, z)\, f(x)\, dy\, dz \\ &+ \sum_{i=1}^{n} \sum_{s=1}^{N} a_s \int_h^H \frac{1}{\prod_{j=1}^{n} \sigma_j(t)^2} \iint K_i(y:\sigma(t))\, K(z:\sigma(t)) \\ &\times \Delta^N(y)\, f(x+sz)\, dy\, dz \frac{\sigma_i'(t)}{\sigma_i(t)}\, dt - f_{\sigma(H)}(x). \end{aligned} \tag{12}$$

If we apply to each term of (12) the operation $\Delta^\mu(\tau)$, where $\tau = (\tau_1, \ldots, \tau_n)$, $0 \leqslant \tau_i \leqslant \sigma(h)$, and express the mixed difference in terms of the integral of the corresponding derivative transfered to the kernel K, we obtain the following estimate in which $R = N|\mu|$:

$$\begin{aligned} \|\Delta^\mu(\tau) f(0)\| \leqslant\; & C \int |K(y)| \sup_{z \in U + R|\sigma(h)|\square} \|\Delta^N(y\sigma(h))\, f(z)\|\, dy \\ &+ C \sum_{i=1}^{n} \int_h^H \frac{\tau^\mu}{\sigma(t)^\mu} \int K_i(y) \sup_{z \in U + R|\sigma(t)|\square} \|\Delta^N(y\sigma(t))\, f(z)\|\, dy \frac{\sigma_i'(t)}{\sigma_i(t)}\, dt \\ &+ C \frac{\tau^\mu}{\sigma(H)^\mu} \sup_{z \in U + R|\sigma(H)|\square} \|f(z)\|. \end{aligned} \tag{13}$$

Under the hypotheses of the theorem, for $N = n(M-1) + 1$ we have

$$\begin{aligned} \Delta^\mu(\tau) f(0)\| \leqslant\; & C \sup_{\substack{|y_i| \leqslant \sigma_i(h) \\ (i=1,\ldots,n)}} \omega^N(y;\, U + l\square;\, f) + C \int_h^H \frac{\tau^\mu}{\sigma(t)^\mu} \\ &\times \sup_{|y_i| \leqslant \sigma_i(t)} \omega^N(y;\, U + l\square;\, f)\, s(t)\, dt + C \frac{\tau^\mu}{H^\mu} \|f\|_{C(B,\, U + l\square)}. \end{aligned} \tag{14}$$

The first difference $\Delta(y)f(x)$ can clearly be represented in the form of a sum of first differences (with shift) of coordinate directions. Hence $\Delta^N(y)f(x)$ can also be represented in terms of a sum of (shifted) differences of coordinate directions $\Delta_i^m(y_i)f(x)$ for $N = n(M-1) + 1$. In view of this we obtain the assertion of the theorem from (14).

REMARK. In estimates (6) of Theorem 1 and (9) of Theorem 2 the differences of the functions or their derivatives on U are estimated in terms of the values of the function on another set ($U + V(\sigma)$ and $U + l\square$, respectively). To circumvent this one can proceed in two ways:

1) Consider an (unbounded) set $E \subset E^n$ with the property $E + V(\sigma) \subset E$ or, respectively, $E + l\Box \subset E$, and apply the derived inequalities (6) and (9) to the estimate

$$\sup_E \| \Delta^\mu(\tau; E) D^\alpha f(x) \| = \sup_z \| \Delta^\mu(\tau; U) D^\alpha f_z(0) \|,$$

where $f_z(x) = f(x + z)$, $U = \{x: 0 \leqslant x_i \leqslant \mu_i \tau_i \ (i = 1, \ldots, n)\}$, and the sup on the right is taken over the z such that $z + U \subset E$.

2) Consider the estimates $\sup_x \|\Delta^\mu(\tau; G) D^\alpha f(x)\|$ for special sets $G \subset E^n$ satisfying the $\sigma(h)$-horn condition or, respectively, the cube condition. Let us define these terms (see [1]).

We say that the set $G \subset E^n$ satisfies the $\sigma(h)$-horn condition if there exists a finite covering $\{G_k\}_1^K$ of G and a collection of $\sigma(h)$-horns $V_k(\sigma)$ such that

$$G = \bigcup_{k=1}^K G_k = \bigcup_{k=1}^K (G_k + V_k(\sigma)) = \bigcup_{k=1}^K G_k^{(d)},$$

where

$$G_k^{(d)} = \{x : x \in G_k, \ \rho(x, \partial G_k \setminus \partial G) > d\}, \ d > 0.$$

We say that $G \subset E^n$ satisfies the cube condition if in the preceding definition the $V_k(\sigma)$ are replaced by the cubes

$$\Box_k(H) = \{x : 0 < \theta_i^{(k)} x_i < H, \ |\theta_i^{(k)}| = 1 \ (i = 1, \ldots, n)\}.$$

Without going into detail (see [1]), we simply note that estimates on sets satisfying the σ-horn condition or the cube condition reduce to a finite combination of estimates such as (6) or, respectively, (9).

Let us apply the results in an example. Let there be open sets $G_0 \subset G_0 + V \subset G \subset E^n$ and $G_0^{(d)} = \{x: x \in G_0, \rho(x, \partial G_0 \backslash \partial G) > d\}$, $d > 0$, and let $V = V(\sigma)$ be the σ-horn (1). Let

$$U = \{x: 0 \leqslant x_i \leqslant \mu_i \tau_i \ (i = 1, \ldots, n)\}.$$

We define the space $C(B_1; U + V)$ as the set of $f(x)$ of the form $f(x) = \varphi(x + z)$, where φ is a scalar function defined on $G_0 + V(\sigma)$, $x \in U + V$ and $z \in G_0(\mu\tau)$, where $G_0(\mu\tau) = \{z: z + U \subset G_0\}$, and

$$\| f(x) \|_{B_1} = \| \varphi(x + \cdot) \|_{L_p(G_0(\mu\tau))}, \quad x \in U + V.$$

For B_2 we take $L_q(G_0^{(d)}(\mu\tau))$, $1 \leqslant p \leqslant q \leqslant \infty$.

Let us verify that (3) holds for $V(\sigma; \Psi) = V$. Relying on the estimate of the convolution for sufficiently small V (depending on d)

$$\left\|\int \Psi(y) f(\sigma(h) y)\, dy\right\|_{B_2} = \left\|\int \Psi(y) \varphi(\cdot + \sigma(h) y)\, dy\right\|_{L_q(G_{(0)}^{(d)}(\mu\tau))}$$
$$\leqslant C_\Psi^{(1)} \prod_{j=1}^{n} \sigma_j(h)^{\frac{1}{q}-\frac{1}{p}} \|\varphi\|_{L_p(G_0^{(d)}(\mu\tau)+V)}$$
$$\leqslant C_\Psi^{(1)} \prod_{j=1}^{n} \sigma_j(h)^{\frac{1}{q}-\frac{1}{p}} \|\varphi\|_{L_p(G_0(\mu\tau))}$$
$$= C_\Psi^{(1)} \prod_{j=1}^{n} \sigma_j(h)^{\frac{1}{q}-\frac{1}{p}} \|f(0)\|_{B_1},$$

we see that (3) does hold.

Now from Theorem 1 we obtain the estimate

$$\|\Delta^\mu(\tau;\, G_0^{(d)}) D^\alpha \varphi\|_{L_q} \leqslant C \left[\frac{\tau}{\sigma(H)}\right]^{\mu-\nu} \sigma(H)^{-\alpha+\beta} \prod_{j=1}^{n} \sigma_j(h)^{\frac{1}{q}-\frac{1}{p}}$$
$$\times \|\Delta^\nu(\tau;\, G_0 + V) D^\beta \varphi\|_{L_p} + C \sum_{i=1}^{n} \left\{\int_0^T + \int_T^H \left[\frac{\tau}{\sigma(h)}\right]^{\mu}\right\}$$
$$\times \sigma(h)^{-\alpha} \prod_{j=1}^{n} \sigma_j(h)^{\frac{1}{q}-\frac{1}{p}}$$
$$\times \int_{\delta_i \varepsilon}^{\delta_i} \|\Delta_i^{m_i}(t\sigma_i(h);\, G_0 + V)\varphi\|_{L_p} \frac{\sigma_i'(h)}{\sigma_i(h)}\, dh.$$

On the right side of this inequality one can obviously take $G \supset G_0 + V$ instead of $G_0 + V$. One can also take G instead of $G_0^{(d)}$ on the left side if G satisfies the $\sigma(h)$-horn condition.

Let us compare these results with the estimates of Brudnyĭ and Šalašov [5] regarding estimates of moduli of continuity in Banach spaces.

Let B be a Banach space and Γ_+ an n-parameter strongly continuous semigroup of linear operators τ_x: $B \longrightarrow B$, $x \in E_+^n = \{x\colon x = (x_1, \ldots, x_n), x_i \geqslant 0\ (i = 1, \ldots, n)\}$, where

a) $\tau_0 = 1_B, \quad \tau_x \cdot \tau_y = \tau_{x+y}, \quad \forall x,\ y \in E_+^n$;

b) $\lim\limits_{x \to y} \|\tau_x u - \tau_y u\| = 0, \quad \forall u \in B.$

Each $u \in B$ generates a (strongly) continuous function with values in B

$$f_u(x) = \tau_x u, \quad \forall x \in E_+^n, \tag{15}$$

which makes it possible to define the analog of a shift in B (M. H. Stone, E. Hille) as well as the analog of the k-modulus of continuity (N. P. Kupcov [9]) and the derivative. For example, the k-modulus of continuity of $u \in B$ in the direction $h \in E_+^n$, $|h| = 1$, is defined as the function

$$\omega_k^{(h)}(u; \rho) = \sup_{\substack{x \in E_+^n \\ |x| \leqslant \rho}} \|(1_B - \tau_x)^k u\|_B, \quad 0 \leqslant \rho < +\infty,$$

and so on.

Brudnyĭ and Šalašov [5] considered an n-parameter strongly continuous semigroup of linear operators with the additional equiboundedness condition

c) $C(\Gamma_+) = \sup_{x \in E_+^n} \|\tau_y\|_{B \to B} < +\infty.$

The set (B, Γ_+) is called here a rigged Banach space.

For fixed $N \geqslant 1$ and $x^{(0)} \in G$ we denote by $\widetilde{C}(B; G)$ the (generally nonlinear) set of functions $f(x)$ in the Banach space $C(B; G)$ for which

$$\|f\|_{C(B;G)} = \sup_G \|f(x)\| \leqslant N_E \|f(x^{(0)})\|_B. \tag{16}$$

The set $\widetilde{C}(B; E_+^n)$ obviously contains for $x^{(0)} = 0$ and $N \geqslant C(\Gamma_+)$ a subspace of functions $f_u(x)$ of the form (15) which is isomorphic to the rigged Banach space (B, Γ_+). Thus the results pertaining to $\widetilde{C}(B; E_+^n)$ (and a fortiori to $C(B; E_+^n)$) contain as a special case the corresponding assertions for the rigged Banach space (B, Γ_+).

Thus Theorems 1 and 2 above for $\sigma(h) = (h^{\alpha_1}, \ldots, h^{\alpha_n})$, $V(\sigma; h) = \{x\colon 0 < x_i < h^{\alpha_i}\ (i = 1, \ldots, n)\}$ and $H = \infty$ encompass the results of Brudnyĭ and Šalašov [5] which deal with moduli of continuity in Banach spaces.

BIBLIOGRAPHY

1. O. V. Besov, *Estimates of moduli of smoothness on domains, and imbedding theorems*, Trudy Mat. Inst. Steklov. **117** (1972), 22–46 = Proc. Steklov Inst. Math. **117** (1972), 25–54.

2. ———, *Théorèmes de plongement des espaces fonctionnels*, Proc. Internat. Congress Math. (Nice, 1970), vol. 2, Gauthier-Villars, Paris, 1971, pp. 467–473.

3. A. F. Timan, *Theory of approximation of functions of a real variable,* Fizmatgiz, Moscow, 1960; English transl., Internat. Ser. Monographs Pure Appl. Math., vol. 34, Macmillan, New York, 1963. MR **22** #8257; **33** #465.

4. Ju. A. Brudnyĭ and V. K. Šalašov, *Lipschitz spaces of functions*, Dokl. Akad. Nauk SSSR **197** (1971), 18–20 = Soviet Math. Dokl. **12** (1971), 383–386. MR **43** #3789.

5. ———, *Lipschitz spaces of functions,* Sibirsk. Mat. Ž. = Siberian Math. J. (to appear).

6. O. V. Besov and V. P. Il'in, *Natural extension of the class of regions in imbedding theorems*, Mat. Sb. **75** (**117**) (1968), 483–495 = Math. USSR Sb. **4** (1968), 445–456. MR **37** #1987.

7. O. V. Besov, *Continuation of functions with conservation of their differential and difference properties in L_p*, Dokl. Akad. Nauk SSSR **150** (1963), 963–966 = Soviet Math. Dokl. **4** (1963), 770–774. MR **28** #3319.

8. ———, *Continuation of functions beyond the boundary of a domain with preservation of differential-difference properties in L_p*, Mat. Sb. **66** (**108**) (1965), 80–96; English transl., Amer. Math. Soc. Transl. (2) **79** (1969), 33–52. MR **30** #5150.

9. N. P. Kupcov, *Direct and inverse theorems of approximation theory and semigroups of operators*, Uspehi Mat. Nauk **23** (1968), no. 4 (142), 117–178 = Russian Math. Surveys **23** (1968), no. 4, 115–177. MR **38** #2507.

Translated by J. MALEK

Trudy Mat. Inst. Steklov.
131 (1974)

Proc. Steklov Inst. Math.
131 (1974)

CONSTRUCTIVE CHARACTERIZATION OF CLASSES OF FUNCTIONS WITH DOMINANT MIXED DERIVATIVE

UDC 517.51

Ja. S. BUGROV

ABSTRACT. This paper investigates the approximation of functions that satisfy multiple Hölder conditions in the L_2 metric, by means of trigonometric sums

$$T(x; N, r) = \sum_{|k|^r \leqslant N} \alpha_k e^{i(k,x)},$$

$$\left(|k|^r = |k_1|^{r_1} \cdots |k_n|^{r_n},\ (k, x) = \sum_{j=1}^{n} k_j x_j,\ r = (r_1, \ldots, r_n)\right),$$

containing a finite number $m = O(N^{1/r_1})$ of terms (under the hypothesis that $r_1 < \min_{j \geqslant 2}\{r_j\}$).

Theorems of the type of the Jackson-Bernšteĭn theorems are obtained for the class $S_2^r H^*$, and the exact order of decrease of the best approximation is found for approximation by the sums $T(x; N, r)$ for the class $S_2^r B^*$.

Bibliography: 9 items.

In the present paper we prove theorems (of the type of the Jackson-Bernšteĭn theorems) on approximating functions that satisfy a multiple Hölder condition, by means of trigonometric sums of a special form that contain a finite number of terms.

§1. Fundamental definitions

Let $x = (x_1, \ldots, x_n)$ $(0 \leqslant x_i \leqslant 2\pi,\ i = 1, \ldots, n)$; $e_n = \{1, \ldots, n\}$; e is an arbitrary subset of e_n, possibly the empty set $\varnothing$; $r = (r_1, \ldots, r_n)$, $r^e = (r_1^e, \ldots, r_n^e)$, where $r_j^e = r_j$ if $j \in e$ and $r_j^e = 0$ if $j \in e_n \backslash e$; $D^{k^e} f = D_{x_1}^{k_1^e} \cdots D_{x_n}^{k_n^e} f$, $D_{x_j}^0 f = f$, where k_i are nonnegative integers; $\Delta_{h_i} f = f(x + \omega^i h_i) - f(x)$, where $\omega^i = (0, \ldots, 0, 1, 0, \ldots, 0)$ and the 1 is in the ith position; $\Delta_{h_i}^{s_i} f = \Delta_{h_i}(\Delta_{h_i}^{s_i - 1} f)$; $\Delta_h^{s^e} f$ are the differences of order $s^e = (s_1^e, \ldots, s_n^e)$, applied successively to the function f with respect to all x_i for which $i \in e$; $h = (h_1, \ldots, h_n)$.

DEFINITION 1. Let $f(x)$ be periodic in each variable x_i with period 2π. We say that $f(x)$ belongs to the class $S_p^r H^*$ $(1 \leqslant p \leqslant \infty,\ r_i > 0)$ if

$$\sup_{h > 0} h^{r^e} \| \Delta_h^{s^e} f \|_p = M_p^{r^e}(f) < \infty, \qquad (s_i > r_i, \quad i = 1, \ldots, n),$$

for all $e \subseteq e_n$. Here $h^\alpha = h_1^{\alpha_1} \ldots h_n^{\alpha_n}$, $\|\psi\|_p = (\int_G |\psi(x)|^p dx)^{1/p}$, $1 \leqslant p < \infty$, $\|\psi\|_\infty = \operatorname{ess\,sup}_{x \in G} |\psi(x)|$, $G = \{0 \leqslant x_i \leqslant 2\pi,\ i = 1, \ldots, n\}$.

In this class we introduce the norm

$$|f, S_p^r H*| = \|f\|_p + \sum_{e \subseteq e_n} M_p^{r^e}(f).$$

DEFINITION 2. We say that $f(x) \in S_p^r B^*$ $(1 \leqslant p < \infty)$, if for all $e \subseteq e_n$ $(e \neq \varnothing)$

AMS (MOS) subject classifications (1970). Primary 42A08, 42A92, 41A25.

$$J_e = \left(\int\limits_{R_n^0(e)} h^{-1-r^e p} \| \Delta_h^{s^e} f \|_p^p \, dh \right)^{\frac{1}{p}} < \infty, \qquad (s_i > r_i, \ i = 1, \ldots, n),$$

$$R_n^0(e) = \{0 \leqslant h_i < \infty, \qquad i \in e\}, \qquad h^{-1-r^e p} = \prod_{j \in e} h_j^{-1-r_j p}.$$

For the space $S_p^r B^*$ we introduce the norm

$$| f, S_p^r B^* | = \| f \|_p + \sum_{e \subseteq e_n} J_e < \infty.$$

Note that when $p = 2$ the class $S_p^r B^*$ coincides with the class $S_2^r W^*$ of functions having all Liouville derivatives $D^{r^e} f \in L_2(G)$.

Let Ω denote the set of all points of n-space with integral coordinates. Let Λ_m be any finite subset of Ω with m points, and let

$$E_{\Lambda_m}(f)_p = \inf_{\alpha_k} \| f - T_{\Lambda_m} \|_p$$

be the best approximation of f by trigonometric sums of the form

$$T_{\Lambda_m}(x) = \sum_{k \in \Lambda_m} \alpha_k e^{i(k,x)}, \qquad \alpha_k = \alpha_{k_1, \ldots, k_n}, \qquad (k, x) = \sum_{i=1}^{n} k_i x_i.$$

Along with Λ_m we consider the set of lattice points whose coordinates satisfy $|k|^r = |k_1|^{r_1} \ldots |k_n|^{r_n} \leqslant N$, where N is fixed. We denote this set by Λ_m^*.

A vector k with integral coordinates, lying on the algebraic surface $|k|^r = N$, for which certain components are zero, can be determined in the following way: in the inequality $|k|^r \leqslant N$ replace the zero coordinates by ones, and determine the rest of the coordinates. If $r_1 < \min_{j \geqslant 2} \{r_j\}$, the number of points in Λ_m^* is precisely of order $m = O(N^{1/r_1})$ (cf. [1], p. 414).

From now on we denote the trigonometric sums $T_{\Lambda_m^*}(x)$ by $T(x; N, r)$, and $E_{\Lambda_m^*}(f)_p$ by $E_N(f; r)_p$.

§2. Equivalent norms of function spaces

Let

$$f(x) \sim \sum_{-\infty}^{\infty} a_k e^{i(k,x)}$$

be the Fourier series of the function $f(x)$ and let a_k be its Fourier coefficients. It is easy to calculate that

$$\| \Delta_h^{s^e} f \|_2^2 = \left\| \sum_{-\infty}^{\infty} a_k e^{i(k,x)} \prod_{j \in e} (e^{ik_j h_j} - 1)^{s_j} \right\|_2^2 = \sum_{-\infty}^{\infty} | a_k |^2 \prod_{j \in e} \left| \sin \frac{k_j h_j}{2} \right|^{s_j}.$$

From this it is clear that $f \in S_2^r H^*$ if and only if

$$\sum_{-\infty}^{\infty} |a_k|^2 \prod_{j\in e} \left|\sin \frac{k_j h_j}{2}\right|^{2s_j} \leqslant c h^{2r^e} \tag{2.1}$$

for all $e \subseteq e_n$.

From (2.1) we can obtain the inequality (cf. [1], p. 415)

$$\sum_{\substack{k_j=-\infty \\ j\in e_n\setminus e}}^{\infty} \sum_{\substack{2^{t_j-1}<|k_j|<2^{t_j} \\ j\in e}} |a_k|^2 \leqslant c 2^{-2(t,r^e)} \tag{2.2}$$

for every $e \subseteq e_n$. Moreover, since

$$J_e = \Big(\int_{R_n^0(e)} h^{-1-2r^e} \| \Delta_h^{s^e} f \|_2^2 \, dh \Big)^{1/2}$$

$$= \Big(\sum_{-\infty}^{\infty} |a_k|^2 \prod_{j\in e} \int_0^{\infty} h_j^{-1-2r_j} \left| \sin \frac{k_j h_j}{2} \right|^{2s_j} dh_j \Big)^{1/2} = c \Big(\sum_{-\infty}^{\infty} |k|^{2r^e} |a_k|^2 \Big)^{1/2},$$

$f(x)$ belongs to $S_2^r B^*$ if and only if the series

$$\sum_{-\infty}^{\infty} |k|^{2r^e} |a_k|^2 < \infty \tag{2.3}$$

converge for all $e \subseteq e_n$.

Let

$$S(f; x, 2^N, r) = \sum_{|k|^r \leqslant 2^N} a_k e^{i(k,x)}$$

be a partial sum of the Fourier series of f, taken over the set Λ_m^* (with N replaced by 2^N).

THEOREM 2.1. *The norm* $|f, S_2^r B^*|$ *is equivalent to*

$$\Lambda = \sum_{N=0}^{\infty} 2^{2^N} \| T(f; x, 2^N, r) \|_2^2,$$

where

$$T(f;\ x,\ 2^N,\ r) = S(f;\ x,\ 2^N,\ r) - S(f;\ x,\ 2^{N-1},\ r),\ (N = 1, 2, \ldots),$$
$$T(f;\ x,\ 2^0,\ r) = S(f;\ x,\ 2^0,\ r).$$

PROOF. 1) Let $f \in S_2^r B^*$. Then by (2.3) we have

$$\Lambda = \sum_{N=0}^{\infty} 2^{2N} \sum_{2^{N-1}<|k|^r \leqslant 2^N} |a_k|^2 = \sum_{N=0}^{\infty} 2^{2N} \sum_{e\subseteq e_n} \sum_{2^{N-1}<|k|^{r^e} \leqslant 2^N} \frac{|k|^{2r^e} |a_k|^2}{|k|^{2r^e}}$$

$$\leqslant c \sum_{N=0}^{\infty} \sum_{e\subseteq e_n} \sum_{2^{N-1}<|k|^{r^e} \leqslant 2^N} |k|^{2r^e} |a_k|^2 \leqslant c \sum_{e\subseteq e_n} \sum_{k=-\infty}^{\infty} |k|^{2r^e} |a_k|^2 \leqslant c\, |f, S_2^r B^*|.$$

2) Let $\Lambda < \infty$. Then, again by (2.3), we obtain

$$|f, S_2^r B*| \leqslant c \sum_{e \subseteq e_n} \sum_{-\infty}^{\infty} |k|^{2re} |a_k|^2 = c \sum_{N=0}^{\infty} \sum_{e \subseteq e_n} \sum_{2^{N-1} < |k|^{re} \leqslant 2^N} |k|^{2r^e} |a_k|^2$$

$$\leqslant c \sum_{e \subseteq e_n} \sum_{N=0}^{\infty} 2^{2N} \sum_{2^{N-1} < |k|^{re} \leqslant 2^N} |a_k|^2 = c \sum_{N=0}^{\infty} 2^{2N} \| T(f; x, 2^N, r)\|_2^2 = c\Lambda,$$

and the theorem is established.

REMARK.

$$\Lambda = \sum_{n=0}^{\infty} 2^{2n} \| T(f; x, 2^n, r)\|_2^2 \sim \sum_{n=0}^{\infty} \| T(f; x, 2^n, r)\|_2^2 \sum_{N=0}^{n} 2^{2N}$$

$$= \sum_{N=0}^{\infty} 2^{2N} \sum_{n=N+1}^{\infty} \| T(f; x, 2^n, r)\|_2^2 = \sum_{N=0}^{\infty} 2^{2N} \sum_{|k|^r \geqslant N} |a_k|^2 = \sum_{N=0}^{\infty} 2^{2N} E_{2^N}(f; r)_2^2 .$$

THEOREM 2.2. *If* $f(x) \in S_2^r H^*$, $r < \min_{j \geqslant 2}\{r_j\}$, *then*

$$\|f(x) - S(f; x, N, \rho)\|_2 \leqslant cN^{-1},$$

where $\rho = (\rho_1, \ldots, \rho_n)$, $\rho_1 = r_1$, $r_1 < \rho_j < \min_{j \geqslant 2}\{r_j\}$, $j = 1, \ldots, n$.

PROOF. By (2.2) we obtain

$$\| f - S(f; x, N, \rho)\|_2^2 = \sum_{|k|^\rho \geqslant N} |a_k|^2$$

$$\leqslant \sum_e \sum_{\substack{k_j = -\infty \\ j \in e_n \setminus e}}^{\infty} \sum_{2^{(t, \rho^e)} \geqslant N} \sum_{2^{t_j - 1} < |k_j| \leqslant 2^{t_j}} |a_k|^2 \leqslant c \sum_e \sum_{2^{(t, \rho^e)} \geqslant N} 2^{-2(t, r^e)} .$$

In the last sum the largest terms are those corresponding to a set e containing a 1. Let $e = e_n$; then ($r_1 = \rho_1$)

$$\sum_{2^{(t, \rho)} \geqslant N} 2^{-2(t, r)} \leqslant N^{-2} \sum_0^{\infty} 2^{2 \sum_{j=2}^{n} t_j(\rho_j - r_j)} \leqslant cN^{-2},$$

since $\rho_j - r_j < 0$ for $j = 2, \ldots, n$. Therefore

$$\|f(x) - S(f; x, N, \rho)\|_2 \leqslant cN^{-1},$$

and the theorem is established.

REMARK 1. If the vector r has several coordinates equal to a minimum, $r_1 = \cdots = r_\nu < \min_{j > \nu}\{r_j\}$, we can obtain an estimate of the form (cf. [1], p. 416)

$$\|f(x) - S(f; x, N, \rho)\|_2 \leqslant cN^{-1} [\ln N]^{\frac{\nu - 1}{2}} .$$

REMARK 2. The number of lattice points on the surface $|k|^\rho = N$ is the same as on the surface $|k|^r = N$, and has precisely order $m = O(N^{1/r_1})$, if the vectors r and ρ are chosen as in the theorem. Therefore

$$\|f(x) - S(f;\ x,\ N,\ \rho)\|_2 \leqslant cm^{-r_1}.$$

REMARK 3. It is shown in our paper [1] (Theorem 3.1) that the approximation by partial sums $S_{\Lambda_m^*}(f)$ is best possible for all sums $S_{\Lambda_m}(f)$ (for the class $S_2^rH^*$ as a whole).

REMARK 4. The problem of the approximation of functions of class SW by trigonometric sums $T(x, N, \rho)$ was considered by Babenko [2], [3], Mitjagin [4], and Teljakovskiĭ [5]. Approximation by quasipolynomials ("angular" approximation) containing infinitely many terms was considered by the author [6] and Potapov [7].

LEMMA. *For the trigonometric sums $T(x, N, r)$ we have the inequality*

$$\| D^{s^e} T(x, N, r)\|_2 \leqslant cN^{\frac{s_j}{r_j}} \| T(x, N, r)\|_2, \tag{2.4}$$

where

$$\frac{s_j}{r_j} = \max_{i \in e} \left\{\frac{s_i}{r_i}\right\}.$$

Inequality (2.4) was proved by B. S. Pozdeev, and can be obtained by using Parseval's equation.

THEOREM 2.3. *If $f(x)$ possesses a sequence of trignometric sums $\{\widetilde{T}(x;\ 2^N, \rho)\}$, $\rho = (\rho_1, \ldots, \rho_n)$, $\rho_1 < \min_{j\geqslant 2}\{\rho_j\}$, such that*

$$\| f(x) - \widetilde{T}(x;\ 2^N, \rho)\|_2 \leqslant c2^{-N},$$

then $f(x)$ belongs to the class $S_2^rH^*$, where $r_j = \rho_j$ if all ρ_j are rational $(j = 1, \ldots, n)$; $r_1 = \rho_1$, $r_j < \rho_j$ $(j = 2, \ldots, n)$ if some of the ρ_j are irrational.

PROOF. By the hypotheses of the theorem, $f(x)$ can be represented as a series

$$f(x) = \sum_{n=0}^{\infty} T(x;\ 2^n, \rho), \tag{2.5}$$

where

$$\begin{aligned} T(x;\ 2^n, \rho) &= \widetilde{T}(x;\ 2^n, \rho) - \widetilde{T}(x;\ 2^{n-1}, \rho), \quad (n = 1, 2, \ldots), \\ T(x;\ 2^0, \rho) &= \widetilde{T}(x;\ 2^0, s), \quad \|T(x;\ 2^n,\ \rho)\|_2 \leqslant c2^{-n}. \end{aligned} \tag{2.6}$$

To prove the theorem it is sufficient to estimate the h-differences of $f(x)$. Let e be any subset of e_n. First consider the case of rational ρ_j. Putting $r_j = \rho_j$, we choose integers $s_j > r_j$ such that

$$\frac{s_1}{r_1} = \ldots = \frac{s_n}{r_n}.$$

Then (2.5) yields

$$\Lambda_e = \| \Delta_h^{s^e} f\|_2 \leqslant \sum_{n=0}^{N} \| \Delta_h^{s^e} T(x, 2^n, r)\|_2 + \sum_{n=N+1}^{\infty} \| T(x;\ 2^n, r)\|_2,$$

where N is fixed for the time being. Note that the numbers N and s_j may be different for each set e.

Using the inequality

$$\| \Delta_h^{s^e} \psi \|_2 \leqslant c h^{s^e} \| D^{s^e} \psi \|_2 \qquad (h_i > 0)$$

and (2.6), we obtain

$$\Lambda_e \leqslant c \sum_{n=0}^{N} h^{s^e} \| D^{s^e} T(x; 2^n, r) \|_2 + c 2^{-(N+1)}.$$

Since all the ratios s_j/r_j $(j = 1, \ldots, n)$ are the same, by (2.4) we have

$$\Lambda_e \leqslant c h^{s^e} \sum_{n=0}^{N} 2^{\frac{s_1}{r_1} n} \| T(x; 2^n, r) \|_2 + c 2^{-(N+1)}$$
$$\leqslant c h^{s^e} \sum_{n=0}^{N} 2^{\left(\frac{s_1}{r_1} - 1\right) n} + c 2^{-(N+1)} \leqslant c h^{s^e} 2^{N\left(\frac{s_1}{r_1} - 1\right)} + c 2^{-(N+1)},$$

since $s_1 > r_1$.

For fixed $h_i > 0$, $i \in e$, we choose N so that

$$2^{-(N+1)} < h^{r^e} < 2^{-N}.$$

Then

$$\Lambda_e \leqslant c h^{s^e} h^{-r^e\left(\frac{s_1}{r_1} - 1\right)} + c h^{r^e} \leqslant c \prod_{i \in e} h_i^{r_i\left[\frac{s_i}{r_i} - \frac{s_1}{r_1} + 1\right]} + c h^{r^e} \leqslant c h^{r^e},$$

since $s_i/r_i = s_1/r_1$ for $i = 1, \ldots, n$.

Hence the theorem is established in this case.

Now let ρ_j be irrational. By hypothesis $r_1 = \rho_1 < \rho_j$ $(j = 2, \ldots, n)$, so that $1/r_1 > 1/\rho_j$. Take rational numbers s_j so that $s_j/r_j \geqslant s_1/r_1 \geqslant s_j/\rho_j$ and $s_j > \rho_j \geqslant r_j$ $(j = 1, \ldots, n)$. This is always possible if $r_j \leqslant \rho_j - r_1/s_1$ $(j = 1, \ldots, n)$.

It is now clear that at the expense of taking s_1 sufficiently large we can take the r_j arbitrarily close to the ρ_j.

Now, as above, if we note that s_1/ρ_1 is the largest of the ratios s_j/ρ_j we obtain

$$\Lambda_e \leqslant c h^{s^e} h^{-r^e\left(\frac{s_1}{\rho_1} - 1\right)} + c h^{r^e} = c \prod_{i \in e} h^{r_i\left[\frac{s_i}{r_i} - \frac{s_1}{r_1} + 1\right]} + c h^{r^e} \leqslant c h^{r^e},$$

since $s_i/r_i \geqslant s_1/r_1 = s_1/\rho_1$, which establishes the theorem.

REMARK 1. The number of lattice points on the surface $|k|^\rho = 2^N$, when ρ is chosen as in the theorem, is $m = O(2^{N/r_1})$. Therefore Theorem 2.3 reduces to Theorem 2.2 if we express it in terms of the number of lattice points (terms) in $T(x; 2^N, \rho)$. It is clear from the proof that the conclusion of the theorem remains true when $\rho_1 = \cdots = \rho_n$, and in fact $f \in S_2^r H^*$, where $r_j = \rho_j$. However, this conclusion does not reduce to Theorem 2.2, since the number of lattice points on the surface $|k|^\rho = 2^N$ $(\rho_1 = \cdots = \rho_n)$ is precisely of order $m = O(2^{N/r_1} N^{n-1})$. Hence our requirement $\|f(x) - \widetilde{T}(x; 2^n, \rho)\|_2 \leqslant c 2^{-N}$ represents an excessively rapid decrease of the best approximation in this case.

REMARK 2. Approximation of functions of classes $S_p^r H^*$ and $S_p^r W^*$ $(1 < p < \infty)$ was considered by Nikol'skaja [9].

REMARK 3. It is possible, by a different method, to prove Theorem 2.3 for irrational ρ_j even in the case $r_j = \rho_j$.

In fact, let

$$\|f(x) - \widetilde{T}(x;\ 2^n,\ r)\|_2 \leqslant c2^{-N},$$

where r is an arbitrary vector. Consider

$$\sigma_e = \Big(\sum_{\substack{k_j = 2^{t_j}+1 \\ j \in e}}^{2^{t_j+1}} |a_k|^2 \Big)^{1/2},$$

where the sum is a multiple sum over all k_j, $j \in e$.

Since $2^{t_j} < k_j \leqslant 2^{t_j+1}$, we have

$$2^{t_j r_j} < k_j^{r_j} \leqslant 2^{(t_j+1) r_j} \text{ and } 2^{(t,r)} < k^r \leqslant 2^{(t,r) + \sum_1^n r_j}.$$

Taking N so that $2^N = 2^{(t,r)}$, we obtain

$$2^N = 2^{\sum_1^n t_j r_j} < k^r \leqslant 2^{\sum_{j=1}^n (t_j+1) r_j} = 2^{N + \sum_1^n r_j}.$$

Hence

$$\sigma_e \leqslant \| \widetilde{T}(x; 2^{N + \sum_1^n r_j}, r) - \widetilde{T}(x; 2^N, r) \|_2 \leqslant c2^{-N} \leqslant c2^{-(t,r)}.$$

Applying Nikol'skiĭ's representation theorem [8], we conclude that $f \in S_2^r H^*$. (The function can be represented in the form

$$f(x) = \sum_e \sum_k Q_{k_e}(x),$$

where $Q_{k_e}(x)$ are trigonometric polynomials of degree 2^{k_j} in x_j $(j \in e)$, satisfying $\|Q_{k_e}\|_2 = \sigma_e$).

Therefore the following theorem follows from Theorems 2.2 and 2.3 and the remarks connected with them.

THEOREM 2.4. *A necessary and sufficient condition for f to belong to $S_2^r H^*$, $r_1 < \min_{j \geqslant 2} \{r_j\}$, is that there is a sequence of trigonometric sums $T_{\Lambda_m^*}(x)$ such that*

$$\| f(x) - T_{\Lambda_m^*}(x) \|_2 \leqslant c m^{-r_1},$$

where m is the number of terms in $T_{\Lambda_m^*}(x)$.

BIBLIOGRAPHY

1. Ja. S. Bugrov, *Approximation of a class of functions with dominant mixed derivative*, Mat. Sb. **64** (**106**) (1964), 410–418. (Russian) MR **29** #2587.

2. K. I. Babenko, *Approximation of periodic functions of many variables by trigonometric polynomials*, Dokl. Akad. Nauk SSSR 132 (1960), 247–250 = Soviet Math. Dokl. **1** (1960), 513–516. MR **22** #12341.

3. ———, *Approximation by trigonometric polynomials in a certain class of periodic functions of several variables*, Dokl. Akad. Nauk SSSR **132** (1960), 982–985 = Soviet Math. Dokl. **1** (1960), 672–675. MR **22** #12342.

4. B. S. Mitjagin, *Approximation of functions in the spaces L^p and C on the torus*, Mat. Sb. **58** (**100**) (1962), 397–414. (Russian) MR **27** #2772.

5. S. A. Teljakovskiĭ, *Some estimates for trigonometric series with quasi-convex coefficients*, Mat. Sb. **63** (**105**) (1964), 426–444; English transl., Amer. Math. Soc. Transl. (2) **86** (1970), 109–131. MR **29** #410; **41** #1462.

6. Ja. S. Burgov, *Approximation of functions of several variables by trigonometric polynomials*, Trudy Naučn. Ob″ed. Prepodavat. Fiz.- Mat. Fak. Ped. Inst. Dal′nego Vostokl., tom 1, Khabarovsk, 1962, pp. 28–49. (Russian)

7. M. K. Potapov, *On "angular" approximation*, Proc. Conf. on Constructive Function Theory (Budapest, 1969), Akad. Kiadó, Budapest, 1972, pp. 371–399.

8. S. M. Nikol′skiĭ, *Functions with dominant mixed derivative satisfying a multiple Hölder condition*, Sibirsk. Mat. Ž. **4** (1963), 1342–1364; English transl., Amer. Math. Soc. Transl. (2) **102** (1973), 27–51. MR **28** #3322.

9. N. S. Nikol′skaja, *Approximation of differentiable functions of several variables by Fourier sums in the L_p metric*, Dokl. Akad. Nauk SSSR **208** (1973), 1282–1285 = Soviet Math. Dokl. **14** (1973), 262–266.

Translated by R. P. BOAS

Trudy Mat. Inst. Steklov.
131 (1974)

Proc. Steklov Inst. Math.
131 (1974)

SOBOLEV'S INTEGRAL REPRESENTATION AND TAYLOR'S FORMULA

UDC 517.518.22

V. I. BURENKOV

ABSTRACT. A connection is established between a well-known integral representation of a function, obtained by S. L. Sobolev, and Taylor's formula. There is given a new derivation of Sobolev's representation, based upon an immediate utilization of Taylor's formula. It is proved that if Ω is a region which is starlike with respect to a sphere K with center at the origin, and if the function $f(x)$ has generalized derivatives in Ω through order r inclusive, then for almost all $x \in \Omega$

$$f(x) = \frac{1}{\varkappa} \sum_{|k|<r} \frac{1}{k!} \int_K D^k f(y)(x-y)^k \omega(y)\, dy + \frac{r}{\varkappa} \sum_{|k|=r} \int_{V_x} D^k f(y)(x-y)^k \frac{w(x, y)}{|x-y|^n}\, dy. \quad (1)$$

Here $\kappa = \int_K \omega(y) dy$, and $\omega(y)$ is a function continuous in K; $V_x = K$ if $x \in \overline{K}$, and otherwise V_x is the region consisting of the sphere K and that part of the cone with vertex at the point x, tangent to the sphere K, which lies between x and K;

$$k = (k_1, \ldots, k_n) \quad |k| = k_1 + \ldots + k_n, \quad k! = k_1! \ldots k_n!,$$
$$(x-y)^k = (x_1 - y_1)^{k_1} \ldots (x_n - y_n)^{k_n};$$
$$w(x, y) = \int_{d_1}^{d_2} \omega\left(x + \rho \frac{y-x}{|y-x|}\right) \rho^{n-1}\, d\rho.$$

d_2 is the length of the segment of the ray which goes from the point x to y and lies in V_x, and $d_1 = \max\{|x-y|, \widetilde{d}_1\}$, where $\widetilde{d}_1$ is the length of the segment of the same ray lying in $V_x - K$.

The second summand in (1) differs only in its written form from the corresponding summand in Sobolev's formula, but the first summand (a polynomial) is clearly expressed (in distinction to Sobolev's formula).

Bibliography: 18 items.

In obtaining inequalities among the norms of the derivatives of a function $f(x)$ of several variables, the representation of this function by integrals containing the function $f(x)$ and its derivatives plays an important role. Such an integral representation was obtained by S. L. Sobolev (see [1], formula (9) or [2], formula (5.7) for functions having ordinary derivatives through order r inclusive; [3], formula (15), or [4], formula (7.13) for functions having generalized derivatives through order r inclusive). A final variant of the integral representation is cited in [5] (formula (7.12)).

Let K be the open sphere of radius h with center at the origin, and let the region Ω be starlike with respect to K. In what follows we shall denote by $\omega(x)$ any infinitely differentiable function which is equal to zero outside of K. Sobolev's integral representation has the following form: if a function $f(x)$ has in the region Ω all generalized derivatives

AMS (MOS) subject classifications (1970). Primary 26A57, 31B10; Secondary 26A93.

$$D^k f = \frac{\partial^{k_1+\dots+k_n} f}{\partial x_1^{k_1} \dots \partial x_n^{k_n}}$$

of order r ($|k| = k_1 + \dots + k_n = r$), then for almost all $x \in \Omega$

$$f(x) = \sum_{|k|<r} x^k \int_K \zeta_k(y) f(y)\, dy + \sum_{|k|=r} \int_{V_x} \frac{w_k(x, y)}{|x-y|^{n-r}} D^k f(y)\, dy. \tag{1}$$

Here $x^k = x_1^{k_1} \dots x_n^{k_n}$; if $x \in \overline{K}$, then $V_x = K$; if $x \in \Omega - \overline{K}$, then V_x is the region formed by K together with that part of the cone with vertex at the point x and tangent to K which lies between x and K. In what follows the region V_x for any $x \in \Omega$ will be called the "cone" for short. Further

$$\zeta_k(y) = \frac{1}{\varkappa} \sum_{\substack{k \leqslant \beta \\ |\beta| \leqslant r-1}} c_{\beta, k} D^\beta [\omega(y) y^{\beta-k}] \tag{2}$$

($k \leqslant \beta$ means that $k_i \leqslant \beta_i$, $i = 1, \dots, n$), where the $c_{\beta,k}$ are constants (binomial coefficients and their combinations), and

$$\varkappa = \int_K \omega(y)\, dy. \tag{3}$$

Finally, if we follow the calculations carried out in [5], it is not difficult to derive that

$$w_k(x, y) = \frac{(-1)^r r!}{\varkappa \cdot k!} \frac{(y-x)^k}{|y-x|^r} \int_{|y-x|}^{\infty} \omega\left(x + \rho \frac{y-x}{|y-x|}\right) \rho^{n-1}\, d\rho \tag{4}$$

is a bounded function which is continuous for $x \neq y$ ($k! = k_1! \dots k_n!$; $(y-x)^k = (y_1 - x_1)^{k_1} \dots (y_n - x_n)^{k_n}$).

In the books of V. I. Smirnov ([6], Chapter IV, formula (188)) and L. A. Ljusternik and V. I. Sobolev ([7], pp. 115 and 116), another form of writing the integral representation (1) was derived:

$$f(x) = \sum_{|k|<r} b_k(x) \int_K \omega(y) D^k f(y)\, dy + \sum_{|k|=r} \int_{V_x} \frac{B_k(x, y)}{|x-y|^{n-r}} D^k f(y)\, dy, \tag{5}$$

where $b_k(x)$ and $B_k(x, y)$ are some functions expressed in terms of $\omega(x)$. For $r = 1$, formula (5) is obtained by integration by parts in the integral $\int_K \omega(y) f(y) dy$, and for $r > 1$ by means of iterated applications of formula (5) for $r = 1$.(1)

In the paper of V. P. Il'in ([8], §1, formula (II')) and in the book of L. V. Kantorovič and G. P. Akilov ([9], Chapter X, §3, formula (1)), the following result was obtained in the case that Ω is a convex region:

$$f(x) = \frac{1}{\operatorname{meas} \Omega} \int_\Omega f(y)\, dy - \frac{1}{\operatorname{meas} \Omega} \sum_{i=1}^{n} \int_\Omega \frac{y_i - x_i}{|y-x|^n} \frac{d^n - |x-y|^n}{n} \frac{\partial f}{\partial x_i}(y)\, dy, \tag{6}$$

(1) We note, by the way, that from (5) it follows that the $b_k(x)$ are essentially polynomials of degree $|k|$. In order to verify this, it suffices to let $f(x) = x^l$ in (5) for successive values $0 \leqslant |l| \leqslant r - 1$.

where $d = d(x, y)$ is the length of the segment of the ray which goes from the point x through the point y to the boundary of the region Ω.

In what follows we use the method of integral representation of functions developed by V. P. Il'in in [10] and [11]. In Il'in's integral representations, the function $f(x)$ is expressed by integrals containing only $f(x)$ and derivatives $\partial^{r_i} f/\partial x_i^{r}$, $i = 1, \ldots, n$ (or derivatives of a fixed order and differences of these derivatives). Similar integral representations are utilized also in the works of Calderón [12], Smith [13], Besov [14], [15], and Besov and Il'in [16].

We further remark that an integral representation for $\Omega = E_n$ was utilized in the paper of Aronszajn, Mulla, and Szeptycki [17]:

$$f(x) = \sum_{l=0}^{r} C_r^l \sum_{|k|=l} \int_{E_n} D_y^k G_{2r}(x-y) \cdot D^k f(y)\, dy, \tag{7}$$

where $G_{2r}(x)$ is the Fourier transform of the function $(2\pi)^{-n/2}(1 + |\xi|^2)^{-r}$ ($G_{2r}(x - y)$ is the fundamental solution of the equation $(1 - \Delta)^r = 0$, where Δ is the Laplacian). Equation (7) is a corollary of the identity

$$1 = \frac{1}{(1+|\xi|^2)^r} \sum_{l=0}^{r} C_r^l \sum_{|k|=l} (-1)^l (i\xi)^k (i\xi)^k.$$

In the book of Nagumo ([18], Russian p. 82), for functions which are finite in Ω the representation reduces to the form

$$f(x) = -\frac{r}{\sigma_n} \sum_{|k|=r} \frac{1}{k!} \int_{\Omega} \frac{(x-y)^k}{|x-y|^n} D^k f(y)\, dy, \tag{8}$$

where σ_n is the surface area of the n-dimensional unit sphere.

In the present work we obtain an integral representation of the type of representation (1) as an immediate consequence of Taylor's formula

$$f(x) = \sum_{|k|<r} \frac{1}{k!} D^k f(y)(x-y)^k + r \sum_{|k|=r} \frac{1}{k!} \int_0^1 (1-t)^{r-1} D^k (y + t(x-y)) \times (x-y)^k\, dt. \tag{9}$$

For this we integrate equation (9) with respect to $y \in K$:

$$f(x) = \frac{1}{\operatorname{meas} K} \Big\{ \sum_{|k|<r} \frac{1}{k!} \int_K D^k f(y)(x-y)^k\, dy + r \sum_{|k|=r} \frac{1}{k!} \int_K \int_0^1 (1-t)^{r-1} D^k f(y + t(x-y))(x-y)^k\, dt\, dy \Big\}. \tag{10}$$

The first summand, as in (1), represents a polynomial, which is exhibited clearly, and in the second summand the values of the derivatives $D^k f(z)$ for $z \in V_x$ are contained in the integral. In what follows we transform the second summand into the form of a potential integral.

Let $x \in \Omega$ and $y \in V_x$. We denote by $d_2 = d_2(x, y)$ the length contained in V_x of the segment of the ray which goes from the point x through the point y, and set $d_1 = d_1(x, y) = \max\{|x - y|, \tilde{d}_1(x, y)\}$, where $\tilde{d}_1(x, y)$ is the length of that same ray which is contained in $V_x - K$. It is not difficult to verify that d_2 is the larger root of the equation

$$d^2 + 2d\left(x, \frac{y-x}{|y-x|}\right) + |x|^2 - h^2 = 0,$$

and $d_1 = \max\{|x-y|, d_1^*\}$, where d_1^* is the smaller root of this equation (the parentheses denote the scalar product).

THEOREM. *Let Ω be a region which is starlike with respect to the sphere K of radius h and center at the origin, and let the function $f(x)$ have generalized derivatives $D^k f$, $|k| \leqslant r$, in Ω. Then for almost all x the following equation (analogous to Taylor's formula) obtains*:

$$f(x) = \frac{1}{\operatorname{meas} K}\left\{ \sum_{|k|<r} \frac{1}{k!} \int_K D^k f(y)(x-y)^k \, dy \right.$$
$$\left. + r \sum_{|k|=r} \frac{1}{k!} \int_{V_x} D^k f(y)(x-y)^k \frac{w(x, y)}{|x-y|^n} \, dy \right\}, \qquad (11)$$

where the function

$$w(x, y) = \frac{d_2^n - d_1^n}{n} \qquad (12)$$

satisfies the inequalities

$$0 \leqslant w(x, y) \leqslant 2h(|x| + h)^{n-1}, \qquad x \in \Omega,\ y \in V_x, \qquad (13)$$

$$0 \leqslant w(x, y) \leqslant \frac{(2h)^n}{n}, \qquad x \in K,\ y \in K. \qquad (14)$$

REMARK. Formula (11) remains valid also in the case that K is not a sphere, but any convex region contained in Ω (in this case $2h$ denotes the diameter of K). For $r = 1$ and $\Omega = K$, (11) coincides with (6).

PROOF. At first let the function $f(x)$ have continuous derivatives in Ω through order r inclusive.

We deduce a formula more general than (11). Let $\omega(x)$ be any measurable function given on E_n which is equal to zero outside of K (for example, $\omega(x) = 1$ in K, $\omega(x) = 0$ outside of K). We further stipulate that the functions $D^k f$, $|k| \leqslant r$, are zero outside of Ω.

We multiply equation (9) by $\omega(y)$ and integrate on E_n (essentially on K). Then

$$f(x) = \frac{1}{\varkappa}\left\{ \sum_{|k|<r} \frac{1}{k!} \int_K D^k f(y)(x-y)^k \omega(y) \, dy \right.$$
$$\left. + r \sum_{|k|=r} \frac{1}{k!} \int_{E_n} \int_0^1 (1-t)^{r-1} D^k f(y + t(x-y))(x-y)^k \omega(y) \, dt \, dy \right\}, \qquad (15)$$

where

$$\varkappa = \int_{E_n} \omega(y) \, dy = \int_K \omega(y) \, dy.$$

We transform the relevant integrals in (15):

$$J_k = \int_0^1 \int_{E_n} (1-t)^{r-1} D^k f(y + t(x-y))(x-y)^k \omega(y) \, dy \, dt,$$

for which we make the change of variables $y + t(x-y) = z$. Taking into account that

$$(x-y)^k = \frac{(x-z)^k}{(1-t)^r},\ dy = \frac{dz}{(1-t)^n},$$

we obtain

$$J_k = \int_{E_n} D^k f(z)(x-z)^k dz \int_0^1 \omega\left(\frac{z-tx}{1-t}\right)\frac{dt}{(1-t)^{n+1}}.$$

Replacing $|x-z|/(1-z)$ by ρ, we have

$$J_k = \int_{E_n} D^k f(z)\frac{(x-z)^k}{|x-z|^n}dz \int_{|x-z|}^{\infty} \omega\left(x+\rho\frac{z-x}{|z-x|}\right)\rho^{n-1}d\rho.$$

We note that for fixed $x \in \Omega$ the function

$$w(x, z) = \int_{|x-z|}^{\infty} \omega\left(x+\rho\frac{z-x}{|z-x|}\right)\rho^{n-1}d\rho = \int_{d_1(x,z)}^{d_2(x,z)} \omega\left(x+\rho\frac{z-x}{|z-x|}\right)\rho^{n-1}d\rho \qquad (16)$$

vanishes outside of the cone V_x, and therefore

$$J_k = \int_{V_x} D^k f(z)\frac{(x-z)^k}{|x-z|^n} w(x, z)\,dz.$$

If $\omega(y) = 1$ in K and $\omega(y) = 0$ outside of K, then

$$w(x, y) = \int_{d_1}^{d_2} \rho^{n-1}d\rho = \frac{d_2^n - d_1^n}{n},$$

whence follows the desired equality (11). Inequality (13) is obtained from (12), since

$$d_2^n - d_1^n \leqslant (|x|+h)^n - (|x|-h)^n \leqslant n(|x|+h)^{n-1}2h.$$

Inequality (14) is obvious.

The passage from functions which are r times continuously differentiable to functions having generalized derivatives is accomplished with the help of mean functions (see [5], proof of formula (7.12)).

Through the proof of the theorem we obtained the following integral representation:

$$f(x) = \frac{1}{\varkappa}\left\{\sum_{|k|<r}\frac{1}{k!}\int_K D^k f(y)(x-y)^k\omega(y)\,dy \right.$$
$$\left. + r\sum_{|k|=r}\frac{1}{k!}\int_{V_x} D^k f(y)(x-y)^k\frac{w(x, y)}{|x-y|^n}dy\right\}; \qquad (17)$$

here the function $w(x, y)$ is defined by (16) (in (16) we have the same integral as in (4)), where $\omega(x)$ is any measurable function which is bounded on the sphere K.

If, finally, the function $\omega(x)$ is $r-1$ times continuously differentiable in K and if it, together with its derivatives through order $r-1$ inclusive, is equal to zero on the boundary, then

$$f(x)=\frac{1}{\varkappa}\Big\{\sum_{|k|<r}\frac{(-1)^{|k|}}{k!}\int_K f(y)\,D^k\,[(x-y)^k\,\omega(y)]\,dy$$
$$+r\sum_{|k|=r}\frac{1}{k!}\int_{V_x}D^kf(y)\,(x-y)^k\,\frac{w(x,y)}{|x-y|^n}\,dy\Big\},\qquad (1')$$

which represents the initial formula (1) with a clearly expressed polynomial.

BIBLIOGRAPHY

1. S. L. Sobolev, *Sur quelques évaluations concernant les familles de fonctions ayant des dérivées à carré intégrable*, C. R. (Dokl.) Acad. Sci. URSS **10** (1936), 279–282.

2. ———, *On a boundary value problem for polyharmonic equations*, Mat. Sb. **2** (**44**) (1937), 467–499; English transl., Amer. Math. Soc. Transl. (2) **33** (1963), 1–40.

3. ———, *Sur une théorème de l'analyse fonctionnelle*, C. R. (Dokl.) Acad. Sci. URSS **20** (1938), 5–9.

4. ———, *On a theorem of functional analysis*, Mat. Sb. **4** (**46**) (1938), 471–497; English transl., Amer. Math. Soc. Transl. (2) **34** (1963), 39–68.

5. ———, *Applications of functional analysis in mathematical physics*, Izdat. Leningrad. Gos. Univ., Leningrad, 1950; reprint, Izdat. Sibirsk. Otd. Akad. Nauk SSSR, Novosibirsk, 1962; English transl., Transl. Math. Monographs, vol. 7, Amer. Math. Soc., Providence, R. I., 1963. MR **14**, 565; **29** #2624.

6. V. I. Smirnov, *A course in higher mathematics*. Vol. 5, rev. ed., Fizmatgiz, Moscow, 1959; English transl., Pergamon Press, Oxford; Addison-Wesley, Reading, Mass., 1964. MR **29** #5964.

7. L. A. Ljusternik and V. I. Sobolev, *Elements of functional analysis*, 2nd rev. ed., "Nauka", Moscow, 1965. MR **35** #698.

8. V. P. Il'in, *Some inequalities in function spaces and their application to the study of the convergence of variational processes*, Trudy Mat. Inst. Steklov. **53** (1953), 64–127; English transl., Amer. Math. Soc. Transl. (2) **81** (1969), 1–66. MR **22** #9738.

9. L. V. Kantorovič and G. P. Akilov, *Functional analysis in normed spaces*, Fizmatgiz, Moscow, 1959; English transl., Internat. Ser. Monographs Pure Appl. Math., vol. 46, Macmillan, New York, 1964. MR **22** #9837; **35** #4699.

10. V. P. Il'in, *The properties of some classes of differentiable functions of several variables defined in an n-dimensional region*, Trudy Mat. Inst. Steklov **66** (1962), 227–363; English transl., Amer. Math. Soc. Transl. (2) **81** (1969), 91–256. MR **27** #3750.

11. V. P. Il'in and V. A. Solonnikov, *On some properties of differentiable functions of several variables*, Trudy Mat. Inst. Steklov. **66** (1962), 205–226; English transl., Amer. Math. Soc. Transl. (2) **81** (1969), 67–90. MR **27** #2768.

12. A. P. Calderón, *Lebesgue spaces of differentiable functions*, Proc. Sympos. Pure Math., vol. 4, Amer. Math. Soc., Providence, R. I., 1961, pp. 33–49. MR **26** #603.

13. K. T. Smith, *Inequalities for formally positive integro-differential forms*, Bull. Amer. Math. Soc. **67** (1961), 368–370. MR **26** #462.

14. O. V. Besov, *The continuation of functions in L_p^l and W_p^l*, Trudy Mat. Inst. Steklov. **89** (1967), 5–17 = Proc. Steklov Inst. Math. **89** (1967), 1–15. MR **35** #5920.

15. ———, *On coercivity in nonisotropic Sobolev spaces*, Mat. Sb. **73** (**115**) (1967), 585–599 = Math. USSR Sb. **2** (1967), 521–534. MR **36** #4120.

16. O. V. Besov and V. P. Il'in, *Natural extension of the class of regions in imbedding theorems*, Mat. Sb. **75** (**117**) (1968), 483–495 = Math. USSR Sb. **4** (1968), 445–456. MR **37** #1987.

17. N. Aronszajn, F. Mulla and P. Szeptycki, *On spaces of potentials connected with L^p classes*, Ann. Inst. Fourier (Grenoble) **13** (1963), 211–306. MR **31** #5076.

18. M. Nagumo, *Lectures on the contemporary theory of partial differential equations*, Kyōritsu Shuppan, Tokyo, 1957. (Japanese); Russian transl., "Mir", Moscow, 1967. MR **38** #6205.

Translated by C. V. and G. A. COMISKY

Trudy Mat. Inst. Steklov.
131 (1974)

Proc. Steklov Inst. Math.
131 (1974)

ON THE DENSITY OF INFINITELY DIFFERENTIABLE FUNCTIONS IN SOBOLEV SPACES FOR AN ARBITRARY OPEN SET

UDC 517.518.22

V. I. BURENKOV

ABSTRACT. The following theorem is proved. Suppose Ω is an open set and $f \in W_p^r(\Omega)$, $1 \leqslant p < \infty$, $r = 1, 2, \ldots$, i.e. f has generalized derivatives $D^k f$ in Ω up to order r inclusively and

$$\|f\|_{W_p^r(\Omega)} = \sum_{|k| \leqslant r} \|D^k f\|_{L_p(\Omega)} < \infty.$$

Then there exists a sequence $\varphi_s(x) \in C^\infty(\Omega)$ (the φ_s depend linearly on f) such that

$$\|f - \varphi_s\|_{W_p^r(\Omega)} \to 0, \quad s \to \infty,$$

and, for $|k| > r$,

$$\|D^k \varphi_s(x)\, \rho(x)^{|k|-r}\|_{L_p(\Omega)} \leqslant c\|f\|_{W_p^r(\Omega)},$$

where c depends only on k while $\rho(x)$ is the distance from $x \in \Omega$ to the boundary $\Gamma(\Omega)$.

Bibliography: 7 items.

§1. Introduction

Let E_n be an n-dimensional space of points $x = (x_1, \ldots, x_n)$ and let Ω be an open set in E_n. A function $f(x)$ belongs to the Sobolev space $W_p^r(\Omega)$ (see [1]), $r = 1, 2, \ldots$, $1 \leqslant p < \infty$, if the generalized partial derivatives

$$D^k f = \frac{\partial^{k_1 + \ldots + k_n} f}{\partial x_1^{k_1} \ldots \partial x_n^{k_n}}, \qquad |k| = k_1 + \ldots + k_n \leqslant r,$$

exist in Ω and the norm

$$\|f\|_{W_p^r(\Omega)} = \sum_{|k| \leqslant r} \|D^k f\|_{L_p(\Omega)} \tag{1}$$

is finite.

In the paper of Deny and Lions ([2], Theorem 2.3; see also the article of Meyers and Serrin [3]) it is proved that the set $C^\infty(\Omega)$ of functions infinitely differentiable in Ω is dense in $W_p^r(\Omega)$ for any open set Ω. Thus, for any open set Ω the spaces obtained by taking the closure of $C^\infty(\Omega)$ with respect to the norm (1) coincide with the spaces $W_p^r(\Omega)$. On the other hand, one can give examples showing that it is not true that the set $C^\infty(\overline{\Omega})$ of functions infinitely differentiable in the closure of Ω is dense in $W_p^r(\Omega)$ for any open set Ω. This result is valid under certain restrictions on Ω (see the papers of Gagliardo [4], Kazarjan [5] and Il'in [6]).

AMS (MOS) subject classifications (1970). Primary 46E35.

In [2] and [3], for any open set Ω and for any function $f \in W_p^r(\Omega)$ there is constructed a sequence $\varphi_s(x) \in C^\infty(\Omega)$ such that $\varphi_s \longrightarrow f$ as $s \longrightarrow \infty$ in the norm of the space $W_p^r(\Omega)$. The construction of the functions φ_s is realized in such a way that their dependence on f is nonlinear. We supplement this result by showing that it is possible to construct a sequence $\varphi_s \in C^\infty(\Omega)$, $\varphi_s \longrightarrow f$ as $s \longrightarrow \infty$, such that (i) the functions φ_s depend linearly on f and, (ii) for $|k| > r$ the derivatives $D^k\varphi_s$ multiplied by a certain power of the distance to the boundary belong to $L_p(\Omega)$. More precisely, we will prove the following theorem.

THEOREM. *Let Ω be an open set in E_n, $1 \leqslant p < \infty$, $r = 1, 2, \ldots$ and $f \in W_p^r(\Omega)$. Then there exists a sequence $\varphi_s(x) \in C^\infty(\Omega)$ (the $\varphi_s(x)$ depend linearly on f) such that*

$$\| f - \varphi_s \|_{W_p^r(\Omega)} \to 0, \qquad s \to \infty,$$

and, for $|k| > r$,

$$\| D^k \varphi_s(x) \rho(x)^{|k|-r} \|_{L_p(\Omega)} \leqslant c_{k,s} \| f \|_{W_p^r(\Omega)}, \tag{2}$$

where $\rho(x)$ is the distance from a point $x \in \Omega$ to the boundary $\Gamma(\Omega)$ and $c_{k,s}$ depends only on k and s.

In order to prove the above theorem, we construct for any function $f \in W_p^r(\Omega)$ a special weighted moving average $f_\delta(x)$ and prove that

$$\| f_\delta - f \|_{W_p^r(\Omega)} \to 0 \quad \text{as} \quad \delta \to 0$$

and that an inequality of type (2) is satisfied for $f_\delta(x)$.

The construction of the weighted moving averages $f_\delta(x)$ is effected in the following way. We first prove (see §3) a lemma on a partition of unity (resembling a partition of unity considered by Whitney [7]). Let Ω_δ denote the set of points $x \in \Omega$ that are at a distance of not less than δ from the boundary $\Gamma(\Omega)$, and let

$$G_{-1} = \varnothing, \qquad G_0 = \Omega_{1/2}, \qquad G_m = \Omega_{1/2^{m+1}} - \Omega_{1/2^m}, \qquad m = 1, 2, \ldots$$

LEMMA 1. *Let Ω be an open set in E_n. There exists a sequence of nonnegative functions $\psi_m(x) \in C^\infty(E_n)$, $m = 0, 1, 2, \ldots$, such that*

1) $\displaystyle\sum_{m=0}^{\infty} \psi_m(x) = \begin{cases} 1, & x \in \Omega, \\ 0, & x \notin \Omega; \end{cases}$

2) $\Omega = \Sigma_0^\infty \operatorname{supp}\psi_m$, *the multiplicity of the covering of Ω by the sets* supp ψ_m *not exceeding two*;

3) $\operatorname{supp} \psi_m \subset G_{m-1} + G_m + G_{m+1}$, $m = 0, 1, 2, \ldots$;

4) *for any vector $k = (k_1, \ldots, k_n)$*

$$| D^k \psi_m(x) | \leqslant c_k 2^{m|k|},$$

where c_k depends only on k.

Let $f \in W_p^r(\Omega)$. We put

$$f_\delta(x) = \sum_{m=0}^{\infty} \psi_m(x) \frac{1}{\delta_m^n} \int_\Omega \omega\left(\frac{x-y}{\delta_m}\right) f(y)\, dy,$$

where $\delta_m = \delta/2^m$ and $\omega(z)$ is a function satisfying the following conditions: $\omega(z) \in C^\infty(E_n)$, $\omega(z) = 0$ for $|z| \geqslant 1$,

$$\int_{|z|\leqslant 1} \omega(z)\, dz = 1, \qquad \int_{|z|\leqslant 1} \omega(z) z^k\, dz = 0, \qquad 0 < |k| \leqslant r-1$$

(here $z^k = z_1^{k_1} \dots z_n^{k_n}$).

We will present some auxiliary assertions in §2, a proof of Lemma 1 in §3, some properties of the weighted moving averages $f_\delta(x)$ in §4, and a proof of the theorem in §5.

§2. Some properties of the sets A_δ and A^δ

Let A be a set in E_n (or, more generally, in some metric space), and let $\delta > 0$. We will use the notation

$$A_\delta = \{x \in A,\ \rho(x, \Gamma(A)) \geqslant \delta\}. \tag{1}$$

Along with the sets A_δ, we will also consider the sets

$$A^\delta = \{x \in A \ \text{ or } \ x \notin A, \text{ but } \rho(x, \Gamma(A)) < \delta\}. \tag{2}$$

The sets A^δ can be equivalently given as follows:

$$A^\delta = \bigcup_{x\in A} K_\delta^x, \tag{3}$$

where K_δ^x is the open ball of radius δ with center at the point x.

In fact, if $x \in \bigcup_{y\in A} K_\delta^y$, then either $x \in A$ or $x \notin A$ but there exists a point $y \in A$ for which $x \in K_\delta^y$, i.e. $\rho(x, \Gamma(A)) = \rho(x, A) \leqslant \rho(x, y) < \delta$. Conversely, if $x \in A$, then $x \in \bigcup_{y\in A} K_\delta^y$, while if $x \notin A$ but $\rho(x, \Gamma(A)) < \delta$, there exists a point $y \in \Gamma(A)$ for which $\rho \equiv \rho(x, y) = \rho(x, \Gamma(A))$ and hence a point $z \in K_{\delta-\rho}^y \cap A$ for which $x \in K_\delta^z$, i.e. $x \in \bigcup_{z\in A} K_\delta^z$.

We note some properties of the sets A_δ and A^δ:

$$\lim_{\delta\to 0} A_\delta = \underline{A}, \qquad \lim_{\delta\to 0} A^\delta = \overline{A} \tag{4}$$

($\overline{A}$ is the closure of A and $\underline{A}$ is the set of all interior points of A),

$$A_\delta = C((CA)^\delta), \qquad A^\delta = C((CA)_\delta) \tag{5}$$

(CX denotes the complement of a set X),

$$(A_{\delta_1})_{\delta_2} = A_{\delta_1+\delta_2}, \qquad (A^{\delta_1})^{\delta_2} = A^{\delta_1+\delta_2} \quad (\delta_1, \delta_2 > 0), \tag{6}$$

$$(A_{\delta_1})^{\delta_2} \subset A_{\delta_1-\delta_2}, \qquad (A^{\delta_1})_{\delta_2} \supset A^{\delta_1-\delta_2} \quad (\delta_1 > \delta_2 > 0), \tag{7}$$

$$(A_\delta)^\delta \subset A \subset (A^\delta)_\delta, \tag{8}$$

$$(A \cup B)_\delta \supset A_\delta \cup B_\delta, \quad (A \cup B)^\delta = A^\delta \cup B^\delta, \tag{9}$$

$$(A \cap B)_\delta = A_\delta \cap B_\delta, \quad (A \cap B)^\delta \subset A^\delta \cap B^\delta, \tag{10}$$

$$(A - B)_\delta = A_\delta - B^\delta, \quad (A - B)^\delta \subset A^\delta - B_\delta, \tag{11}$$

and if $\overline{B} \subset \underline{A}$, then

$$(A - B)^\delta = A^\delta - B_\delta. \tag{12}$$

Properties (4) are obvious. Properties (5) can be rewritten as follows:

$$CA_\delta = (CA)^\delta, \quad CA^\delta = (CA)_\delta. \tag{5'}$$

We prove, for example, the second of them: $x \in CA^\delta$ means that $x \notin A$ and $\rho(x, \Gamma(A)) \geqslant \delta$ or, in other notation, $x \in CA$ and $\rho(x, \Gamma(CA)) \geqslant \delta$, i.e. $x \in (CA)_\delta$.

We now prove the second of equalities (6) by making use of formula (3). We have $(A^{\delta_1})^{\delta_2} = \bigcup_{x \in A^{\delta_1}} K^x_{\delta_2}$. For any $x \in A^{\delta_1}$, there exists a $y \in A$ for which $x \in K^y_{\delta_1}$. But then $K^x_{\delta_2} \subset K^y_{\delta_1+\delta_2}$, and hence $(A^{\delta_1})^{\delta_2} \subset A^{\delta_1+\delta_2}$. Conversely, for any $y \in A^{\delta_1+\delta_2}$ there exists an $x \in A$ for which $y \in K^x_{\delta_1+\delta_2}$. Suppose $y \in K^x_{\delta_1}$. Then $y \in A^{\delta_1} \subset (A^{\delta_1})^{\delta_2}$. Suppose $y \notin K^x_{\delta_1}$ and consider the point $z \in [x, y]$ such that

$$\rho(x, z) = \delta_1 - \frac{\delta_1 + \delta_2 - \rho(x, y)}{2}.$$

Then $z \in K^x_{\delta_1} \subset A^{\delta_1}$ and

$$\rho(y, z) = \rho(x, y) - \rho(x, z) = \delta_2 - \frac{\delta_1 + \delta_2 - \rho(x, y)}{2} < \delta_2,$$

i.e. $y \in K^z_{\delta_2}$. Thus $y \in (A^{\delta_1})^{\delta_2}$. The first of equalities (6) follows from the second and properties (5) and (5'):

$$(A_{\delta_1})_{\delta_2} = C((CA_{\delta_1})^{\delta_2}) = C(((CA)^{\delta_1})^{\delta_2}) = C((CA)^{\delta_1+\delta_2}) = A_{\delta_1+\delta_2}. \tag{13}$$

The first of properties (7) is proved as follows. Suppose $x \in (A_{\delta_1})^{\delta_2}$. Then either $x \in A_{\delta_1} \subset A_{\delta_1-\delta_2}$, or $x \notin A_{\delta_1}$ but $\rho(x, \Gamma(A_{\delta_1})) < \delta_2$. In the second case we choose a point $y \in \Gamma(A_{\delta_1})$ such that $\rho(x, y) = \rho(x, \Gamma(A_{\delta_1}))$. If $x \notin A$, then the segment $[y, x]$ contains a point $z \in \Gamma(A)$, and

$$\begin{aligned}\rho(x, \Gamma(A)) = \rho(x, z) &\geqslant \rho(y, z) - \rho(x, y) \\ &\geqslant \rho(y, \Gamma(A)) - \rho(x, \Gamma(A_{\delta_1})) > \delta_1 - \delta_2.\end{aligned}$$

which is a contradiction. Hence $x \in A$. We further choose a point $v \in \Gamma(A)$ such that $\rho(x, v) = \rho(x, \Gamma(A))$. Then

$$\begin{aligned}\rho(x, \Gamma(A)) = \rho(x, v) &\geqslant \rho(y, v) - \rho(x, y) \\ &\geqslant \rho(y, \Gamma(A)) - \rho(x, \Gamma(A_{\delta_1})) > \delta_1 - \delta_2,\end{aligned}$$

i.e. $x \in A_{\delta_1-\delta_2}$. The second of properties (7) is proved in the same way as the first of equalities (6) (see (13)), starting from the first of properties (7) and properties (5) and (5').

The inclusions (8) are obtained from (7) by passing to the limit; for example,

$$(A_\delta)^\delta = (\bar{A}_\delta)^\delta = \lim_{\delta_1\to\delta} (A_\delta)^{\delta_1} = \bigcup_{\delta_1>\delta} (A_\delta)^{\delta_1} \subset \bigcup_{\delta_1>\delta} A_{\delta_1-\delta} \subset A.$$

The second of properties (9) is obvious:

$$(A\cup B)^\delta = \bigcup_{x\in A\cup B} K_\delta^x = (\bigcup_{x\in A} K_\delta^x)\cup(\bigcup_{x\in B} K_\delta^x) = A^\delta \cup B^\delta.$$

The second of properties (10) is also obvious: if $x \in (A \cap B)^\delta$, then $x \in K_\delta^y$, where $y \in A \cap B$, i.e. $x \in A^\delta$ and $x \in B^\delta$. The first of formulas (9) and (10) are proved by starting from the second and using relations (5) and (5′); for example,

$$(A\cap B)_\delta = C((C(A\cap B))^\delta) = C((CA\cup CB)^\delta)$$
$$= C((CA)^\delta \cup (CB)^\delta) = C((CA)^\delta)\cap C((CB)^\delta) = A_\delta \cap B_\delta.$$

Properties (11) are proved as follows:

$$(A-B)_\delta = (A\cap CB)_\delta = A_\delta\cap(CB)_\delta = A_\delta \cap C(B^\delta) = A_\delta - B^\delta,$$
$$(A-B)^\delta = (A\cap CB)^\delta \subset A^\delta\cap(CB)^\delta = A^\delta\cap C(B_\delta) = A^\delta - B_\delta.$$

It suffices for the proof of equality (12) to establish that $A^\delta - B_\delta \subset (A-B)^\delta$. Let $x \in A^\delta - B_\delta$ and consider the ball K_δ^x. We will show that $K_\delta^x \cap (A-B) \neq 0$. In fact, we would otherwise get by taking into account the inclusion $\bar{B} \subset \underline{A}$ that either $K_\delta^x \cap A = \varnothing$ or $K_\delta^x \subset B$. In the first case $\rho(x, \Gamma(A)) \geqslant \delta$, which contradicts the fact that $x \in A^\delta$; in the second case $\rho(x, \Gamma(B)) \geqslant \delta$, which contradicts the fact that $x \notin B$. We now choose a point $y \in K_\delta^x \cap (A-B)$. Then $x \in K_\delta^y$, i.e. $x \in (A-B)^\delta$.

We conclude this section by noting that it is not difficult to give examples showing that the inclusion signs in formulas (7)–(11) cannot be replaced by equality signs.

§3. Proof of Lemma 1

Consider the sets

$$\tilde{G}_m = \Omega_{\frac{1}{2^{m+1}}-\frac{1}{2^{m+3}}} - \Omega_{\frac{1}{2^m}+\frac{1}{2^{m+3}}}$$

and let

$$\varphi_m(x) = \frac{1}{\delta_m^n}\int_{\tilde{G}_m} \omega\left(\frac{x-y}{\delta_m}\right) dy, \tag{1}$$

where $\delta_m = 1/2^{m+4}$ and the function $\omega(z)$ is such that $\omega(z) = 0$ for $|z| \geqslant 1$, $\omega(z) \geqslant 0$, $\omega(z) \in C^\infty(E_n)$ and $\int_{|z|\leqslant 1}\omega(z)\,dz = 1$.

Thus the function $\varphi_m(x)$ is an average of the characteristic function of the set $\tilde{G}_m$ with averaging radius δ_m. From the properties of mean functions (see [1]) it follows that

1) $\varphi_m(x) \in C^\infty(E_n)$;
2) $|D^k\varphi_m(x)| \leqslant c_k 2^{m|k|}$;
3) $\operatorname{supp}\psi_m \subset \overline{\Omega_{\frac{1}{2^{m+1}}-\frac{1}{2^{m+3}}-\frac{1}{2^{m+4}}} - \Omega_{\frac{1}{2^m}+\frac{1}{2^{m+3}}+\frac{1}{2^{m+4}}}}$;
4) $0 \leqslant \varphi_m(x) \leqslant 1$, $\quad \varphi_m(x) = 1 \quad$ for $\quad x \in G_m$.

Let us explain property 3). Property (1) implies that $\varphi_m(x)$ can be different from zero only for those x for which $K^x_{\delta_m} \cap \widetilde{G}_m \neq \varnothing$, i.e. for $x \in (\widetilde{G}_m)^{\delta_m}$ in the notation of §2. Further, according to properties (12), (7), and (6) of §2, we get

$$(\widetilde{G}_m)^{\delta_m} = (\Omega_{\frac{1}{2^{m+1}} - \frac{1}{2^{m+3}}} - \Omega_{\frac{1}{2^m} + \frac{1}{2^{m+3}}})^{\frac{1}{2^{m+4}}} = (\Omega_{\frac{1}{2^{m+1}} - \frac{1}{2^{m+3}}})^{\frac{1}{2^{m+4}}}$$
$$- (\Omega_{\frac{1}{2^m} + \frac{1}{2^{m+3}}})_{\frac{1}{2^{m+4}}} = \Omega_{\frac{1}{2^{m+1}} - \frac{1}{2^{m+3}} - \frac{1}{2^{m+4}}} - \Omega_{\frac{1}{2^m} + \frac{1}{2^{m+3}} + \frac{1}{2^{m+4}}},$$

which implies property 3).

Property 3) in turn implies

5) $\operatorname{supp} \varphi_m \cap \operatorname{supp} \varphi_{m+j} = \varnothing, \quad j \geqslant 2;$

6) $\operatorname{supp} \varphi_m \subset G_{m-1} + G_m + G_{m+1}.$

From property 5) it follows that the multiplicity of the covering of Ω by the sets $\operatorname{supp} \varphi_m$ does not exceed two. As a result, we get by taking into account property 4) that

$$1 \leqslant \sum_{l=0}^{\infty} \varphi_l(x) \leqslant 2. \tag{2}$$

for $x \in \Omega$. The function $\Sigma_0^\infty \varphi_l(x)$ is infinitely differentiable for $x \in \Omega$, with

$$D^k\left(\sum_{l=0}^{\infty} \varphi_l(x)\right) = D^k\varphi_{m-1}(x) + D^k\varphi_m(x) + D^k\varphi_{m+1}(x) = D^k\varphi_{m-1}(x) + D^k\varphi_{m+1}(x)$$

for $x \in G_m$. Here, depending on $x \in G_m$, either $D^k\varphi_{m-1}(x) = 0$ or $D^k\varphi_{m+1}(x) = 0$ by virtue of 5). Therefore

$$\left| D^k\left(\sum_{l=0}^{\infty} \varphi_l(x)\right)\right| \leqslant \widetilde{c}_k\, 2^{m|k|}. \tag{3}$$

Now let

$$\psi_m(x) = \begin{cases} \dfrac{\varphi_m(x)}{\sum_{l=0}^{\infty} \varphi_l(x)}, & x \in \Omega, \\ 0, & x \notin \Omega. \end{cases}$$

The above properties of the functions $\varphi_m(x)$ imply the fulfilment of properties 1)–3) of Lemma 1. Further, for $x \in \operatorname{supp} \psi_m$

$$D^k\psi_m(x) = \sum_{\alpha \leqslant k} \frac{k!}{\alpha!\,(k-\alpha)!} D^{k-\alpha}\varphi_m D^\alpha\left(\frac{1}{\sum_{l=0}^{\infty} \varphi_l(x)}\right). \tag{4}$$

Let $\Phi(x) = \Sigma_0^\infty \varphi_l(x)$. By virtue of the rules of differentiation we have

$$D^\alpha\left(\frac{1}{\Phi(x)}\right) = \frac{\Sigma}{\Phi^{|\alpha|+1}},$$

where the sum Σ consists of a finite number of terms, taken with certain coefficients, of the form $D^{\gamma_1}\Phi, D^{\gamma_2}\Phi, \ldots, D^{\gamma_{|\alpha|}}\Phi$, with $\gamma_1 + \gamma_2 + \ldots + \gamma_{|\alpha|} = |\alpha|$. Taking into account this fact, property 2) of the functions φ_m, and inequalities (2) and (3), we get from (4) that the functions φ_m also have property 4) of the lemma.

Lemma 1 is completely proved.

§4. Some properties of averages

Suppose now $f(x)$ is a function from $L_p(\Omega)$, $1 \leqslant p < \infty$. With the use of the functions $\psi_m(x)$ of Lemma 1, we construct the following weighted moving average of $f(x)$ with "averaging radius δ":

$$f_\delta(x) = \sum_{m=0}^{\infty} \psi_m(x) \frac{1}{\delta_m^n} \int_\Omega \omega\left(\frac{x-y}{\delta_m}\right) f(y)\, dy, \tag{1}$$

where $\delta_m = \delta/2^m$. Here $\omega(z)$ is a function satisfying the following conditions: $\omega(z) \in C^\infty(E_n)$, $\omega(z) = 0$ for $|z| \geqslant 1$, and

$$\int_{E_n} \omega(z)\, dz = 1. \tag{2}$$

We will prove two lemmas for the weighted moving averages constructed above that are analogous to corresponding lemmas for the ordinary averages

$$f_\delta(x) = \frac{1}{\delta^n} \int_\Omega \omega\left(\frac{x-y}{\delta}\right) f(y)\, dy.$$

LEMMA 2. *Suppose Ω is an open set and $f \in L_p(\Omega)$, $1 \leqslant p < \infty$. Then for $0 < \delta < 1/8$*

$$\| f_\delta \|_{L_p(\Omega)} \leqslant c_1 \| f \|_{L_p(\Omega)}, \tag{3}$$

where $c_1 = 9 \int_{|z| \leqslant 1} |\omega(z)|\, dz$.

PROOF. According to condition 2) of Lemma 1, not more than two of the sets $\operatorname{supp} \psi_m(x)$ cover any point $x \in \Omega$. Therefore not more than two summands of the sum (1) are different from zero for any $x \in \Omega$. Consequently

$$| f_\delta(x) |^p \leqslant 2^{p-1} \sum_{m=0}^{\infty} | \psi_m(x) \frac{1}{\delta_m^n} \int_\Omega \omega\left(\frac{x-y}{\delta_m}\right) f(y)\, dy |^p.$$

Further, taking into account condition 3) of Lemma 1, we have

$$\begin{aligned}
\int_\Omega | f_\delta(x) |^p\, dx &= \sum_{s=0}^{\infty} \int_{G_s} | f_\delta(x) |^p\, dx \\
&\leqslant 2^{p-1} \sum_{s=0}^{\infty} \int_{G_s} \sum_{m=0}^{\infty} \left| \psi_m(x) \frac{1}{\delta_m^n} \int_\Omega \omega\left(\frac{x-y}{\delta_m}\right) f(y)\, dy \right|^p dx \\
&= 2^{p-1} \sum_{s=0}^{\infty} \int_{G_s} \sum_{m=s-1}^{s+1} \left| \psi_m(x) \frac{1}{\delta_m^n} \int_\Omega \omega\left(\frac{x-y}{\delta_m}\right) f(y)\, dy \right|^p dx \\
&\leqslant 2^{p-1} \sum_{s=0}^{\infty} \sum_{m=s-1}^{s+1} \int_{G_s} \left| \frac{1}{\delta_m^n} \int_\Omega \omega\left(\frac{x-y}{\delta_m}\right) f(y)\, dy \right|^p dx \\
&= 2^{p-1} \sum_{s=0}^{\infty} \sum_{m=s-1}^{s+1} \int_{G_s} \left| \int_{|z| \leqslant 1} \omega(z) f(x - z\delta_m)\, dz \right|^p dx.
\end{aligned}$$

Applying Minkowski's inequality, we get for $\delta < 1/8$ that

$$\int_{G_s} \Big| \int_{|z|\leqslant 1} \omega(z) f(x - z\delta_m)\, dz \Big|^p dx \leqslant \Big(\int_{|z|\leqslant 1} |\omega(z)| \cdot \| f(x - z\delta_m) \|_{L_{p,\,x}(G_s)}\, dz \Big)^p$$

$$\leqslant \Big(\int_{|z|\leqslant 1} |\omega(z)| \cdot \| f(y) \|_{L_p\left((G_s)^{\frac{1}{2^{m+3}}}\right)} dz \Big)^p$$

$$= \Big(\int_{|z|\leqslant 1} |\omega(z)|\, dz \Big)^p \int_{(G_s)^{\frac{1}{2^{m+3}}}} |f(y)|^p\, dy$$

$$\leqslant \Big(\int_{|z|\leqslant 1} |\omega(z)|\, dz \Big)^p \Big(\int_{G_{s-1}} + \int_{G_s} + \int_{G_{s+1}} \Big) |f(y)|^p\, dy,$$

since, according to properties (12), (7) and (6) of §2, for $m = s - 1, s, s + 1$

$$(G_s)^{\frac{1}{2^{m+3}}} \subset (G_s)^{\frac{1}{2^{s+2}}} = (\Omega_{\frac{1}{2^{s+1}}} - \Omega_{\frac{1}{2^s}})^{\frac{1}{2^{s+2}}} = (\Omega_{\frac{1}{2^{s+1}}})^{\frac{1}{2^{s+2}}} - (\Omega_{\frac{1}{2^s}})_{\frac{1}{2^{s+2}}}$$

$$= \Omega_{\frac{1}{2^{s+1}} - \frac{1}{2^{s+2}}} - \Omega_{\frac{1}{2^s} + \frac{1}{2^{s+2}}} \subset \Omega_{\frac{1}{2^{s+2}}} - \Omega_{\frac{1}{2^{s-1}}} = G_{s-1} + G_s + G_{s+1}.$$

Thus for $\delta < 1/8$

$$\int_\Omega |f_\delta(x)|^p\, dx \leqslant 3 \cdot 2^{p-1} \Big(\int_{|z|\leqslant 1} |\omega(z)|\, dz \Big)^p \sum_{s=0}^{\infty} \Big(\int_{G_{s-1}} + \int_{G_s} + \int_{G_{s+1}} \Big) |f(y)|^p\, dy$$

$$\leqslant 9 \cdot 2^{p-1} \Big(\int_{|z|\leqslant 1} |\omega(z)|\, dz \Big)^p \sum_{s=0}^{\infty} \int_{G_s} |f(y)|^p\, dy = 9 \cdot 2^{p-1} \Big(\int_{|z|\leqslant 1} |\omega(z)|\, dz \Big)^p \int_\Omega |f(y)|^p dy,$$

Q.E.D.

COROLLARY. *Suppose $f \in L_p(\Omega)$. Then for $\delta < 1/8$*

$$\Big\| \sum_{m=l}^{\infty} \psi_m(x) \frac{1}{\delta_m^n} \int_\Omega \omega\Big(\frac{x-y}{\delta_m}\Big) f(y)\, dy \Big\|_{L_p(\Omega)} \leqslant c_1 \| f \|_{L_p\left(\Omega - \Omega_{\frac{1}{2^{l-2}}}\right)}. \tag{4}$$

PROOF. Arguing in the same way as in the proof of Lemma 2, we get for $\delta < 1/8$ that

$$\int_\Omega \Big| \sum_{m=l}^{\infty} \psi_m(x) \frac{1}{\delta_m^n} \int_\Omega \omega\Big(\frac{x-y}{\delta_m}\Big) f(y)\, dy \Big|^p dx$$

$$= \sum_{s=l-1}^{\infty} \int_{G_s} \Big| \sum_{m=l}^{\infty} \psi_m(x) \frac{1}{\delta_m^n} \int_\Omega \omega\Big(\frac{x-y}{\delta_m}\Big) f(y)\, dy \Big|^p dx$$

$$\leqslant 3 \cdot 2^{p-1} \int_{|z|\leqslant 1} |\omega(z)|\, dz \sum_{s=l-1}^{\infty} \Big(\int_{G_{s-1}} + \int_{G_s} + \int_{G_{s+1}} \Big) |f(y)|^p\, dy$$

$$\leqslant 9 \cdot 2^{p-1} \int_{|z|\leqslant 1} |\omega(z)|\, dz \sum_{s=l-2}^{\infty} \int_{G_s} |f(y)|^p\, dy$$

$$= 9 \cdot 2^{p-1} \int_{|z|\leqslant 1} |\omega(z)|\, dz \int_{\Omega - \Omega_{\frac{1}{2^{l-2}}}} |f(y)|^p\, dy,$$

which implies (4).

LEMMA 3. *Suppose* $f \in L_p(\Omega)$. *Then as* $\delta \to 0$

$$\| f_\delta - f \|_{L_p(\Omega)} \to 0.$$

PROOF. Consider the difference

$$f_\delta(x) - f(x) = \sum_{m=0}^{\infty} \psi_m(x) \frac{1}{\delta_m^n} \int_\Omega \omega\left(\frac{x-y}{\delta_m}\right) [f(y) - f(x)]\, dy \equiv \sum_{m=0}^{\infty} a_m(x, \delta). \tag{5}$$

We will prove that the series (5) coverges uniformly relative to δ in the metric of $L_p(\Omega)$. In fact, from the inequality

$$\left| \sum_{m=l}^{\infty} a_m(x, \delta) \right| \leqslant \sum_{m=l}^{\infty} \psi_m(x) \frac{1}{\delta_m^n} \int_\Omega \left| \omega\left(\frac{x-y}{\delta_m}\right) f(y) \right| dy + |f(x)| \sum_{m=l}^{\infty} \psi_m(x),$$

the properties of the functions ψ_m and inequality (4) we get

$$\left\| \sum_{m=l}^{\infty} a_m(x, \delta) \right\|_{L_p(\Omega)} \leqslant \left\| \sum_{m=l}^{\infty} \psi_m(x) \frac{1}{\delta_m^n} \int_\Omega \left| \omega\left(\frac{x-y}{\delta_m}\right) f(y) \right| dy \right\|_{L_p\left(\sum_{s=l-1}^{\infty} G_s\right)}$$

$$+ \| f \|_{L_p\left(\sum_{s=l-1}^{\infty} G_s\right)} \leqslant (1 + c_1) \| f \|_{L_p\left(\Omega - \Omega_{\frac{1}{2^{l+2}}}\right)},$$

which implies the uniform convergence of the series (5).

Since each summand $a_m(x, \delta) \to 0$ in $L_p(\Omega)$ as $\delta \to 0$ (this follows from the properties of mean functions), we get (lim is taken in $L_p(\Omega)$)

$$\lim_{\delta \to 0} [f_\delta(x) - f(x)] = \sum_{m=0}^{\infty} \lim_{\delta \to 0} a_m(x, \delta) = 0.$$

Q.E.D.

§5. Proof of the theorem

1. Since for any $x \in \Omega$ there exists a ball with center at x that contains not more than two of the sets supp ψ_m, we can differentiate the equality

$$f_\delta(x) = \sum_{m=0}^{\infty} \psi_m(x) \frac{1}{\delta_m^n} \int_\Omega \omega\left(\frac{x-y}{\delta_m}\right) f(y)\, dy \tag{1}$$

term by term for $x \in \Omega$, getting

$$D^k f_\delta(x) = \sum_{m=0}^{\infty} D^k \left[\psi_m(x) \frac{1}{\delta_m^n} \int_\Omega \omega\left(\frac{x-y}{\delta_m}\right) f(y)\, dy \right]$$

$$= \sum_{m=0}^{\infty} \sum_{\alpha \leqslant k} \frac{k!}{\alpha!\,(k-\alpha)!} D^\alpha \psi_m(x) \frac{1}{\delta_m^n} \int_\Omega D_x^{k-\alpha} \left[\omega\left(\frac{x-y}{\delta_m}\right) \right] f(y)\, dy. \tag{2}$$

Here we have made use of the well-known fact that the function

$$\frac{1}{\delta_m^n}\int_\Omega \omega\left(\frac{x-y}{\delta_m}\right) f(y)\,dy$$

is infinitely differentiable everywhere in E_n, it being permissible to differentiate under the integral sign.

Thus the functions $f_\delta(x)$ are infinitely differentiable in Ω and formula (2) is valid.

Suppose now $|k| \leqslant r$. If $x \in \operatorname{supp} \psi_m$, then

$$x \subset G_{m-1} + G_m + G_{m+1} = \Omega_{\frac{1}{2^{m+2}}} - \Omega_{\frac{1}{2^{m-1}}},$$

so that when $\delta < 1/8$ the ball of radius $\delta_m = \delta/2^m$ with center at x lies strictly interior to Ω. Consequently, for $x \in \operatorname{supp} \psi_m$ the support of the function $\omega((x-y)/\delta_m)$ as a function of y lies in Ω, and

$$\int_\Omega D_x^{k-\alpha}\left[\omega\left(\frac{x-y}{\delta_m}\right)\right] f(y)\,dy = (-1)^{|k-\alpha|}\int_\Omega D_y^{k-\alpha}\left[\omega\left(\frac{x-y}{\delta_m}\right)\right] f(y)\,dy$$
$$= \int_\Omega \omega\left(\frac{x-y}{\delta_m}\right) D^{k-\alpha} f(y)\,dy.$$

Thus, for $x \in \Omega$

$$D^k f_\delta(x) = \sum_{m=0}^{\infty}\sum_{\alpha\leqslant k}\frac{k!}{\alpha!\,(k-\alpha)!} D^\alpha\psi_m(x)\frac{1}{\delta_m^n}\int_\Omega \omega\left(\frac{x-y}{\delta_m}\right) D^{k-\alpha} f(y)\,dy$$
$$= \sum_{\alpha\leqslant k}\frac{k!}{\alpha!\,(k-\alpha)!}\sum_{m=0}^{\infty} D^\alpha\psi_m(x)\int_{|z|\leqslant 1}\omega(z)\,D^{k-\alpha} f(x-z\delta_m)\,dz \tag{3}$$

(here we have again taken into account the fact that when $|z| \leqslant 1$ and $x \in \operatorname{supp} \psi_m$ the point $x - z\delta_m \in \Omega$ if $\delta < 1/8$).

Further, for $x \in \Omega$

$$D^k f_\delta(x) - D^k f(x) = \sum_{\alpha\leqslant k}\frac{k!}{\alpha!(k-\alpha)!}\sum_{m=0}^{\infty} D^\alpha\psi_m(x)\int_{|z|\leqslant 1}\omega(z)$$
$$\times\,[D^{k-\alpha}f(x-z\delta_m) - D^{k-\alpha}f(x)]\,dz. \tag{4}$$

We have taken into consideration the fact that when $\alpha = 0$

$$\int_{|z|\leqslant 1}\omega(z)\,dz = 1,\quad \sum_{m=0}^{\infty}\psi_m(x) = 1,$$

and when $|\alpha| > 0$

$$\sum_{m=0}^{\infty} D^\alpha\psi_m(x) = D^\alpha\left(\sum_{m=0}^{\infty}\psi_m(x)\right) = 0.$$

When $\alpha = 0$ the single sum in the right side of (4) tends to zero in $L_p(\Omega)$ as $\delta \to 0$; this follows from Lemma 3. We will prove below that the sums in the right side of (4) also tend to zero in $L_p(\Omega)$ as $\delta \to 0$ when $|\alpha| > 0$.

2. Let us prove that if

$$\int\limits_{|z|\leqslant 1} \omega(z)\,dz = 1; \quad \int\limits_{|z|\leqslant 1} \omega(z)\, z^k dz = 0, \quad 0 < |k| \leqslant r-1, \tag{5}$$

then for $l \leqslant r$

$$\Big\| \int\limits_{|z|\leqslant 1} \omega(z)\,[\Phi(x - z\delta_m) - \Phi(x)]\,dz \Big\|_{L_p(G)} \leqslant c_1 \delta_m^l \|\Phi\|_{W_p^l(G^{\delta_m})} . \tag{6}$$

In fact, by making use of Taylor's formula and taking into account (5), we get

$$\int\limits_{|z|\leqslant 1} \omega(z)\,[\Phi(x - z\delta_m) - \Phi(x)]\,dz = \int\limits_{|z|\leqslant 1} \omega(z) \sum_{0<|k|<l} \frac{D^k\Phi(x)}{k!} (-z\delta_m)^k\,dz$$

$$+ l \int\limits_{|z|\leqslant 1} \Big(\omega(z) \sum_{|k|=l} \frac{(-z\delta_m)^k}{k!} \int\limits_0^1 (1-t)^{l-1} D^k\Phi(x - tz\delta_m)\,dt\Big)\,dz$$

$$= (-1)^l \delta_m^l l \int\limits_{|z|\leqslant 1} \Big(\omega(z) \sum_{|k|=l} \frac{z^k}{k!} \int\limits_0^1 (1-t)^{l-1} D^k\Phi(x - tz\delta_m)\,dt\Big)\,dz.$$

Applying Minkowski's inequality, we have

$$\Big\| \int\limits_{|z|\leqslant 1} \omega(z)\,[\Phi(x - z\delta_m) - \Phi(x)]\,dz \Big\|_{L_p(G)}$$

$$\leqslant l\delta_m^l \int\limits_{|z|\leqslant 1} \Big(|\omega(z)| \sum_{|k|=l} \frac{1}{k!} \int\limits_0^1 (1-t)^{l-1} \|D^k\Phi(x - tz\delta_m)\|_{L_p(G)}\,dt\Big)\,dz$$

$$\leqslant l\delta_m^l \int\limits_{|z|\leqslant 1} \Big(|\omega(z)| \sum_{|k|=l} \frac{1}{k!} \int\limits_0^1 (1-t)^{l-1} \|D^k\Phi(y)\|_{L_p(G^{t|z|\delta_m})}\,dt\Big)\,dz$$

$$\leqslant l\delta_m^l \int\limits_{|z|\leqslant 1} |\omega(z)|\,dz \sum_{|k|=l} \|D^k\Phi\|_{L_p(G^{\delta_m})},$$

which implies (6).

3. We will now prove that the sums in the right side of (4) tend to zero in $L_p(\Omega)$ as $\delta \longrightarrow 0$ when $|\alpha| > 0$.

According to condition 2) of Lemma 1, not more than two of the sets supp ψ_m contain any point $x \in \Omega$, so that not more than two of the summands in the sums of (4) are different from zero for any $x \in \Omega$. Consequently, by carrying out arguments similar to those in the proof of Lemma 2, we get

$$J_\alpha = \int\limits_\Omega \Big| \sum_{m=0}^{\infty} D^\alpha\psi_m(x) \int\limits_{|z|\leqslant 1} \omega(z)\,[D^{k-\alpha} f(x - z\delta_m) - D^{k-\alpha} f(x)]\,dz\Big|^p dx$$

$$\leqslant 2^{p-1} \int\limits_\Omega \sum_{m=0}^{\infty} \Big| D^\alpha\psi_m(x) \int\limits_{|z|\leqslant 1} \omega(z)\,[D^{k-\alpha} f(x - z\delta_m) - D^{k-\alpha} f(x)]\,dz\Big|^p dx$$

$$= 2^{p-1} \sum_{s=0}^{\infty} \int\limits_{G_s} \sum_{m=s-1}^{s+1} \Big| D^\alpha\psi_m(x) \int\limits_{|z|\leqslant 1} \omega(z)\,[D^{k-\alpha} f(x - z\delta_m) - D^{k-\alpha} f(x)]\,dz\Big|^p dx$$

$$\leqslant c_1 \sum_{s=0}^{\infty} 2^{s|\alpha|p} \sum_{m=s-1}^{s+1} \int\limits_{G_s} \Big| \int\limits_{|z|\leqslant 1} \omega(z)\,[D^{k-\alpha} f(x - z\delta_m) - D^{k-\alpha} f(x)]\,dz\Big|^p dx.$$

By making use of inequality (6) for $l = |\alpha|$ and $\Phi = D^{k-\alpha}f$, we get

$$J_\alpha \leqslant c_2 \sum_{s=0}^{\infty} 2^{s|\alpha|p} \sum_{m=s-1}^{s+1} \delta_m^{|\alpha|p} \sum_{|i|\leqslant r} \int_{(G_s)^{\delta_m}} |D^i f(y)|^p \, dy$$

$$\leqslant c_3 \delta^{|\alpha|p} \sum_{s=0}^{\infty} 2^{s|\alpha|p} \sum_{m=s-1}^{s+1} \frac{1}{2^{m|\alpha|p}} \sum_{|i|\leqslant r} \int_{(G_s)^{\frac{\delta}{2^m}}} |D^i f(y)|^p \, dy$$

$$\leqslant c_4 \delta^{|\alpha|p} \sum_{s=0}^{\infty} \sum_{|i|\leqslant r} \int_{(G_s)^{\frac{\delta}{2^{s-1}}}} |D^i f(y)|^p \, dy.$$

When $\delta < 1/8$ we have

$$(G_s)^{\frac{\delta}{2^{s-1}}} \subset (G_s)^{\frac{1}{2^{s+2}}} = (\Omega_{\frac{1}{2^{s+1}}} - \Omega_{\frac{1}{2^s}})^{\frac{1}{2^{s+2}}} \subset \Omega_{\frac{1}{2^{s+2}}} - \Omega_{\frac{1}{2^{s-1}}} = G_{s-1} + G_s + G_{s+1},$$

so that

$$J_\alpha \leqslant c_4 \delta^{|\alpha|p} \sum_{|i|\leqslant r} \sum_{s=0}^{\infty} \left(\int_{G_{s-1}} + \int_{G_s} + \int_{G_{s+1}} \right) |D^i f(y)|^p \, dy$$

$$\leqslant 3c_4 \delta^{|\alpha|p} \sum_{|i|\leqslant r} \sum_{s=0}^{\infty} \int_{G_s} |D^i f(y)|^p \, dy \leqslant c_5 \delta^{|\alpha|p} \|f\|^p_{W_p^r(\Omega)}.$$

Thus $J_\alpha \longrightarrow 0$ in $L_p(\Omega)$ as $\delta \longrightarrow 0$. Q.E.D.

4. Suppose now $|k| > r$. We use (2). If $|k| - |\alpha| > r$, then for $x \in \operatorname{supp} \psi_m$

$$\frac{1}{\delta_m^n} \int_\Omega D_x^{k-\alpha} \left[\omega\left(\frac{x-y}{\delta_m}\right) \right] f(y) \, dy = \frac{1}{\delta_m^n} \int_\Omega D_x^{k-\alpha-j} \left[\omega\left(\frac{x-y}{\delta_m}\right) \right] D^j f(y) \, dy$$

$$= \frac{1}{\delta_m^{n+|k|-|\alpha|-r}} \int_\Omega (D^{k-\alpha-j}\omega)\left(\frac{x-y}{\delta_m}\right) D^j f(y) \, dy$$

$$= \frac{1}{\delta_m^{|k|-|\alpha|-r}} \int_{|z|\leqslant 1} D^{k-\alpha-j}\omega(z) \, D^j f(x - z\delta_m) \, dz$$

(here j is a vector such that $|j| = r$ and $k - \alpha - j \geqslant 0$). It follows that

$$D^k f_\delta(x) \rho(x)^{|k|-r}$$

$$= \sum_{\substack{\alpha\leqslant k \\ |\alpha|\geqslant|k|-r}} \frac{k!}{\alpha!(k-\alpha)!} \sum_{m=0}^{\infty} D^\alpha \psi_m(x) \rho(x)^{|k|-r} \int_{|z|\leqslant 1} \omega(z) D^{k-\alpha} f \; (x - z\delta_m) \, dz$$

$$+ \sum_{\substack{\alpha\leqslant k \\ |\alpha|<|k|-r}} \frac{k!}{\alpha!(k-\alpha)!} \sum_{m=0}^{\infty} \frac{D^\alpha \psi_m(x) \rho(x)^{|k|-r}}{\delta_m^{|k|-|\alpha|-r}} \int_{|z|\leqslant 1} D^{k-\alpha-j}\omega(z) \; D^j f \; (x - z\delta_m) \, dz.$$

When $x \in \operatorname{supp} \psi_m$ the inequality $\rho(x) \leqslant 1/2^{m-1}$ is satisfied, and hence

$$|D^\alpha \psi_m(x)| \rho(x)^{|k|-r} \leqslant c_\alpha 2^{m(|\alpha|+r-|k|)}.$$

Moreover, as in subsection 3, under the sign of the integral with respect to z in the first sum it is possible to subtract the necessary segment of the Taylor series of the function $D^{k-\alpha} f$, inasmuch as $|\alpha| \geqslant |k| - r > 0$ and $\Sigma_{m=0}^{\infty} D^\alpha \psi_m(x) = 0$. Thus, taking into account inequality (6), we get that

$$\| D^k f_\delta \rho^{|k|-r} \|_{L_p(\Omega)} \leqslant \frac{c_k}{\delta^{|k|-r}} \| f \|_{W_p^r(\Omega)},$$

where c_k depends only on k ($\delta < 1/8$). Q.E.D.

BIBLIOGRAPHY

1. S. L. Sobolev, *Applications of functional analysis in mathematical physics,* Izdat. Leningrad. Gos. Univ., Leningrad, 1950; English transl., Transl. Math. Monographs, vol. 7, Amer. Math. Soc., Providence, R. I., 1963. MR **14**, 565; **29** #2624.

2. J. Deny and J.-L. Lions, *Les espaces du type de Beppo Levi,* Ann. Inst. Fourier (Grenoble) **5** (1953/54), 305–370 (1955). MR **17**, 646.

3. N. G. Meyers and J. Serrin, $H = W$, Proc. Nat. Acad. Sci. U. S. A. **51** (1964), 1055–1056. MR **29** #1551.

4. E. Gagliardo, *Proprietà di alcune classi di funzioni in più variabili,* Ricerche Mat. **7** (1958), 102–137. MR **21** #1526.

5. G. G. Kazarjan, *The density in* $\mathring{W}_p^{(r)}(\Omega)$ *of smooth functions with compact support,* Mat. Zametki **2** (1967), 45–52 = Math. Notes **2** (1967), 514–517. MR **35** #3431.

6. V. P. Il'in, *On approximation of functions in the spaces* $\widetilde{W}_p^{(1)}(D)$ *and* $W_p^{(1)}(D)$ *by continuously differentiable functions,* Trudy Mat. Inst. Steklov. **64** (1961), 61–78; English transl., Amer. Math. Soc. Transl. (2) **44** (1965), 89–107. MR **25** #447.

7. H. Whitney, *Analytic extensions of differentiable functions defined in closed sets,* Trans. Amer. Math. Soc. **36** (1934), 63–89.

Translated by S. SMITH

Trudy Mat. Inst. Steklov.
131 (1974)

Proc. Steklov Inst. Math.
131 (1974)

ON THE APPROXIMATION OF FUNCTIONS IN THE SPACE $W_p^r(\Omega)$ BY FUNCTIONS WITH COMPACT SUPPORT FOR AN ARBITRARY OPEN SET Ω(1)

UDC 517.518.22

V. I. BURENKOV

ABSTRACT. Let $W_p^r(\Omega)$ be the Sobolev space of functions f with finite norm

$$\|f\|_{W_p^r(\Omega)}=\sum_{|k|\leqslant r}\|D^kf\|_{L_p(\Omega)}, \tag{1}$$

where $\Omega\subset E_n$ is an arbitrary open set. A number of equivalent necessary and sufficient conditions are obtained for $p>n$ that show in what sense the functions $D^kf(x)$, $|k|\leqslant r-1$, must vanish on the boundary $\Gamma(\Omega)$ in order for it to be possible to approximate a function $f\in W_p^r(\Omega)$ with arbitrary accuracy in the norm (1) by infinitely differentiable functions with compact support, two examples of which are the condition that

$$\|f\|_{L_p(\Omega-\Omega_\delta)}=o(\delta)^r,$$

where Ω_δ is the set of points $x\in\Omega$ that are at a distance of not less than δ from the boundary $\Gamma(\Omega)$, and the condition that the function f be equivalent to a function $\bar f$ such that for any $x\in\Gamma(\Omega)$

$$\lim_{\substack{y\to x\\ y\in\Omega}}D^k\bar f(y)=0,\qquad |k|\leqslant r-1.$$

The connection between the possibility of approximating a function $f\in W_p^r(\Omega)$ with any degree of accuracy by functions with compact support and the possibility of extending a function f by zero with preservation of class is investigated.

Bibliography: 6 items.

Let E_n be an n-dimensional Euclidean space of points $x=(x_1,\ldots,x_n)$ and let Ω be an arbitrary open set in E_n. A function $f(x)$ belongs to the Sobolev space $W_p^r(\Omega)$, $r=1,2,\ldots$, $1\leqslant p<\infty$, if the generalized (see [1]) partial derivatives

$$D^kf=\frac{\partial^{k_1+\ldots+k_n}f}{\partial x_1^{k_1}\ldots\partial x_n^{k_n}},\quad |k|=k_1+\ldots+k_n\leqslant r,$$

exist in Ω and the norm

$$\|f\|_{W_p^r(\Omega)}=\sum_{|k|\leqslant r}\|D^kf\|_{L_p(\Omega)}$$

is finite ($D^kf=f$ for $k=0$). In many questions it is necessary to consider in conjunction with the spaces $W_p^r(\Omega)$ the spaces obtained by taking the closure of the set $C_0^\infty(\Omega)$ (the set of infinitely differentiable functions with compact support contained in Ω) in the norm of

AMS (MOS) subject classifications (1970). Primary 46E35, 41A30.

(1) The results of the present article were given without proof in the author's note [6].

the space $W_p^r(\Omega)$. It is known that if Ω is a domain with a sufficiently "good" boundary, then these spaces consist of the functions $f \in W_p^r(\Omega)$ that, together with their derivatives of order up to $r-1$ inclusively, vanish in a certain sense on the boundary $\Gamma(\Omega)$. A similar result is obtained in the article [2] of G. G. Kazarjan for the class of domains admitting a "local translation". Namely, it is proved that if

$$\Phi(x) = \begin{cases} f(x), & x \in \Omega \\ 0, & x \notin \Omega \end{cases} \in W_p^r(E_n),$$

where the domain Ω admits a "local translation", then the function f can be approximated with any degree of accuracy by functions from $C_0^\infty(\Omega)$ (see, in this connection, Theorem 6, where necessary and sufficient conditions on Ω, under which this assertion is valid, are given for $p > n$). The article [2] makes use of a modification of a method of approximating a function by an average of it with translation, which was previously used by E. Gagliardo in [3].

Necessary and sufficient conditions have been obtained by the author [4], [5] for an arbitrary open set Ω under which a function from the space $C^r(\Omega)$ of r times continuously differentiable functions with finite norm

$$\| f \|_{C^r(\Omega)} = \max_{|k| \leqslant r} \sup_{x \in \Omega} | D^k f(x) |$$

can be approximated with arbitrary accuracy by functions from $C_0^\infty(\Omega)$. These conditions have the form: for any $x \in \Gamma(\Omega)$

$$\lim_{\substack{y \to x \\ y \in \Omega}} D^k f(y) = 0, \qquad |k| \leqslant r,$$

and, in the case of an unbounded open set Ω,

$$\lim_{\substack{y \to \infty \\ y \in \Omega}} D^k f(y) = 0, \qquad |k| \leqslant r.$$

It is also proved in [4] and [5] that, when Ω is a bounded open set for which

$$\Gamma(\Omega) = \Gamma(\overline{\Omega}),$$

these conditions are equivalent to the fact that the function $\Phi(x) \in C^r(E_n)$. Also necessary and sufficient conditions are obtained on Ω under which the indicated equivalence holds for any $f \in C^r(\Omega)$.

In the present article we investigate similar questions for the Sobolev spaces $W_p^r(\Omega)$ when $p > n$. We obtain a number of equivalent necessary and sufficient conditions showing in what sense the functions $D^k f(x)$, $|k| \leqslant r-1$, must vanish on the boundary $\Gamma(\Omega)$ of an arbitrary open set Ω in order for it to be possible to approximate a function $f \in W_p^r(\Omega)$ with arbitrary accuracy by functions from $C_0^\infty(\Omega)$ (Theorems 2–5). As in the case of the spaces $C^r(\Omega)$, we prove that if the open set Ω is such that $\Gamma(\Omega) = \Gamma(\overline{\Omega})$, then these conditions are equivalent to the fact that $\Phi(x) \in W_p^r(E_n)$. We obtain necessary and sufficient conditions on Ω under which such an equivalence holds for any function $f \in W_p^r(\Omega)$ when $p > n$ (Theorem 6). These conditions have a different form than in the case of the spaces $C^r(\Omega)$ (see [4] and [5]).

LEMMA 1. *Suppose $p > n$, Ω is a star-shaped domain with respect to a measurable set $S \subset \Omega$ of positive measure and d is the diameter of Ω. Then for any function $f \in W_p^1(\Omega)$*

$$\| f \|_{L_p(\Omega)} \leqslant (\text{meas } \Omega)^{1/p} (\text{meas } S)^{-1/q} \| f \|_{L_q(S)} + \frac{d}{1 - n/p} \sum_{i=1}^{n} \left\| \frac{\partial f}{\partial x_i} \right\|_{L_p(\Omega)}, \tag{1}$$

where $1 \leqslant q \leqslant \infty$.

PROOF. Suppose first $f \in C^1(\Omega)$. Then for $x \in S$ and $y \in \Omega$

$$f(y) = f(x) + \sum_{i=1}^{n} \int_0^1 \frac{\partial f}{\partial x_i} (x + t(y - x)) (y_i - x_i)\, dt. \tag{2}$$

Applying Minkowski's inequality with respect to $y \in \Omega$, we get

$$\| f \|_{L_p(\Omega)} \leqslant | f(x) | (\text{ meas } \Omega)^{1/p} + d \sum_{i=1}^{n} \int_0^1 \left\| \frac{\partial f}{\partial x_i} (x + t(y - x)) \right\|_{L_p(\Omega)} dt.$$

We replace the variables y_i by z_i with the use of the formula $x_i + t(y_i - x_i) = z_i$. Taking into account the fact that $z \in \Omega$ if $y \in \Omega$, we get

$$\| f \|_{L_p(\Omega)} \leqslant | f(x) | (\text{meas } \Omega)^{1/p} + d \int_0^1 t^{-n/p}\, dt \cdot \sum_{i=1}^{n} \left\| \frac{\partial f}{\partial x_i} (z) \right\|_{L_p(\Omega)}$$

$$= | f(x) | (\text{meas } \Omega)^{1/p} + \frac{d}{1 - \frac{n}{p}} \sum_{i=1}^{n} \left\| \frac{\partial f}{\partial x_i} \right\|_{L_p(\Omega)}.$$

Taking the L_q-norm with respect to $x \in S$ of both sides of this inequality, we obtain the desired inequality for $f \in C^1(\Omega)$.

Suppose now $f \in W_p^1(\Omega)$. Then $f \in L_q(\Omega)$ for any q, $1 \leqslant q \leqslant \infty$. Let f_n denote the sequence of functions from $C^1(\Omega)$ such that

$$\| f_n - f \|_{W_p^r(\Omega)} \to 0, \qquad \| f_n - f \|_{L_q(\Omega)} \to 0$$

(in regard to the construction of such a sequence, see [3], §2). Writing inequality (1) for the functions f_n and passing to the limit for $n \longrightarrow \infty$, we obtain inequality (1) for $f \in W_p^1(\Omega)$.

COROLLARY 1. *Under the assumptions of Lemma 1, for any function $f \in W_p^r(\Omega)$*

$$\| f \|_{L_p(\Omega)} \leqslant c_1 \Big[(1 + d^{r-1}) (\text{meas } \Omega)^{1/p} (\text{meas } S)^{-1/q} \| f \|_{W_q^{r-1}(S)} + d^r \sum_{|k|=r} \| D^k f \|_{L_p(\Omega)} \Big], \tag{3}$$

where $1 \leqslant q \leqslant \infty$ and c_1 depends only on r, p and n.

COROLLARY 2. *If it is known in addition that $f(x) \equiv 0$ for $x \in S$, then*

$$\| f \|_{L_p(\Omega)} \leqslant c_1 d^r \sum_{|k|=r} \| D^k f \|_{L_p(\Omega)}. \tag{4}$$

We now give an example showing for $r = 1$ that inequality (4) is not satisfied when $p < n$. Suppose Ω is the ball of diameter d with center at the origin while $S = K_h$ is the ball of radius h with center at the origin. We put $f(x) = \varphi(|x|/h)$, where $\varphi(u)$ is an element of $C^\infty(-\infty, \infty)$ such that $\varphi(u) \equiv 0$ for $u \leqslant 1$ and $\varphi(u) \equiv 1$ for $u \geqslant 2$. Then

$$\|f\|_{L_p(\Omega)} \geqslant c_2 [(d/2)^n - (2h)^n]^{1/p},$$

while

$$\left\|\frac{\partial f}{\partial x_i}\right\|_{L_p(\Omega)} = \left\|\frac{1}{h}\varphi'\left(\frac{|x|}{h}\right)\frac{x_i}{|x|}\right\|_{L_p(K_{2h})} \leqslant c_3 h^{n/p-1},$$

where c_2 and c_3 do not depend on h or d. If (4) were valid, it would follow that

$$c_2 [(d/2)^n - (2h)^n]^{1/p} \leqslant c_1 dnc_3 h^{n/p-1},$$

which is impossible (for $h \to 0$ or $d \to \infty$).

THEOREM 1. *Suppose $p > n$ and $\Omega \neq E_n$ is an arbitrary open set. For any function $f \in C_0^\infty(\Omega)$*

$$\|f\|_{L_p(\Omega-\Omega_\delta)} \leqslant c_4 \delta^r \sum_{|k|=r} \|D^k f\|_{L_p(\Omega-\Omega_{2\delta})},\text{(2)} \tag{5}$$

where c_4 depends only on r, p and n.

PROOF. We complete the definition of a function $f \in C_0^\infty(\Omega)$ by setting it equal to zero outside Ω, and we denote the resultant function again by f. We cover the whole space E_n by open balls of radius $\delta/2$ with centers at the points $(\delta l_1/2, \ldots, \delta l_n/2)$, where the l_i are arbitrary integers. Consider those of the indicated balls that intersect the set $\Omega - \Omega_\delta$. We denote them by L_i and their centers by $x^{(i)}$ (if Ω is bounded, there are a finite number of them, depending on δ; if Ω is unbounded, there are a countable number of them). Suppose, further, $y^{(i)}$ is a point of the boundary Γ such that $\rho(x^{(i)}, y^{(i)}) = \rho(x^{(i)}, \Gamma)$. We denote by M_i the ball of radius 2δ with center at $y^{(i)}$. Since $L_i \subset M_i$, we have

$$\int\limits_{\Omega-\Omega_\delta} |f(x)|^p\,dx \leqslant \sum_i \int\limits_{L_i} |f(x)|^p\,dx \leqslant \sum_i \int\limits_{M_i} |f(x)|^p\,dx.$$

But $f(x) = 0$ in $\Omega - \Omega_h$ for sufficiently small h, and hence M_i contains a ball in which $f(x) \equiv 0$. Therefore, by virtue of (4),

$$\int\limits_{M_i} |f|^p\,dx \leqslant c_5 (2\delta)^{pr} \sum_{|k|=r} \int\limits_{M_i} |D^k f|^p\,dx. \tag{6}$$

Thus

$$\int\limits_{\Omega-\Omega_\delta} |f|^p\,dx \leqslant c_6 \delta^{pr} \sum_{|k|=r} \sum_i \int\limits_{\Omega M_i} |D^k f|^p\,dx. \tag{7}$$

Suppose now $x \in \Sigma_i \Omega M_i$. We estimate the number of balls M_i containing the point x. Any ball M_i containing x is obtained from a ball L_i such that

$$|x - x^{(i)}| \leqslant |x - y^{(i)}| + |y^{(i)} - x^{(i)}| < 4\delta.$$

Consequently the number of such balls does not exceed the number of points of the form $(\delta l_1/2, \ldots, \delta l_n/2)$ lying in the ball of radius 4δ with center at x, i.e. the number of integral points lying in an n-dimensional open ball of radius 8, which in turn never exceeds

(2) Ω_δ denotes the set of those $x \in \Omega$ that are at a distance of not less than δ from the boundary $\Gamma(\Omega)$.

16^n. It follows that

$$\sum_i \int_{\Omega M_i} |\Phi|\, dx \leq 16^n \int_{\sum_i \Omega M_i} |\Phi|\, dx. \tag{8}$$

In fact, we represent the sets $S_i = QM_i$ in the form of a union of nonintersecting parts: $S_i = \Sigma_j S_{ij}$, the sets S_{ij} and S_{kl}, $i \neq k$, either being disjoint or coinciding. This can be done, for example, in the following manner:

$$S_i = \prod_m S_m + \sum_{m_1 \neq i} \Big(\prod_{m \neq m_1} S_m - S_{m_1} \Big) + \sum_{\substack{m_1, m_2 \neq i \\ m_1 < m_2}} \Big(\prod_{m \neq m_1, m_2} S_m - (S_{m_1} + S_{m_2}) \Big) + \ldots \equiv \sum_j S_{ij}.$$

In this connection,

$$\sum_i S_i = \prod_m S_m + \sum_{m_1} \Big(\prod_{m \neq m_1} S_m - S_{m_1} \Big) + \sum_{m_1 < m_2} \Big(\prod_{m \neq m_1, m_2} S_m - (S_{m_1} + S_{m_2}) \Big) + \ldots .$$

By virtue of the above remarks, not more than 16^n terms can coincide in the unions $\Sigma_j S_{ij}$ and $\Sigma_j S_{kj}$, $i \neq k$. Therefore

$$\sum_i \int_{S_i} |\Phi|\, dx = \sum_{i,j} \int_{S_{ij}} |\Phi|\, dx \leq 16^n \int_{\sum_{i,j} S_{ij}} |\Phi|\, dx = 16^n \int_{\sum_i S_i} |\Phi|\, dx,$$

which implies (8).

From (7), (8) and the fact that $\Sigma_i \Omega M_i \subset \Omega - \Omega_{2\delta}$ we get

$$\int_{\Omega - \Omega_\delta} |f|^p\, dx \leq c_6 16^n \delta^{pr} \sum_{|k|=r} \int_{\Omega - \Omega_{2\delta}} |D^k f|^p\, dx.$$

Q.E.D.

Arguments analogous to those developed in the proof of Theorem 1 lead to the following assertion.

Let $\Gamma'(\Omega)$ denote the set of those points $x \in \Gamma(\Omega)$ that can be touched by some ball (depending on x) lying in Ω.

LEMMA 2. *Suppose $\Omega \neq E_n$ is an open set and $f \in W_p^r(\Omega)$, $1 \leq p < \infty$. If for any $x \in \Gamma'(\Omega)$ and any $\delta > 0$*

$$\|f\|_{L_p(\Omega K_\delta^x)} \leq c_7 \delta^r \sum_{|k|=r} \|D^k f\|_{L_p(\Omega K_\delta^x)}, \tag{9}$$

where K_δ^x is the ball of radius δ with center at x and c_7 depends only on r, p and n, then inequality (5) *with a constant depending only on r, p and n is valid for any $\delta > 0$.*

PROOF. We construct balls L_i and M_i in the same way as in the proof of Theorem 1. We note that the centers of the balls M_i are the points belonging to $\Gamma'(\Omega)$. It follows from (9) that

$$\int_{\Omega M_i} |f|^p\,dx \leqslant \bar{c}_7 (2\delta)^{pr} \sum_{|k|=r} \int_{\Omega M_i} |D^k f|^p\,dx,$$

which implies (7). The subsequent arguments are the same as in the proof of Theorem 1.

LEMMA 3. *Suppose $f \in W_p^r(\Omega)$, where Ω is an open set and $1 \leqslant p < \infty$. If for some compactum S*

$$\|f\|_{L_1[S(\Omega-\Omega_\delta)]} = o(\delta^r),\ \ \delta \to 0,$$

then

$$\Phi(x) = \begin{cases} f(x), & x \in \Omega \\ 0, & x \notin \Omega \end{cases} \in W_p^r(E_n). \tag{10}$$

PROOF. It suffices to prove that $\Phi(x)$ has generalized derivatives $D^k\Phi$ of order $|k| \leqslant r$ on E_n, with

$$D^k\Phi(x) = \begin{cases} D^k f(x), & x \in \Omega \\ 0, & x \notin \Omega \end{cases}.$$

Suppose $\phi \in C_0^{|k|}(E_n)$, with $\phi(x) = 0$ outside the compactum S. We must prove that

$$\int_{E_n} \Phi D^k \varphi dx = (-1)^{|k|} \int_{E_n} D^k \Phi \varphi dx,$$

i.e.

$$\int_{\Omega S} f D^k \varphi dx = (-1)^{|k|} \int_{\Omega S} D^k f \varphi dx. \tag{11}$$

If $\Omega S = \varnothing$ or $S \subset \Omega$, this equality is obvious (since $D^k f$ is a generalized derivative of f in Ω). Suppose now $\Omega S \neq \varnothing$ and $S \subset \Omega$. Let χ_δ denote the moving average with step $\delta/2$ of the characteristic function of $\Omega_{3\delta/2}$. We have $\chi_\delta(x) \in C_0^\infty(\Omega)$, with $\chi_\delta(x) = 0$ outside Ω_δ and $\chi_\delta(x) = 1$ in $\Omega_{2\delta}$; moveover,

$$|D^\alpha \chi_\delta(x)| \leqslant \frac{c_\alpha}{\delta^{|\alpha|}}. \tag{12}$$

Further,

$$\int_{\Omega S} f D^k \varphi dx = \int_{\Omega S} f D^k (\varphi \chi_\delta)\,dx + \int_{\Omega S} f D^k [\varphi(1-\chi_\delta)]\,dx. \tag{13}$$

Since $\varphi\chi_\delta \in C_0^{|k|}(\Omega)$, we have

$$\int_{\Omega S} f D^k (\varphi\chi_\delta)\,dx = (-1)^{|k|} \int_{\Omega S} D^k f \varphi \chi_\delta dx. \tag{14}$$

Taking into account the fact that

$$D^k[\varphi(1-\chi_\delta)] = (D^k\varphi)(1-\chi_\delta) - \sum_{\substack{\alpha \leqslant k \\ |\alpha| > 0}} C_k^\alpha D^{k-\alpha}\varphi \cdot D^\alpha \chi_\delta,$$

we get from (13) and (14) that

$$\int_{\Omega S} fD^k\varphi dx = (-1)^{|k|} \int_{\Omega S} D^k f\cdot \varphi\chi_\delta dx + \int_{(\Omega-\Omega_{2\delta})S} fD^k\varphi\cdot(1-\chi_\delta)\,dx - \sum_{\substack{\alpha\leqslant K\\ |\alpha|>0}} C_k^\alpha \int_{(\Omega_\delta-\Omega_{2\delta})S} fD^{k-\alpha}\varphi\cdot D^\alpha\chi_\delta dx. \tag{15}$$

Let us pass to the limit for $\delta \longrightarrow 0$. Then

$$\int_{\Omega S} D^k f\cdot\varphi\chi_\delta dx \to \int_{\Omega S} D^k f\cdot\varphi dx,$$

since

$$\left|\int_{\Omega S} D^k f\cdot\varphi(1-\chi_\delta)\,dx\right| \leqslant \int_{(\Omega-\Omega_{2\delta})S} |D^k f|\cdot|\varphi|\,dx \leqslant c_1 \int_{(\Omega-\Omega_{2\delta})S} |D^k f|\,dx \to 0.$$

It is proved analogously that

$$\int_{(\Omega-\Omega_{2\delta})S} f\cdot D^k\varphi\cdot(1-\chi_\delta)\,dx \to 0.$$

Further, taking (12) into consideration, we get ($|\alpha| \leqslant |k| \leqslant r$)

$$\left|\int_{(\Omega_\delta-\Omega_{2\delta})S} fD^{k-\alpha}\varphi\cdot D^\alpha\chi_\delta dx\right| \leqslant \frac{c_2}{\delta^{|\alpha|}} \int_{(\Omega_\delta-\Omega_{2\delta})S} |f|\,dx \to 0,\ \delta\to 0.$$

Thus (11) follows from (15) when $\delta \longrightarrow 0$. Q.E.D.

LEMMA 4. *Suppose $f \in W_p^r(\Omega)$, $1 \leqslant p < \infty$, where Ω is an arbitrary open set in E_n. If there exists a sequence φ_s such that*

$$\varphi_s \in C_0^\infty(\Omega),\ \lim_{s\to\infty} \|f-\varphi_s\|_{W_p^r(\Omega)} = 0, \tag{16}$$

then $\Phi(x) \in W_p^r(E_n)$.

PROOF. We have

$$\|\varphi_s-\varphi_t\|_{W_p^r(E_n)} = \|\varphi_s-\varphi_t\|_{W_p^r(\Omega)} \leqslant \|\varphi_s-f\|_{W_p^r(\Omega)} + \|f-\varphi_t\|_{W_p^r(\Omega)} \to 0,$$

By virtue of the completeness of the space $W_p^r(E_n)$ there exists a function $F(x) \in W_p^r(E_n)$ such that

$$\lim_{s\to\infty} \|F-\varphi_s\|_{W_p^r(E_n)} = 0;$$

in particular,

$$\lim_{s\to\infty} \|F-\varphi_s\|_{L_p(E_n)} = 0,$$

which implies that

$$\|F\|_{L_p(E_n-\Omega)} = 0, \qquad \|F-f\|_{L_p(\Omega)} = 0,$$

i.e. $F(x) = \Phi(x)$ almost everywhere. Q.E.D.

THEOREM 2. *Suppose $f \in W_p^r(\Omega)$, $r = 1, 2, \ldots, p > n$, where Ω is an arbitrary open* *set in E_n. In order for there to exist a sequence φ_s such that* (16) *holds it is necessary and sufficient that*

$$\|f\|_{L_p(\Omega-\Omega_\delta)} = o(\delta^r), \quad \delta \to 0.^{(3)} \tag{17}$$

PROOF. The necessity follows from Theorem 1. Suppose first Ω is a bounded open set and condition (17) is satisfied. Let (as in [4])

$$f_\delta(x) = \frac{1}{\delta^n} \int\limits_{\Omega_{2\delta}} \omega\left(\frac{x-y}{\delta}\right) f(y)\, dy \in C_0^\infty(\Omega),$$

where $\omega(z) \in C^\infty(E_n)$, $\omega(z) = 0$ for $|z| \geqslant 1$ and $\int_{E_n} \omega(z)\,dz = 1$, and let us prove that

$$\lim_{\delta \to 0} \|f - f_\delta\|_{W_p^r(\Omega)} = 0.$$

We have

$$\|f - f_\delta\|_{W_p^r(\Omega)} \leqslant \|f - f_\delta\|_{W_p^r(\Omega_{3\delta})} + \|f\|_{W_p^r(\Omega-\Omega_{3\delta})} + \|f_\delta\|_{W_p^r(\Omega-\Omega_{3\delta})}.$$

Since for $x \in \Omega_{3\delta}$ the function $f_\delta(x)$ is a weighted moving average with radius δ of $f(x)$ or $\Phi(x)$ (see (10)), i.e. $f_\delta(x) = [f(x)]_\delta = [\Phi(x)]_\delta$, $x \in \Omega_{3\delta}$, and $\Phi(x) \in W_p^r(E_n)$ (Lemma 3), it follows on the basis of the properties of mean functions (see [1]) that

$$\|f - f_\delta\|_{W_p^r(\Omega_{3\delta})} = \|\Phi - [\Phi]_\delta\|_{W_p^r(\Omega_{3\delta})} \leqslant \|\Phi - [\Phi]_\delta\|_{W_p^r(E_n)} \to 0.$$

As $\delta \to 0$

$$\|f\|_{W_p^r(\Omega-\Omega_{3\delta})} \to 0.$$

Let us now estimate $\|f_\delta\|_{W_p^r(\Omega-\Omega_{3\delta})} = \|f_\delta\|_{W_p^r(\Omega_\delta-\Omega_{3\delta})}$. Since

$$D^k f_\delta(x) = \frac{1}{\delta^{n+|k|}} \int\limits_{\substack{y \in \Omega_{2\delta} \\ |x-y| \leqslant \delta}} D^k\omega\left(\frac{x-y}{\delta}\right) f(y)\, dy = \frac{1}{\delta^{n+|k|}} \int\limits_{\substack{x-z \in \Omega_{2\delta} \\ |z| \leqslant \delta}} D^k\omega\left(\frac{z}{\delta}\right) f(x-z)\,dz,$$

it follows that

$$\begin{aligned}
\|D^k f_\delta\|_{L_p(\Omega_\delta-\Omega_{3\delta})} &\leqslant \frac{1}{\delta^{n+|k|}} \left\| \int\limits_{|z| \leqslant \delta} \left| D^k\omega\left(\frac{z}{\delta}\right)\right| \cdot |f(x-z)|\,dz \right\|_{L_p(\Omega_\delta-\Omega_{3\delta})} \\
&\leqslant \frac{1}{\delta^{n+|k|}} \int\limits_{|z| \leqslant \delta} \left| D^k\omega\left(\frac{z}{\delta}\right)\right| \cdot \|f(x-z)\|_{L_{p,x}(\Omega_\delta-\Omega_{3\delta})}\,dz \\
&\leqslant \frac{1}{\delta^{|k|}} \|f\|_{L_p(\Omega-\Omega_{4\delta})} \int\limits_{|z| \leqslant 1} |D^k\omega(z)|\,dz \leqslant \frac{c_k}{\delta^r} \|f\|_{L_p(\Omega-\Omega_{4\delta})} \to 0
\end{aligned}$$

as $\delta \to 0$. If Ω is an unbounded open set, we must consider in place of $f_\delta(x)$ the function $f_\delta(x)\chi_\delta(x)$, where $\chi_\delta(x) = \chi(\delta x)$, $\chi(u) \in C^\infty(E_n)$, $\chi(u) = 1$ for $|u| \leqslant 1$ and $\chi(u) = 0$ for $|u| \geqslant 2$.

REMARK. The sufficiency of condition (17) has been proved for any $p \geqslant 1$.

COROLLARY 1. *When $f \in W_p^r(\Omega)$ and $p > n$, condition* (17) *is equivalent to the condition*

---

(3) If $\Omega = E_n$, we will consider $(E_n)_\delta = E_n$, so that condition (17) is trivially satisfied for any f.

$$\| D^k f \|_{L_p(\Omega-\Omega_\delta)} = o(\delta), \qquad |k| \leq r-1, \delta \to 0. \tag{18}$$

PROOF. From (18) it follows that $\|f\|_{L_p(\Omega-\Omega_\delta)} = o(\delta)$. Hence, by virtue of Theorem 2, (16) is satisfied for $r = 1$. From (16) and Theorem 1 we get the validity of inequality (5) for $r = 1$. From (5) and (18) it follows that $\|f\|_{L_p(\Omega-\Omega_\delta)} = o(\delta^2)$, and so on. Thus (18) implies (17). The converse implication is obtained by applying Theorem 2 to the functions $D^k f \in W_p^{r-|k|}(\Omega)$.

COROLLARY 2. *When $f \in W_p^r(\Omega)$ and $p > n$, condition* (17) *is equivalent to the following condition*: *inequality* (9) *is satisfied for any $x \in \Gamma'(\Omega)$.*

PROOF. Suppose (17) and hence (16) are satisfied. The functions $\varphi_s \in C_0^\infty(\Omega)$ satisfy (9) for any $x \in \Gamma$ by virtue of (4). By passing to the limit, we obtain (9) for $f \in W_p^r(\Omega)$. Conversely, inequality (5) and with it (17) follow from (9) by virtue of Lemma 2.

Let us consider condition (18) in more detail for $n = 1, p = 1$ and $\Omega = (a, b)$.(4) In this case it takes the form

$$\lim_{\delta\to 0} \frac{\int_a^{a+\delta} |f^{(m)}|\,dx}{\delta} = \lim_{\delta\to 0} \frac{\int_{b-\delta}^{b} |f^{(m)}|\,dx}{\delta} = 0, \qquad m = 0, 1, \ldots, r-1. \tag{19}$$

We will obtain conditions of type (19) below for $n = 1, p > 1$ and for $n > 1$.

THEOREM 3. *Suppose $f \in W_p^r(\Omega)$, $p > n$, where Ω is an arbitrary open set in E_n. In order for there to exist a sequence φ_s such that* (16) *holds it is necessary and sufficient that*

$$\forall x \in \Gamma(\Omega) \qquad \lim_{\delta\to 0} \left(\frac{\int_{\Omega K_\delta^x} |D^k f|^q\,dx}{\operatorname{meas} \Omega K_\delta^x} \right)^{1/q} = 0, \quad |k| \leq r-1. \tag{20}$$

The conditions (20) *for different $q \in [1, \infty]$ are equivalent. Further, the condition* (20′) *obtained by replacing the set $\Gamma(\Omega)$ in* (20) *by the set $\Gamma'(\Omega)$ is equivalent to condition* (20).

When $q = \infty$, condition (20) means that

$$\lim_{\delta\to 0} \operatorname*{ess\,sup}_{\Omega K_\delta^x} |D^k f| = 0, \quad |k| \leq r-1. \tag{21}$$

If we take into account the fact that $p > n$, we can rewrite condition (20) in the following manner: the function $f \in W_p^r(\Omega)$ is equivalent to a function $\bar{f}$ that is $r-1$ times continuously differentiable in Ω and satisfies the condition

$$\lim_{\substack{y\to x\\ y\in\Omega}} D^k \bar{f}(y) = 0, \quad |k| \leq r-1, \tag{22}$$

for any $x \in \Gamma(\Omega)$. Condition (22) is similar to the corresponding condition for the spaces $C^r(\Omega)$ (see the beginning of the article).

---

(4) Theorem 2 remains true for $p = n = 1$. This follows from the fact that in the one-dimensional case it is not difficult to obtain the following inequality in place of inequality (1) for any $p \geqslant 1$:

$$\| f \|_{L_p(a,b)} \leqslant (b-a)^{1/p} (\beta - \alpha)^{-1/q} \| f \|_{L_q(\alpha,\beta)} + (b-a) \| f' \|_{L_p(a,b)}$$

(starting from the equality $f(y) = f(x) + \int_x^y f'(\xi)\,d\xi$), where $1 \leqslant q \leqslant \infty$.

Thus (22) is the "strongest" of the conditions mentioned in Theorem 3, while the "weakest" is the condition that for any $x \in \Gamma'(\Omega)$

$$\lim_{\delta \to 0} \frac{\int\limits_{\Omega K_\delta^x} |D^k f| \, dx}{\operatorname{meas} \Omega K_\delta^x} = 0, \qquad |k| \leqslant r-1. \tag{23}$$

PROOF. It suffices to prove that (16) implies (22) while (23) implies (16).

1. Suppose (16) is satisfied. Then by virtue of Lemma 4

$$\Phi(x) = \begin{cases} f(x), & x \in \Omega \\ 0, & x \notin \Omega \end{cases} \in W_p^r(E_n).$$

Since $p > n$, it follows from the imbedding theorem $W_p^r(E_n) \to C^{r-1}(E_n)$ that

$$\begin{aligned} \|\bar{f} - \varphi_s\|_{C^{r-1}(\Omega)} &= \|\bar{\Phi} - \varphi_s\|_{C^{r-1}(E_n)} \\ &\leqslant c\|\Phi - \varphi_s\|_{W_p^r(E_n)} = c\|f - \varphi_s\|_{W_p^r(\Omega)} \to 0, \qquad s \to \infty, \end{aligned}$$

where $\bar{\Phi}$ is a function from $C^{r-1}(E_n)$ that is equivalent to Φ and $\bar{f}$ is the restriction of $\bar{\Phi}$ to Ω. Since

$$\lim_{s\to\infty} \|\bar{f} - \varphi_s\|_{C^{r-1}(\Omega)} = 0, \qquad s \to \infty,$$

(22) is satisfied for any $x \in \Gamma(\Omega)$ (see the beginning of the article).

2. Suppose (23) is satisfied. As in the proof of Theorem 1, we cover the whole space E_n by open balls of radius $\delta/2$ with centers at the points $(\delta l_1/2, \ldots, \delta l_n/2)$, where the l_i are arbitrary integers. Consider those of the indicated balls that intersect the set $\Omega_{\delta/2} - \Omega_\delta$. We denote these balls by L_i and their centers by $x^{(i)}$. Further (in contrast to the proof of Theorem 1), we denote by M_i the ball with center at $x^{(i)}$ of radius $\rho_i = \rho(x^{(i)}, \Gamma) \leqslant 2\delta$. Let $y^{(i)}$ be a point of the boundary such that $\rho(x^{(i)}, y^{(i)}) = \rho_i$, and let K_h be the ball with center at $y^{(i)}$ of radius h. By virtue of Corollary 1 to Lemma 1

$$f\|_{L_p(M_i)} \leqslant c_1 \left[(1 + (2\delta)^{r-1}) (\operatorname{meas} M_i)^{1/p} \frac{\|f\|_{W_1^{r-1}(K_L M_i)}}{\operatorname{meas} K_h M_i} + (2\delta)^r \sum_{|k|=r} \|D^k f\|_{L_p(M_i)} \right],$$

where c_1 depends only on r, p and n. Since

$$\operatorname{meas} K_h M_i \geqslant v_n (h/2)^n \geqslant 2^{-n} \operatorname{meas} \Omega K_h$$

for $h \leqslant 2\delta$, it follows by virtue of (23) that

$$\frac{\|f\|_{W_1^{r-1}(K_h M_i)}}{\operatorname{meas} K_h M_i} \leqslant 2^n \frac{\|f\|_{W_1^{r-1}(\Omega K_h)}}{\operatorname{meas} \Omega K_h} \to 0$$

as $h \to 0$. Consequently

$$\|f\|_{L_p(M_i)} \leqslant c_1 (2\delta)^r \sum_{|k|=r} \|D^k f\|_{L_p(M_i)}.$$

Continuing the argument in the same way as in the proof of Theorem 1, we get

$$\|f\|_{L_p(\Omega_{\delta/2}-\Omega_\delta)} \leqslant c_2\delta^r \sum_{|k|=r} \|D^k f\|_{L_p(\Omega-\Omega_{3\delta})}.$$

Further,

$$\|f\|_{L_p(\Omega-\Omega_\delta)} \leqslant \sum_{m=1}^{\infty} \|f\|_{L_p(\Omega_{\delta/2^{m+1}}-\Omega_{\delta/2^m})}$$

$$\leqslant \sum_{m=1}^{\infty} c_2(\delta/2^m)^r \sum_{|k|=r} \|D^k f\|_{L_p(\Omega-\Omega_{3\delta})} \leqslant 2c_2\delta^r \sum_{|k|=r} \|D^k f\|_{L_p(\Omega-\Omega_{3\delta})},$$

i.e. $\|f\|_{L_p(\Omega-\Omega_\delta)} = o(\delta^r)$ for $\delta \to 0$. Q.E.D.

In Theorem 3 one is concerned with the "pointwise" convergence to zero of a function and its derivatives on the boundary $\Gamma(\Omega)$. Below we give a theorem in which the condition equivalent to (16) is expressed in terms of convergence in the mean with respect to a "boundary strip".

THEOREM 4. *Suppose $f \in W_p^r(\Omega)$, $p > n$, where Ω is an open set in E_n satisfying the condition*

$$\text{meas}\,(\Omega - \Omega_\delta) \leqslant c_8\delta, \tag{24}$$

in which c_8 does not depend on δ. In order for (16) *to be satisfied it is necessary and sufficient that for any ball K with center at any point $x \in \Gamma(\Omega)$*

$$\lim_{\delta\to 0}\left(\frac{\int_{K(\Omega-\Omega_\delta)} |D^k f|^q\,dx}{\text{meas}\,K(\Omega-\Omega_\delta)}\right)^{1/q} = 0, \qquad |k| \leqslant r-1, \tag{25}$$

where $q \in [1, \infty]$. The conditions (25) *for different q are equivalent to each other.*

When $q = \infty$, condition (25) means that

$$\lim_{\delta\to 0}\ \operatorname*{ess\,sup}_{K(\Omega-\Omega_\delta)} |D^k f| = 0, \qquad |k| \leqslant r-1. \tag{26}$$

PROOF. *Necessity.* Condition (16) implies (22) by virtue of Theorem 3. According to Lemma 2 of [5] it follows from (22) that

$$\lim_{\delta\to 0}\|D^k \bar f\|_{C(\Omega-\Omega_\delta)} = 0, \quad |k| \leqslant r-1.$$

Hence (26) is satisfied for any K. But (26) implies (25) for any $q \in [1, \infty]$ and for any ball K with center at any point $x \in \Gamma(\Omega)$.

Sufficiency. From Hölder's inequality it follows that

$$\frac{\|D^k f\|_{L_1[K(\Omega-\Omega_\delta)]}}{\text{meas}\,K(\Omega-\Omega_\delta)} \leqslant \frac{\|D^k f\|_{L_q[K(\Omega-\Omega_\delta)]}}{[\text{meas}\,K(\Omega-\Omega_\delta)]^{1/q}},$$

which implies, in particular, by virtue of (25) for any q and (24) that for any compactum S

$$\|D^k f\|_{L_1[S(\Omega-\Omega_\delta)]} = o(\delta), \qquad |k| \leqslant r-1. \tag{27}$$

It follows from this condition for $k = 0$ that the function Φ defined by (1) belongs to $W_p^1(E_n)$. Again applying Lemma 3 and condition (27) for $|k| = 1$, we get that $\Phi \in W_p^2(E_n)$, and so on. Thus $\Phi \in W_p^r(E_n)$, with

$$D^k\Phi(x) = \begin{cases} D^k f(x), & x \in \Omega \\ 0, & x \notin \Omega \end{cases}$$

for $|k| \leqslant r$.

Making use of Corollary 1 to Lemma 1 for the function Φ, we get that for any $x \in \Gamma(\Omega)$

$$\|f\|_{L_p(\Omega K_\delta^x)} = \|\Phi\|_{L_p(K_\delta^x)} \leqslant \tilde{c}_1 \Big\{(1 + \delta^{r-1})\, \delta^{n/p}\, [\text{meas } K_\delta^x(\Omega - \Omega_h)]^{-1/q}$$

$$\times \|\Phi\|_{W_q^{r-1}[K_\delta^x(\Omega-\Omega_h)]} + \delta^r \sum_{|k|=r} \|D^k\Phi\|_{L_p(K_\delta^x)}\Big\}$$

$$= \tilde{c}_1 \Big\{(1 + \delta^{r-1})\, \delta^{n/p}\, [\text{meas } K_\delta^x(\Omega - \Omega_h)]^{-1/q} \|f\|_{W_q^{r-1}[K_\delta^x(\Omega-\Omega_h)]}$$

$$+ \delta^r \sum_{|k|=r} \|D^k f\|_{L\,(\Omega K_\delta^x)}\Big\}.$$

Passing to the limit for $h \to 0$ and taking (25) into account, we get that for $x \in \Gamma(\Omega)$

$$\|f\|_{L_p(\Omega K_\delta^x)} \leqslant \tilde{c}_1 \delta^r \sum_{|k|=r} \|D^k f\|_{L_p(\Omega K_\delta^x)},$$

which implies (16) (see Corollary 2 to Theorem 2).

REMARK 1. Condition (24) cannot be weakened, i.e. it cannot be replaced by the condition

$$\text{meas } (\Omega - \Omega_\delta) \leqslant c_8 \varphi(\delta)\, \delta,$$

where $\varphi(\delta) > 0$ and $\varphi(\delta) \to \infty$ as $\delta \to 0$, as the following example shows. It suffices to consider a function $\varphi(\delta)$ having a continuous derivative $\varphi'(\delta)$ such that $\varphi'(\delta) \geqslant -1/\delta$, inasmuch as for any function $\varphi(\delta) \to \infty$ as $\delta \to 0$ it is possible to construct a function $\varphi_1(\delta)$, $0 < \varphi_1(\delta) \leqslant \varphi(\delta)$, satisfying the above condition.

Suppose $n = 1$ and $\Omega = (-1, 0) + \Sigma_{k=1}^\infty (a_k, b_k)$, where $b_k - a_k = 2\varphi^{-1}(k)$ and $a_k \to 0$ as $k \to \infty$. For $\delta < 1/2$

$$\text{meas}\,(\Omega - \Omega_\delta) = 2\delta + \sum_{b_k - a_k \geqslant 2\delta} 2\delta + \sum_{b_k - a_k < 2\delta} (b_k - a_k)$$

$$= 2\delta + \sum_1^{[\varphi(\delta)]} 2\delta + \sum_{[\varphi(\delta)]+1}^\infty 2\varphi^{-1}(k) \leqslant 2\delta\,(1 + \varphi(\delta)) + 2\varphi^{-1}([\varphi(\delta)] + 1)$$

$$+ 2\int_{[\varphi(\delta)]+1}^\infty \varphi^{-1}(x)\, dx \leqslant 2\delta\,(2 + \varphi(\delta)) + 2\int_{\varphi(\delta)}^\infty \varphi^{-1}(x)\, dx$$

$$= 2\delta\,(2 + \varphi(\delta)) + 2\int_0^\delta y\,|\varphi'(y)|\, dy \leqslant 2\delta\,(3 + \varphi(\delta)).$$

Thus

$$\text{meas } (\Omega - \Omega_\delta) \sim 2\delta\varphi(\delta)$$

for $\delta \to 0$.

Suppose now that

$$f(x)=\begin{cases} x+1, & x\in(-1,0);\\ 0, & x\in\sum_{k=1}^{\infty}(a_k,b_k).\end{cases}$$

Condition (22) is not satisfied for this function $f\in W_p^1(\Omega)$ at $x=0$, and hence (16) is not satisfied. On the other hand, condition (25) is satisfied for $1\leqslant q\leqslant\infty$ (if $x=0$, then $\int_{K(\Omega-\Omega_\delta)}|f|^q\,dx\leqslant\delta$, while meas $K(\Omega-\Omega_\delta)\sim 2\delta\varphi(\delta)$ for $\delta\to 0$).

REMARK 2. From the proof of Theorem 4 we see that $\Gamma(\Omega)$ can be replaced by $\Gamma'(\Omega)$ in the statement of condition (25).

Suppose now Ω is a bounded open set such that for any ball K with center at any point $x\in\Gamma(\Omega)$

$$c_4\delta\leqslant \text{meas}\,K(\Omega-\Omega_\delta)\leqslant c_5\delta, \tag{28}$$

where c_4 and c_5 depend only on Ω and K (this is essentially an assumption on the "uniform $(n-1)$-dimensionality" of the boundary $\Gamma(\Omega)$). From Theorem 4 we obtain the following theorem.

THEOREM 5. *Suppose $f\in W_p^r(\Omega)$, $p>n$, where Ω is a bounded open set in E_n satisfying condition* (28). *In order for condition* (16) *to be satisfied it is necessary and sufficient that*

$$\lim_{\delta\to 0}\left(\frac{\int_{\Omega-\Omega_\delta}|D^k f|^q\,dx}{\text{meas}\,(\Omega-\Omega_\delta)}\right)^{1/q}=0,\qquad |k|\leqslant r-1. \tag{29}$$

The conditions (29) *for different $q\in[1,\infty]$ are equivalent.*

In conclusion, we prove the following theorem on the connection between the density in $W_p^r(\Omega)$ of infinitely differentiable functions with compact support and the possibility of extending a function $f\in W_p^r(\Omega)$ by zero to the whole space with preservation of class.

THEOREM 6. *Suppose Ω is an open set in E_n. In order for condition* (16) *to be equivalent to condition* (10) *for any function $f\in W_p^r(\Omega)$ it is necessary and sufficient that for any neighborhood K of any point $x\in\Gamma(\Omega)$*

$$\text{meas}(K-\Omega)>0. \tag{30}$$

PROOF. *Sufficiency.* Suppose $\Phi\in W_p^r(E_n)$, $x\in\Gamma(\Omega)$ and K_δ^x is the ball of radius δ with center at x. Making use of inequality (4) (the role of S in the present case is played by the set $K_\delta^x-\Omega$, $\text{meas}(K_\delta^x-\Omega)>0$, in which $\Phi(x)\equiv 0$), we get

$$\|f\|_{L_p(\Omega K_\delta^x)}=\|\Phi\|_{L_p(K_\delta^x)}\leqslant \bar c_1\delta^r\sum_{|k|=r}\|D^k\Phi\|_{L_p(K_\delta^x)}=\bar c_1\delta^r\sum_{|k|=r}\|D^k f\|_{L_p(\Omega K_\delta^x)},$$

which implies (16) by virtue of Corollary 2 to Theorem 2.

Necessity. Suppose condition (30) is not satisfied: suppose there exists a point $x_0\in\Gamma(\Omega)$ and a ball K_δ with center at x_0 such that $\text{meas}(K_\delta-\Omega)=0$, and consider a function $\rho\in C_0^\infty(E_n)$ that is equal to one in $K_{\delta/3}$ and to zero outside $K_{2\delta/3}$. We put $f(x)=\rho(x)$ for $x\in\Omega$. The function $\Phi(x)$ belongs to $W_p^r(E_n)$ since it coincides almost everywhere with

$\rho(x)$. On the other hand, (22) at x_0 and hence (16) are not satisfied for $f(x)$, i.e. in this case conditions (16) and (10) are not equivalent.

Condition (30) means that in a certain sense the set $E_n - \Omega$ has "uniform dimension" n. In this context, it is interesting to compare condition (30) with condition (25) of Theorem 5 of [5], in which the spaces $C^r(\Omega)$ are considered in place of the spaces $W_p^r(\Omega)$.

If the open set Ω satisfies the condition

$$\Gamma(\Omega) = \Gamma(\overline{\Omega}), \tag{31}$$

then condition (29) is satisfied. This follows from the fact that condition (31) is equivalent to the following condition: for any $x \in \Gamma(\Omega)$ and any ball K_δ^x there exists a ball K such that $K \subset K_\delta^x$ and $K\Omega = \varnothing$ (see [5]).

We cite in addition a typical example of a set Ω for which (30), but not (31), is satisfied. Let S be a perfect nowhere dense set of positive (one-dimensional) measure on the interval $[-½, ½]$. The set

$$K - (\underbrace{S \times S \times \ldots \times S}_{n}),$$

where K is the unit open ball in E_n, can be taken as Ω.

BIBLIOGRAPHY

1. S. L. Sobolev, *Applications of functional analysis in mathematical physics*, Izdat. Leningrad. Gos. Univ., Leningrad, 1950; reprint, Izdat. Sibirsk. Otd. Akad. Nauk SSSR, Novosibirsk, 1962; English transl., Transl. Math. Monographs, vol. 7, Amer. Math. Soc., Providence, R. I., 1963. MR **14**, 565; **29** #2624.

2. G. G. Kazarjan, *The density in* $\overset{\circ}{W}{}_p^r(\Omega)$ *of smooth functions with compact support*, Mat. Zametki **2** (1967), 45–52 = Math. Notes **2** (1967), 514–517. MR **35** #3431.

3. E. Gagliardo, *Proprietà di alcune classi di funzioni in più variabili*, Ricerche Mat. **7** (1958), 102–137. MR **21** #1526.

4. V. I. Burenkov, *On the approximation of functions in* $C^r(\Omega)$, *where* Ω *is an arbitrary open set*, Dokl. Akad. Nauk SSSR **202** (1972), 16–18 = Soviet Math. Dokl. **13** (1972), 10–12. MR **45** #2462.

5. ———, *On approximating functions of the space* $C^r(\Omega)$ *by functions with compact support, for an arbitrary open set* Ω, Trudy Mat. Inst. Steklov. **117** (1972), 62–74 = Proc. Steklov Inst. Math. **117** (1972), 73–88.

6. ———, *On approximation of functions in Sobolev spaces by functions with compact support on an arbitrary open set*, Dokl. Akad. Nauk SSSR **202** (1972), 259–262 = Soviet Math. Dokl. **13** (1972), 60–64. MR **45** #2463.

Translated by S. SMITH

Trudy Mat. Inst. Steklov.
131 (1974)

Proc. Steklov Inst. Math.
131 (1974)

ON THE SEARCH FOR THE MAXIMUM OF A FUNCTION AND ON THE APPROXIMATE GLOBAL SOLUTION OF A SYSTEM OF NONLINEAR EQUATIONS

E. A. VOLKOV

UDC 518: 519,3

ABSTRACT. A method is considered for obtaining an upper estimate of the maximum of a twice continuously differentiable function on an n-dimensional closed cube with an error not greater than a given $\epsilon > 0$. The estimate is expressed in terms of (i) the values of the function at a finite number of points of the cube, which are chosen according to a well defined rule, and (ii) uniform lower estimates of the pure second derivatives of the function. Conditions are indicated under which the number of elementary operations used to obtain the estimate is of order $O(\ln \epsilon^{-1})$ while the number of intermediate quantities requiring storage is uniformly bounded with respect to $\epsilon > 0$.

A method is proposed and investigated for the approximate global solution of systems of n nonlinear equations with n unknowns on an arbitrary bounded closed domain $\overline{\Omega} \subset G \subset E^n$. Under the assumption that the functions giving the equations have continuous and bounded second derivatives on G, the method permits one to separate each solution on $\overline{\Omega}$ at which the Jacobian of the system of functions giving the equations is not equal to zero (to completely define a sphere containing it and no other solution) after a finite number of tests of certain elementary criteria. A separated solution can be found with any accuracy by an iterative method. If the system of equations has only a finite number of solutions on $\overline{\Omega}$ that can be separated by the given method, a sequence of closed sets is constructed that contracts to the set $\overline{\omega'} \subset \overline{\Omega}$ of nonseparable solutions.

Bibliography: 4 items.

Introduction

In the present paper we consider a method for obtaining an upper estimate of the maximum of a function with any given accuracy $\epsilon > 0$ on a closed n-dimensional cube and a method for the approximate solution of a system of n nonlinear equations with n unknowns on a bounded closed domain $\overline{\Omega}$ of n-dimensional Euclidean space. These methods have certain elements in common; namely, they are constructed with the use of a sequence of increasingly finer uniform cubic partitions of the space and are based on analogous assumptions concerning the twice continuous differentiability of the functions in question.

In §1 we prove auxiliary propositions on upper and lower estimates of the maximum of a function on the elementary cubes of cubic partitions, and we describe the method of finding an upper estimate of the maximum of a function on the basic unit cube with a given admissible error $\epsilon > 0$. The estimate of the maximum is expressed in terms of (i) certain values of the function at the vertices of the cubic partitions, and (ii) uniform lower estimates of the pure second derivatives of the function on the given unit cube.

AMS (MOS) subject classifications (1970). Primary 65H10; Secondary 65D99.

The method possesses, in particular, the following properties. If the function takes its maximal value at a finite number of points of the closed unit cube, and at each such point the second differential of the function with respect to the group of variables with respect to which the first partial derivatives vanish at the point in question is a definite quadratic form, then the desired estimate of the maximum is found with the use of $O(\ln \epsilon^{-1})$ elementary operations while the number of intermediate quantities requiring storage is uniformly bounded with respect to $\epsilon > 0$ (Theorem 1.1). But if the maximum is achieved only at certain vertices of the unit cube at which all of the first partial derivatives are not equal to zero, then the considered procedure is generally finite and the upper estimate coincides with the maximum of the function itself for sufficiently small ϵ.

A description is given in [1] of a method of searching for the maximum of a function that is constructed on other principles and for which the above-noted properties are not established.

In §2 we consider the system of nonlinear equations

$$f_k(x) = 0, \quad k = 1, 2, \ldots, n, \tag{1}$$

where the functions $f_k(x)$ have continuous and bounded second derivatives on a certain domain $G \supset \overline{\Omega}$.

Theorem 2.1 establishes, in terms of (i) the values of the functions f_k, $k = 1, \ldots, n$, at certain points of an elementary cube and (ii) uniform estimates of the second derivatives of these functions on G, a sufficient criterion for the existence of an isolated solution of system (1) that is unique in a completely defined sphere whose center coincides with the center of the elementary cube.

Theorem 2.2 guarantees, in the case when the Jacobian of the system of functions $\{f_k\}_1^n$ is different from zero at a point that is a solution of the system of equations (1), the fulfilment of this criterion for a sufficiently fine partition for all elementary cubes containing the given solution.

Theorem 2.3, which follows directly from the fixed point theorem for contractions, substantiates the convergence of the method of successive approximations for a separated isolated solution of system (1).

In §3 we construct and investigate a method for the approximate global solution of the system of nonlinear equations (1) on a closed domain $\overline{\Omega} \subset G$, for the substantiation of which we mainly use the results of the preceding section. We consider a sequence of uniform cubic partitions, each successive partition being effected by elementary cubes with edge length decreased by one half. For the elementary cubes of a given partition we test a series of criteria. Three cases are possible: 1) an isolated solution of system (1) is discovered that is unique within a certain sphere containing, in particular, the elementary cube in question, i.e. there occurs a separation of an isolated solution of system (1), which together with a sphere of uniqueness is "entered" in a special list; 2) the elementary cube is discarded; 3) the elementary cube is retained for subsequent consideration under a finer partition.

This method possesses the following basic properties. If the procedure of testing the criteria turns out to be finite, then the set of solutions of system (1) on $\overline{\Omega}$ consists of not more than a finite number of isolated points (Theorem 3.1). Every isolated solution of system (1) on $\overline{\Omega}$ for which the Jacobian of the system of functions $\{f_k\}_1^n$ is different from zero is separated, independently of the properties of the set of solutions of system (1) in the large, by means of a finite number of elementary tests (Theorem 3.2). When the set of solutions of system (1) on $\overline{\Omega}$ consists of a finite number of isolated solutions with Jacobian different from zero, the procedure of testing the criteria in the large is finite (Theorem 3.3). If the procedure of testing the criteria is infinite, then the sequence of residual sets (the unions of the undiscarded elementary cubes) possesses the property that the distance from the farthest point of a residual set to the set of solutions of system (1) on $\overline{\Omega}$ (from which any finite number of isolated solutions with Jacobian different from zero can be discarded) tends (with respect to the number of partitions) to zero (Theorems 3.4 and 3.5). It is guaranteed that at any stage every solution of system (1) that is situated on $\overline{\Omega}$ is either separated and determined in a special way or contained in the residual set. In this sense, taking into account the above-mentioned properties established by Theorems 3.4 and 3.5, the considered method is a global approximation method of solving a system of nonlinear equations.

According to Theorem 2.3, all of the separated solutions of system (1) can be calculated by an iterative method that converges in geometric progression, with any preassigned accuracy.

§1. Method of searching for a maximum of a function on a cube

Suppose there is defined on the n-dimensional cube $\overline{R} = \{x = (x_1, \ldots, x_n): 0 \leqslant x_i \leqslant 1, i = 1, \ldots, n\}$ a twice continuously differentiable function $f(x)$, with

$$-\infty < F_{jj} \leqslant \min_{\overline{R}} \frac{\partial^2 f}{\partial x_j^2}, \qquad j = 1, 2, \ldots, n, \tag{1.1}$$

where the F_{jj} are certain numbers, and it is required to find an estimate from above of the maximal value of f on $\overline{R}$ with an error not exceeding a given number $\epsilon > 0$.

LEMMA 1.1. *Let $\varphi(t)$ and $\psi(t)$ be twice continuously differentiable functions on an interval $[a, b]$, with $\psi(a) \leqslant \varphi(a)$, $\psi(b) \leqslant \varphi(b)$ and $\psi''(t) \geqslant \varphi''(t)$ $(a \leqslant t \leqslant b)$. Then $\psi(t) \leqslant \varphi(t)$ $(a \leqslant t \leqslant b)$.*

Lemma 1.1 is a corollary to Lemma 1 of [2].

LEMMA 1.2. *Let m and k be integers such that $m \geqslant 0$ and $0 \leqslant k \leqslant 2^m - 1$, and let $\varphi(t)$ be a twice continuously differentiable function on the interval $[0, 1]$, with $\varphi''(t) \geqslant \Phi_2$ $(0 \leqslant t \leqslant 1)$. Then*

$$\max_{[k2^{-m},\,(k+1)2^{-m}]} \varphi(t) \leqslant \Xi^m\left(\varphi\left(\frac{k}{2^m}\right), \varphi\left(\frac{k+1}{2^m}\right), \Phi_2\right),$$

where

$$\Xi^m(\alpha, \beta, \gamma) = \begin{cases} \max\{\alpha, \beta\}, & |\alpha - \beta| \geqslant -2^{-1-2m}\gamma, \\ \tilde{r}(\tau^*), & |\alpha - \beta| < -2^{-1-2m}\gamma, \end{cases} \tag{1.2}$$

$$\tilde{r}(\tau) = \alpha(1 - \tau) + \beta\tau + 2^{-1-2m}((\tau - 2^{-1})^2 - 2^{-2})\gamma,$$

$$\tau^* = 2^{-1} + 2^{2m}(\alpha - \beta)/\gamma.$$

PROOF. Let

$$P_k^m(t) = \varphi\left(\frac{k}{2^m}\right)\frac{(k+1)2^{-m}-t}{2^{-m}} + \varphi\left(\frac{k+1}{2^m}\right)\frac{t-k2^{-m}}{2^{-m}} + \frac{\Phi_2}{2}\left((t-(k+2^{-1})2^{-m})^2 - 2^{-2-2m}\right).$$

By Lemma 1.1,

$$\max_{[k2^{-m},\,(k+1)2^{-m}]} \varphi(t) \leqslant \max_{[k2^{-m},\,(k+1)2^{-m}]} P_k^m(t) = \Xi^m\left(\varphi\left(\frac{k}{2^m}\right), \varphi\left(\frac{k+1}{2^m}\right), \Phi_2\right).$$

Lemma 1.2 is proved.

Let m be a nonnegative integer, let $p = (p_1, \ldots, p_n)$, where the p_i are integers such that $0 \leqslant p_i \leqslant 2^m - 1$, $i = 1, \ldots, n$, let

$$M(m) = \{p : 0 \leqslant p_i \leqslant 2^m - 1, \quad i = 1,2, \ldots, n\},$$
$$\bar{\sigma}_p^m = \{x : p_i 2^{-m} \leqslant x_i \leqslant (p_i + 1)2^{-m}, \quad i = 1,2, \ldots, n\},$$
$$\delta^k = \{\delta_1^k, \ldots, \delta_n^k\}, \ \delta_i^k \text{ is the Kronecker delta,}$$

$$\lambda_0^m[f; p] = f\left(\frac{p_1}{2^m}, \ldots, \frac{p_n}{2^m}\right),$$
$$\lambda_k^m[f; p] = \Xi^m(\lambda_{k-1}^m[f; p + \delta^k], \lambda_{k-1}^m[f; p], F_{kk}), \qquad k = 1, 2, \ldots, n,$$
$$\Lambda_p^m[f; F_{11}, \ldots, F_{nn}] = \lambda_n^m[f; p], \tag{1.3}$$

let e_k^m be the set of vertices of the elementary cube $\bar{\sigma}_p^m$, and let

$$\tilde{F}_p^m = \max_{e_p^m} f(x), \quad \bar{F}_p^m = \Lambda_p^m[f; F_{11}, \ldots, F_{nn}]. \tag{1.4}$$

LEMMA 1.3. *The following inequalities hold*:

$$\tilde{F}_p^m \leqslant \max_{\bar{\sigma}_p^m} f(x) \leqslant \bar{F}_p^m, \tag{1.5}$$

$$0 \leqslant \bar{F}_p^m - \tilde{F}_p^m \leqslant 2^{-2m-3}\sum_{j=1}^n \max\{0, -F_{jj}\}. \tag{1.6}$$

PROOF. The left inequality of (1.5) is obvious. The right inequality of (1.5) is a consequence of (1.1) and Lemma 1.2, by virtue of which the quantity $\lambda_k^m[f; p]$, $k = 1, \ldots, n-1$, serves as an estimate from above of the maximum of the function f on the face of the cube $\bar{\sigma}_p^m$ of dimension k containing the point $(2^{-m}p_1, \ldots, 2^{-m}p_n)$ and orthogonal to the axes of coordinates with indices $j = k+1, k+2, \ldots, n$, and the quantity (1.3) is correspondingly an upper estimate of the maximum of f on the closed elementary cube $\bar{\sigma}_p^m$. Inequalties (1.6) follows from (1.2), (1.4) and (1.5). Lemma 1.3 is proved.

The right side of (1.6) does not depend on p. Consequently, on the basis of (1.5) and (1.6),

$$\max_{p \in M(m_*)} \tilde{F}_p^{m_*} - \varepsilon \leqslant \max_{\bar{R}} f(x) \leqslant \max_{p \in M(m_*)} \bar{F}_p^{m_*}, \tag{1.7}$$

where

$$m_* = \max\left\{0, -\left[2^{-1}\log_2\left(\varepsilon \Big/ \sum_{j=1}^{n} \max\{0, -F_{jj}\}\right)\right] - 1\right\}, \tag{1.8}$$

in which [ ] denotes the integral part; that is, the right part of (1.7) is the desired estimate of the maximum of f on $\overline{R}$. But, in order to obtain the given estimate, one must estimate the maximum of f on each elementary cube $\overline{\sigma}_p^{m_*} \subset \overline{R}$ of the cubic lattice of rank m_*; this requires $O(\epsilon^{-n/2})$ elementary operations.(1)

We will consider a somewhat modified method of obtaining an estimate of a maximum, which leads to the same estimate but, as a rule, is essentially more economical in regard to the number of elementary operations required for its realization.

Let $\Pi_0 = (0, \ldots, 0)$, let $q = (q_1, \ldots, q_n)$, where the q_i, $i = 1, 2, \ldots, n$, are integers, and for $m = 0, 1, \ldots$ let

$$\widetilde{F}^m = \max_{p \in \Pi_m} \widetilde{F}_p^m, \quad \overline{F}^m = \max_{p \in \Pi_m} \overline{F}_p^m, \tag{1.9}$$

$$\varepsilon_m = \overline{F}^m - \widetilde{F}^m, \tag{1.10}$$

$$\pi_m = \{p : \overline{F}_p^m \geqslant \widetilde{F}^m, \ p \in \Pi_m\}, \tag{1.11}$$

$$\Pi_{m+1} = \{q : 2p_i \leqslant q_i \leqslant 2p_i + 1, \ i = 1, 2, \ldots, n; \ p \in \pi_m\}. \tag{1.12}$$

Clearly, the quantity $\widetilde{F}^m$ is the maximum achieved by the function f at the vertices of the whole lattice formed by the cubic lattice of rank m on the cube $\overline{R}$, while

$$\overline{F}^m = \max_{p \in M(m)} \overline{F}_p^m,$$

i.e. $\overline{F}^{m_*}$ coincides with the global upper estimate considered above of the maximum of f on $\overline{R}$. The set π_m contains information only on those elementary cubes of the cubic lattice of rank m in which the upper estimate of the maximum of the function is not less than the maximum achieved by $\widetilde{F}^m$ on the whole lattice. The remaining elementary cubes of the cubic lattice of rank m, being obviously devoid of information on the maximum of the function, are discarded.

The cyclic process (1.9)–(1.12) terminates as soon as the inequality $\epsilon_m \leqslant \epsilon$ is satisfied for some $m = m'$. The quantity $\overline{F}^{m'}$ found in this way is an estimate from above of the maximum of f on $\overline{R}$ with an error not greater than ϵ and with, clearly, $m' \leqslant m_*$. It is also clear that the function f can take a maximal value only at the points of the set

$$\overline{Q}' = \bigcup_{p \in \pi_{m'}} \overline{\sigma}_p^{m'}.$$

When necessary, it is not difficult to specify a point (belonging to $\overline{Q}'$) at which the value of the function is equal to $\widetilde{F}^{m'}$, i.e. differs on the underside from the maximal value by not more than ϵ.

We note some characteristic cases and cite corresponding estimates of the number N of elementary operations used to find an estimate from above, depending on ϵ, of the maximum of a function on $\overline{R}$ by the method (1.9)–(1.12).

a) The function f takes its maximal value on a set $E \subset \overline{R}$ of positive n-dimensional

---

(1) By *elementary* we mean the arithmetic operations, the operation of calculating the value of a function f at a point and the operation of comparing two numbers.

measure. In this case, as in the case of estimate (1.7), $N = O(\epsilon^{-n/2})$.

b) $\min_{1 \leqslant j \leqslant n} F_{jj} \geqslant 0$, in which case $\overline{F}^0 = \widetilde{F}^0 = \max_{\overline{R}} f(x)$, i.e. $N = O(1)$.

c) The function f takes its maximal value only at one or several vertices of the cube $\overline{R}$, the following condition being satisfied at all of the indicated vertices:

$$\frac{\partial f}{\partial x_i} \neq 0, \quad i = 1, 2, \ldots, n. \tag{1.13}$$

As follows from estimate (1.4) upon taking (1.2) and (1.13) into account, the equality $\overline{F}^{m_*} = \widetilde{F}^{m_*}$ is satisfied for some finite m^* depending on f, i.e. $N = O(1)$.

d) $f(x) = -\sum_{i=1}^n (x_i - a_i)^{2\nu}$, where $\nu \geqslant 1$ is an integer and $a_i \in [0, 1]$ for $i = 1, \ldots, n$.

Let $F_{jj} = -4\nu$, $j = 1, \ldots, n$. We have $0 \geqslant \widetilde{F}^m \geqslant -n2^{-2\nu(m+1)}$. It is easily verified that if

$$\bar{\sigma}_p^m \cap \bar{R}_m = \varnothing, \quad p \in M(m),$$

where

$$\bar{R}_m = \{x : |x_i - a_i| \leqslant \sqrt[2\nu]{n(\nu+1)}\, 2^{-\frac{2m+1}{2\nu}}, \quad i = 1, 2, \ldots, n\},$$

then $\overline{F}_p^m < \widetilde{F}^m$. Hence, as simple calculations show, the number of elements of the set π_m does not exceed the quantity

$$C_m = (\sqrt[2\nu]{n(\nu+1)}\, 2^{m+1-\frac{2m+1}{2\nu}} + 2)^n. \tag{1.14}$$

From (1.8) and (1.14) it follows, in particular, that when $\nu = 1$ the number of elements of π_m (requiring storage for the practical realization of the method) is bounded from above by a number not depending on m, i.e. on ϵ, while $N = O(\ln \epsilon^{-1})$.

A more general assertion can be established analogously, as in case d).

THEOREM 1.1. *Suppose a twice continuously differentiable function f on the closed cube $\overline{R}$ takes its absolute maximal value at a finite number of points of $\overline{R}$, with* (1.1) *being satisfied. If at each such point the second differential of f with respect to the group of variables with respect to which the first partial derivatives vanish is a definite quadratic form, then an estimate from above of the maximum of f on $\overline{R}$ with an error not greater than a given $\epsilon > 0$ can be found by the method* (1.9)–(1.12) *with the use of $O(\ln \epsilon^{-1})$ elementary operations. In this connection, the number of elements of the intermediate sets requiring memorization is uniformly bounded with respect to $\epsilon > 0$.*

PROOF. We assume for the sake of simplicity that the point of absolute maximum of the function f on $\overline{R}$ is unique; we denote it by x^*. Under the conditions of Theorem 1.1 there exists an open sphere S^* of radius $\rho^* > 0$ with center at x^* and a number $\mu^* > 0$ such that on $\overline{R} \cap S^*$

$$f(x) \leqslant f(x^*) - \mu^* \rho^2(x, x^*), \tag{1.15}$$

where $\rho(x, x^*)$ is the distance between the points x and x^*.

Let S^{**} be the open sphere of radius $\rho^*/2$ with center at x^*, and let

$$\vartheta_m = \{p : \bar{\sigma}_p^m \subset \bar{R} \setminus S^{**}, p \in M(m)\}.$$

Clearly, $\lim_{m\to\infty}\widetilde{F}^m = f(x^*)$ and

$$\lim_{m\to\infty}\max_{p\in\vartheta_m}\bar{F}_p^m = \max_{\bar{R}\setminus S^{**}} f(x) < f(x^*),$$

and hence (see (1.11) and (1.12)) there exists an m' such that the condition

$$\bigcup_{p\in\Pi_m}\bar{\sigma}_p^m \subset \bar{R}\cap S^*$$

is satisfied for all $m \geqslant m'$. From this result, upon taking (1.1) and (1.15) into account and reproducing the estimates completely analogous to case d), we arrive at the conclusion that $N = O(\ln \epsilon^{-1})$ while the number of elements of the intermediate sets π_m, $m = 0, 1, \ldots$, requiring storage is uniformly bounded with respect to $\epsilon > 0$.

The extension of the above proof to the case of any finite number of points of absolute maximum if obvious. Theorem 1.1 is proved.

Theorem 1.1 covers a sufficiently general and the most natural case.

§2. Criteria for the existence of an isolated solution of a system of nonlinear equations

Let $f_k(x)$, $k = 1, \ldots, n$, be twice continuously differentiable functions on a bounded domain $G \subset E^n$, with

$$-\infty < \Phi_{ij}^{k1} \leqslant \inf_G \frac{\partial^2 f_k}{\partial x_i \partial x_j} \leqslant \sup_G \frac{\partial^2 f_k}{\partial x_i \partial x_j} \leqslant \Phi_{ij}^{k0} < \infty, \tag{2.1}$$

where the $\Phi_{ij}^{k\nu}$, $1 \leqslant i, j, k \leqslant n$ and $0 \leqslant \nu \leqslant 1$, are certain numbers.

We will consider the system of generally nonlinear equations

$$f_k(x) = 0, \qquad k = 1, 2, \ldots, n, \tag{2.2}$$

in a domain $\Omega \subset \bar{\Omega} \subset G$. Let

$$\bar{\omega} = \left\{x : \sum_{k=1}^n f_k^2(x) = 0, x \in \bar{\Omega}\right\}$$

denote the set of solutions of equations (2.2) on $\bar{\Omega}$. It will be assumed without loss of generality that $\Omega \subset \bar{R}$, where $\bar{R}$ is the closed unit cube defined in §1. Let m_0 be a natural number such that any closed cube $\bar{\sigma}_p^{m_0}$, $p \in M(m_0)$, with edge length 2^{-m_0} and having a nonempty intersection with $\bar{\Omega}$ entirely belongs to the domain G.

Suppose m is an integer not less than m_0, $\bar{\sigma}_p^m \subset G$, $k = 1, \ldots, n$ and

$$\bar{F}_{kp}^{m\nu} = (-1)^{1-\nu}\Lambda_p^m[f_k; \check{\alpha}_{k1}, \ldots, \check{\alpha}_{kn}], \tag{2.3}$$

where $\alpha_{ki}^\nu = (-1)^{1-\nu}\Phi_{ii}^{k\nu}$ and Λ_p^m is the quantity (1.3). On the basis of (2.1) and Lemma 1.3 we have

$$\begin{gathered} 0 \leqslant \min_{\bar{\sigma}_p^m} f_k - \bar{F}_{kp}^{m0} \leqslant 2^{-2m-3}\sum_{\mu=1}^n |\Phi_{\mu\mu}^{k0}|, \\ 0 \leqslant \bar{F}_{kp}^{m1} - \max_{\bar{\sigma}_p^m} f_k \leqslant 2^{-2m-3}\sum_{\mu=1}^n |\Phi_{\mu\mu}^{k1}|. \end{gathered} \tag{2.4}$$

Suppose also $\rho(x, y)$ is the distance between the points x and y of the space E^n, $S(x, r) = \{y: \rho(x, y) < r\}$, $\bar{S}(x, r) = \{y: \rho(x, y) \leqslant r\}$, $x^m_{(p)}$ is the center of the cube $\bar{\sigma}^m_p$, ρ^m_p is the distance from the point $x^m_{(p)}$ to the boundary ∂G of G and $A^m_p = \{a^m_{pij}\}^n_{i,j=1}$ is the square matrix with elements

$$a^m_{pij} = 2^{-m}[f_i(x^m_{(p)} + 2^{-1}\delta^j) - f_i(x^m_{(p)} - 2^{-1}\delta^j)]. \tag{2.5}$$

If $\det A^m_p \neq 0$, we denote the inverse of A^m_p by $A^{m,-1}_p = \{a^{m,-1}_{pij}\}^n_{i,j=1}$ and consider the system of equations

$$x = x - A^{m,-1}_p f(x) \equiv \psi(x) \tag{2.6}$$

(x and $f(x)$ are the columns with elements $x_1, \ldots, x_n$ and $f_1(x), \ldots, f_n(x)$ respectively), which is clearly equivalent to the given system of equations (2.2).(2) Let

$$\Phi^k_{ij} = \max\{|\Phi^{k0}_{ij}|, |\Phi^{k1}_{ij}|\},$$

$$\varkappa_j = \left(\sum_{\mu=1}^n (\Phi^j_{\mu\mu})^2\right)^{1/2}, \qquad \tau_j = \left(\sum_{\mu=1}^n \sum_{\nu=1}^n (\Phi^j_{\mu\nu})^2\right)^{1/2},$$

$$a^m_p = \left(\sum_{i=1}^n \left(\sum_{j=1}^n a^{m,-1}_{pij} f_j(x^m_{(p)})\right)^2\right)^{1/2},$$

$$b^m_p = 2^{-m-1}\left(\sum_{i=1}^n \left(\sum_{j=1}^n |a^{m,-1}_{pij}|\varkappa_j\right)^2\right)^{1/2},$$

$$c^m_p = \left(\sum_{i=1}^n \left(\sum_{j=1}^n |a^{m,-1}_{pij}|\tau_j\right)^2\right)^{1/2},$$

$$r^{m\nu}_p = \frac{1 - b^m_p - (-1)^\nu\sqrt{(1 - b^m_p)^2 - 4a^m_p c^m_p}}{2c^m_p} \tag{2.7}$$

$$(c^m_p > 0, \qquad \nu = 0, 1).$$

For real r^{m0}_p and r^{m1}_p we introduce the notation

$$l^{m0}_p = r^{m0}_p, \qquad l^{m1}_p = \min\{r^{m1}_p, \rho^m_p\}. \tag{2.8}$$

REMARK 2.1. The quantity c^m_p can vanish only in the uninteresting (to us) case when the system of equations (2.2) is linear on G. It will therefore be assumed in the sequel that $\Sigma^n_1\tau_j > 0$, i.e. $c^m_p > 0$ and $l^{m0}_p \geqslant 0$.

THEOREM 2.1. *Suppose* $\bar{\sigma}^m_p \subset G$, $\det A^m_p \neq 0$, *and*

$$b^m_p < 1, \qquad a^m_p < (1 - b^m_p)^2/4c^m_p \tag{2.9}$$

and, in addition, $l^{m0}_p < l^{m1}_p$. *Then the system of equations* (2.2) *has a unique solution on* $\bar{S}(x^m_{(p)}, l^{m0}_p)$ *and no solutions on the spherical annulus*

$$S(x^m_{(p)}, l^{m1}_p) \setminus \bar{S}(x^m_{(p)}, l^{m0}_p).$$

PROOF. Let x and y be arbitrary points of $S(x^m_{(p)}, l^{m1}_p) \subset G$, and let

(2) For the sake of simplicity we will use the same notation for points (of n-dimensional space) and columns (of dimension n) whose respective coordinates and elements with the same index are equal.

$$r = \max \{\rho (x, x_{(p)}^m),\ \rho (y, x_{(p)}^m)\}.$$

We have

$$\rho (x_{(p)}^m, \psi (x_{(p)}^m)) = a_p^m. \tag{2.10}$$

Moreover, on the basis of the theorem on finite increments and the triangle inequality

$$\begin{aligned}
&\rho (\psi (x), \psi (y)) \\
&= \left(\sum_{i=1}^n \left(\sum_{j=1}^n a_{pij}^{m,-1} \sum_{\mu=1}^n \frac{\partial f_j (\xi^j)}{\partial x_\mu} (y_\mu - x_\mu) - (y_i - x_i)\right)^2\right)^{1/2} \\
&= \left(\sum_{i=1}^n \left(\sum_{j=1}^n a_{pij}^{m,-1} \sum_{\mu=1}^n \left(\frac{\partial f_j (\xi^j)}{\partial x_\mu} - a_{pj\mu}^m\right) (y_\mu - x_\mu)\right)^2\right)^{1/2} \\
&\leqslant \left(\sum_{i=1}^n \left(\sum_{j=1}^n a_{pij}^{m,-1} \sum_{\mu=1}^n \left(\frac{\partial f_j (\xi^j)}{\partial x_\mu} - \frac{\partial f_j (x_{(p)}^m)}{\partial x_\mu}\right) (y_\mu - x_\mu)\right)^2\right)^{1/2} \\
&+ \left(\sum_{i=1}^n \left(\sum_{j=1}^n a_{pij}^{m,-1} \sum_{\mu=1}^n \left(\frac{\partial f_j (x_{(p)}^m)}{\partial x_\mu} - a_{pj\mu}^m\right) (y_\mu - x_\mu)\right)^2\right)^{1/2},
\end{aligned} \tag{2.11}$$

where the ξ^j, $j = 1, \ldots, n$, belong to the line segment joining the points x and y.

Since

$$\left|\frac{\partial f_j (x_{(p)}^m)}{\partial x_\mu} - a_{pj\mu}^m\right| \leqslant 2^{-m-1}\Phi_{\mu\mu}^j, \qquad 1 \leqslant j,\ \mu \leqslant n, \tag{2.12}$$

we get with the use of Cauchy's inequality that

$$\left|\sum_{\mu=1}^n \left(\frac{\partial f_j (x_{(p)}^m)}{\partial x_\mu} - a_{pj\mu}^m\right) (y_\mu - x_\mu)\right| \leqslant 2^{-m-1}\varkappa_j\rho (x, y), \qquad 1 \leqslant j \leqslant n. \tag{2.13}$$

Applying the theorem on finite increments and Cauchy's inequality, we find

$$\begin{aligned}
&\left|\sum_{\mu=1}^n \left(\frac{\partial f_j (\xi^j)}{\partial x_\mu} - \frac{\partial f_j (x_{(p)}^m)}{\partial x_\mu}\right) (y_\mu - x_\mu)\right| \\
&\leqslant \sum_{\mu=1}^n \sum_{\nu=1}^n \left|\frac{\partial^2 f_j (\eta^{j\mu})}{\partial x_\mu \partial x_\nu} \left(\xi_\nu^j - \frac{2p_\nu + 1}{2^{m+1}}\right) (y_\mu - x_\mu)\right| \leqslant \tau_j r\rho (x, y), \quad 1 \leqslant j \leqslant n,
\end{aligned} \tag{2.14}$$

where the $\eta^{j\mu}$, $\mu = 1, \ldots, n$, belong to the line segment joining the points $x_{(p)}^m$ and ξ^j.

From (2.11), (2.13) and (2.14) it follows that

$$\rho (\psi (x),\ \psi (y)) \leqslant (b_p^m + c_p^m r)\ \rho (x, y). \tag{2.15}$$

By virtue of (2.7)–(2.10) and (2.15), for any l satisfying the inequalities $l_p^{m\,0} \leqslant l < l_p^{m\,1}$ we have

$$\begin{aligned}
&\rho (\psi (x), \psi (y)) \leqslant \alpha_l\rho (x, y), \qquad x, y \in \bar{S} (x_{(p)}^m, l), \\
&\rho (x_{(p)}^m, \psi (x_{(p)}^m)) = (1 - b_p^m - c_p^m r_p^{m0})\, r_p^{m0} \\
&= (1 - b_p^m - c_p^m r_p^{m1})\, r_p^{m1} \leqslant (1 - \alpha_l)\, l,
\end{aligned} \tag{2.16}$$

where $\alpha_l = b_p^m + c_p^m l < b_p^m + c_p^m r_p^{m\,1} \leqslant 1$.

It follows on the basis of the fixed point theorem for contractions ([3], Chapter 7, §4, Theorem 2) that the system of equations (2.6) and hence system (2.2) have a unique solution on the closed sphere $\overline{S}(x^m_{(p)}, l)$, where $l \in [l^{m0}_p, l^{m1}_p)$. Theorem 2.1 is proved.

THEOREM 2.2. *Let $x^* \in \overline{\Omega}$ be a solution of the system of equations* (2.2) *at which*

$$\left.\frac{\partial(f_1, \ldots, f_n)}{\partial(x_1, \ldots, x_n)}\right|_{x=x^*} \neq 0. \tag{2.17}$$

Then the point x^* is an isolated solution on G of system (2.2) *and, in addition, there exists an integer $m^* \geqslant m_0$ such that the following four conditions are satisfied for any $m \geqslant m^*$ and any $p \in M(m)$ such that $x^* \in \overline{\sigma}^m_p$: $\overline{\sigma}^m_p \in G$, $\det A^m_p \neq 0$, (2.9) is satisfied and*

$$l^{m1}_p > 2l^{m0}_p + 2^{-m-1}\sqrt{n}. \tag{2.18}$$

PROOF. The isolatedness of the solution x^* of system (2.2) follows from the implicit function theorem (see, for example, [4], §41.5). The fulfilment of the first of the four conditions in the assertion of Theorem 2.2 is guaranteed by the choice of m_0.

Let $\gamma^m_* = \{p: x^* \in \overline{\sigma}^m_p, p \in M(m)\}$. On the basis of (2.1) and (2.12) we have

$$\lim_{m\to\infty} \max_{p\in\gamma^m_*} \max_{1\leqslant i, j\leqslant n} \left|\frac{\partial f_i(x^*)}{\partial x_j} - a^m_{pij}\right| = 0. \tag{2.19}$$

Relations (2.17) and (2.19) imply the existence of the limits

$$\lim_{m\to\infty} \max_{p\in\gamma^m_*} a^m_p = \lim_{m\to\infty} \max_{p\in\gamma^m_*} b^m_p = 0, \tag{2.20}$$

$$\lim_{m\to\infty} \min_{p\in\gamma^m_*} c^m_p = \lim_{m\to\infty} \max_{p\in\gamma^m_*} c^m_p = c^* > 0, \tag{2.21}$$

in which the positiveness of c^* follows from the assumption adopted in Remark 2.1.

The remaining assertions of Theorem 2.2 readily follow from (2.7), (2.8) and (2.19)–(2.21).

REMARK 2.2. From (2.7), (2.8), (2.20) and (2.21) we have

$$\lim_{m\to\infty} \max_{p\in\gamma^m_*} l^{m0}_p = 0, \tag{2.22}$$

$$\lim_{m\to\infty} \min_{p\in\gamma^m_*} l^{m1}_p = \lim_{m\to\infty} \max_{p\in\gamma^m_*} l^{m1}_p = \min\left\{\frac{1}{c^*}, \rho_*\right\}, \tag{2.23}$$

where ρ_* is the distance from the point x^* to ∂G.

THEOREM 2.3. *Suppose the conditions of Theorem 2.1 are satisfied and suppose $x^{(0)} = x^m_{(p)}$ and*

$$x^{(k+1)} = x^{(k)} - A^{m,-1}_p f(x^{(k)}), \qquad k = 0, 1, \ldots. \tag{2.24}$$

Then $x^{(k)} \in \overline{S}(x^m_{(p)}, l^{m0}_p) \subset G$, $k = 0, 1, \ldots$, the following limit exists:

$$\lim_{k\to\infty} x^{(k)} = x^* \in \overline{S}(x^m_{(p)}, l^{m0}_p), \tag{2.25}$$

in which x^* *is the unique solution of system* (2.2) *on* $S(x^m_{(p)}, l^{m_1}_p)$, *and in addition,*

$$\rho(x^{(k)}, x^*) \leqslant l_p^{m_0}\alpha^k, \tag{2.26}$$

$$\rho(x^{(k+1)}, x^*) \leqslant \frac{\alpha}{1-\alpha}\rho(x^{(k+1)}, x^{(k)}), \tag{2.27}$$

where $k = 0, 1, \ldots$ *and* $\alpha = b^m_p + c^m_p l^{m_0}_p < 1$.

PROOF. All of the assertions of Theorem 2.3 except (2.26) and (2.27) follow from the equivalence of the systems of equations (2.2) and (2.6), the proof of Theorem 2.1 and the fixed point theorem for contractions ([3], Chapter 7, §4.1, the first two theorems). Taking into account the fact that $x^* = \psi(x^*)$, we get from (2.16), (2.24) and (2.25) that

$$\rho(x^{(k+1)}, x^*) = \rho(\psi(x^{(k)}), \psi(x^*)) \leqslant \alpha\rho(x^{(k)}, x^*), \qquad k = 0, 1, \ldots, \tag{2.28}$$

where $\alpha = b^m_p + c^m_p l^{m_0}_p < b^m_p + c^m_p r^{m_1}_p \leqslant 1$. In addition, $\rho(x^{(0)}, x^*) \leqslant l^{m_0}_p$. This result together with (2.28) implies (2.26). Finally, by virtue of (2.28)

$$\rho(x^{(k)}, x^*) \leqslant \rho(x^{(k)}, x^{(k+1)}) + \rho(x^{(k+1)}, x^*) \leqslant \rho(x^{(k+1)}, x^{(k)}) + \alpha\rho(x^{(k)}, x^*),$$

where $k = 0, 1, \ldots$, i.e.

$$\rho(x^{(k)}, x^*) \leqslant \frac{1}{1-\alpha}\rho(x^{(k+1)}, x^{(k)}),$$

which together with (2.28) implies (2.27). Theorem 2.3 is proved.

§3. Method for the approximate global solution of a system of nonlinear equations

In this section we describe a method for the approximate global solution of the system of nonlinear equations (2.2) on a closed domain $\overline{\Omega} \subset G$, where G is a bounded open domain of n-dimensional Euclidean space. To substantiate the method, we rely mainly on the results of the preceding section. The notation introduced earlier is preserved.

We will consider a recurrence process of forming certain sets. Let $m = m_0, m_0 + 1, \ldots$, let B_m and D^k_m, $k = 0, 1, \ldots, N_m$, be finite sets of points of E^n with nonnegative integral coordinates, let N_m be the number of elements of B_m, and let

$$B_{m_0} = \{p : \overline{\sigma}^{m_0}_p \cap \overline{\Omega} \neq \varnothing, \quad p \in M(m_0)\} \neq \varnothing, \qquad D^0_{m_0} = \varnothing.$$

Further, let H^k_m, $m = m_0, m_0 + 1, \ldots$ and $k = 0, 1, \ldots, N_m$, be finite sets of 4-tuples of the form $\{\mu, q, r, l\}$, where μ is a natural number, q is a point of E^n with nonnegative integral coordinates and r and l are nonnegative numbers such that $2r < l$, and let $H^0_{m_0} = \varnothing$. To each element $\{\mu, q, r, l\}$ of a set H^k_m (if it is not empty) there corresponds the pair of spheres $\overline{S}(x^\mu_{(q)}, r)$ and $S(x^\mu_{(q)}, l)$, where $x^\mu_{(q)}$ is the center of the cube $\overline{\sigma}^\mu_q$.

Running ahead, we note that $S(x^\mu_{(q)}, l) \subset G$, while the closed sphere $\overline{S}(x^\mu_{(q)}, r)$ contains a unique solution of the sustem of equations (2.2), with the spherical annulus $S(x^\mu_{(q)}, l) \backslash \overline{S}(x^\mu_{(q)}, r)$ containing no solutions of system (2.2), i.e. the solution existing in $\overline{S}(x^\mu_{(q)}, r)$ is an isolated solution. In addition, for any two distinct elements $\{\mu, q, r, l\}$ and $\{\mu^*, q^*, r^*, l^*\}$ of a set H^k_m (m and k are restricted only by the requirement that H^k_m have at least two elements), the spheres $\overline{S}(x^\mu_{(q)}, r)$ and $\overline{S}(x^{\mu^*}_{(q^*)}, r^*)$ do not have any points in common (all of what has been said follows from the subsequent constructions). This means that system (2.2) has at least as many isolated solutions as the number of elements in the set H^k_m in question. From what has been said it also follows that

$$\bar{S}(x^{\mu}_{(q)}, r) \setminus S(x^{\mu^*}_{(q^*)}, l^*) \neq \varnothing,$$
$$\bar{S}(x^{\mu^*}_{(q^*)}, r^*) \setminus S(x^{\mu}_{(q)}, l) \neq \varnothing.$$

Let us assume that the sets B_m, D^{k-1}_m and H^{k-1}_m are given for some $m \in [m_0, \infty)$ and some $k \in [1, N_m]$, and that the elements of B_m have been indexed in an arbitrary order.

Consider the kth element p of B_m (the index of p is omitted for the sake of simplicity). Only three cases are possible.

Case 1. At least one of the following three conditions is satisfied.

1a) $\bar{\sigma}^m_p \cap \bar{\Omega} = \varnothing$.

1b) $\max_{1 \leqslant i \leqslant n} \bar{F}^{m\,0}_{ip} \bar{F}^{m\,1}_{ip} > 0$. (3.1)

1c) There exists an element $(\mu, q, r, l) \in H^{k-1}_m$ such that

$$\bar{S}(x^m_{(p)}, 2^{-m-1}\sqrt{n}) \subset S(x^{\mu}_{(q)}, l),$$

i.e.

$$\rho(x^m_{(p)}, x^{\mu}_{(q)}) < l - 2^{-m-1}\sqrt{n}.$$

We then put $H^k_m = H^{k-1}_m$ and $D^k_m = D^{k-1}_m$.

Case 2. None of the conditions 1a)–1c) is satisfied but $\det A^m_p \neq 0$, inequalities (2.9) and (2.18) are valid and the condition

$$\bar{S}(x^m_{(p)}, l^{m0}_p) \cap \bar{S}(x^{\mu}_{(q)}, r) = \varnothing \tag{3.2}$$

is satisfied for any element $(\mu, q, r, l) \in H^{k-1}_m$, i.e. $\rho(x^m_{(p)}, x^{\mu}_{(q)}) > l^{m\,0}_p + r$. Then

$$H^k_m = H^{k-1}_m \cup (m, p, l^{m0}_p, l^{m1}_p),$$
$$D^k_m = \{q : \rho(x^m_{(q)}, x^m_{(p)}) \geqslant l^{m1}_p - 2^{-m-1}\sqrt{n};\ q \in D^{k-1}_m\}.$$

Case 3. None of the condition 1a)–1c) is satisfied and not all of the additional conditions of Case 2 are satisfied. In this case we put $H^k_m = H^{k-1}_m$ and $D^k_m = D^{k-1}_m \cup p$.

If $k < N_m$, then the sets H^{k-1}_m and D^{k-1}_m are constructed analogously from the $(k+1)$th element of B_m and the sets H^k_m and D^k_m, and so on. When $k = N_m$ we put

$$H^0_{m+1} = H^{N_m}_m,\ D^0_{m+1} = \varnothing,$$
$$B_{m+1} = \{q : 2p_i \leqslant q_i \leqslant 2p_i + 1,\ i = 1, 2, \ldots, n;\ p \in D^{N_m}_m\}.$$

For the sake of simplicity the sets $H^{N_m}_m$ and $D^{N_m}_m$ will also be denoted by H_m and D_m respectively.

If $B_{m+1} = \varnothing$, i.e. $D_m = \varnothing$, then the process of constructing sets effectively terminates and we formally put $H_{m+\nu} = H_m$, $B_{m+\nu} = \varnothing$, $D_{m+\nu} = \varnothing$, $\nu = 1, 2, \ldots$. If, on the other hand, B_{m+1} is not empty, then the elements of this set are indexed in an arbitrary order and the subsequent process is continued by the described method.

Inasmuch as the process of constructing the sets B_m, D^k_m and H^k_m, $m = m_0, m_0 + 1, \ldots, k = 0, 1, \ldots, N_m$, was described purely formally, we will explain in somewhat greater detail the meaning of the procedures realized in this connection.

If condition 1a) is satisfied, the cube $\bar{\sigma}_p^m$ is situated outside $\bar{\Omega}$. When condition 1b) is satisfied it follows by virtue of (2.4) that at least one of the functions of the system $\{f_i\}_1^n$ does not vanish on $\bar{\sigma}_p^m$, and consequently the system of equations (2.2) does not have any solutions on this cube. Under the fulfilment of condition 1c) the cube $\bar{\sigma}_p^m$ is obviously situated in a sphere $S(x_{(q)}^\mu, l)$, where $(\mu, q, r, l) \in H_m^{k-1}$, and therefore, even if system (2.2) has a solution on the cube $\bar{\sigma}_p^m$, it is already determined in the form of the element (μ, q, r, l) of the set H_m^{k-1}. Thus in Case 1 the cube $\bar{\sigma}_p^m$ is naturally discarded and does not participate in subsequent considerations, i.e. one puts $H_m^k = H_m^{k-1}$ and $D_m^k = D_m^{k-1}$.

When all of the conditions corresponding to Case 2 (including the conditions of Theorem 2.1) are satisfied, there is discovered (separated) a new isolated solution of the system of equations (2.2), which is the only solution of (2.2) in the sphere $S(x_{(p)}^m, l_p^{m1})$ and, in particular, on the closed sphere $\bar{S}(x_{(p)}^m, l_p^{m0})$. This fact is determined in forming the set H_m^k by adding the element $(m, p, l_p^{m0}, l_p^{m1})$ to the set H_m^{k-1}. Here again the cube $\bar{\sigma}_p^m$ drops out of the subsequent considerations, since, by virtue of the fulfilment of condition (2.18) and Theorem 2.1, it cannot contain additional solutions of the system of equations (2.2) except, possibly, one previously determined. In forming the set D_m^k one discards certain elements $q \in D_m^{k-1}$ for which the corresponding elementary cube $\bar{\sigma}_p^m$ is known to be contained in the sphere of uniqueness $S(x_{(p)}^m, l_p^{m1})$ of the solution just determined, and, consequently, does not carry additional information on the solutions of the system of equations (2.2).

In Case 3 the question of whether or not the system of equations (2.2) has solutions on the cube $\bar{\sigma}_p^m$ remains unanswered. We therefore put $H_m^k = H_m^{k-1}$ and add the element p to the set D_m^{k-1} to form a set $D_m^k = D_m^{k-1} \cup p$ carrying the information, at the kth step $(1 \leqslant k \leqslant N_m)$ of consideration of B_m on the cubes, concerning the cubic lattice of rank m that require additional investigation, in particular when $k = N_m$, with the use of a finer partition (cubic lattice of higher rank).

After all of the elements of the set B_m have been dealt with, i.e. when $k = N_m$, we arrive at the sets H_m and D_m, which characterize in a certain way the set of solutions of the system (2.2). More precisely, it follows from the above constructions upon taking into account Theorem 2.1 that system (2.2) has at least as many isolated solutions as the number of elements of H_m. In addition, as has already been mentioned above, each element $(\mu, q, r, l) \in H_m$ specifies two spheres: the sphere $\bar{S}(x_{(q)}^\mu, r)$ of existence of a solution of system (2.2) and the sphere $S(x_{(q)}^\mu, l) \subset G$ of uniqueness of the solution, with $l > 2r$.

The set of indicated isolated (separated) solutions of the system of equations (2.2) corresponding to the set H_m is denoted by $\bar{\omega}_m^0$.

From our constructions it is clear that the set $\bar{\omega}\backslash\bar{\omega}_m^0$, where $\bar{\omega}$ is the set of all solutions of system (2.2) on $\bar{\Omega}$, is contained in the closed set

$$\bar{Q}_m = \bigcup_{p \in D_m} \bar{\sigma}_p^m,$$

such that, obviously, $\bar{Q}_{m_0} \supset \bar{Q}_{m_0+1} \supset \cdots$.

The above remarks imply, in particular, the following assertion.

THEOREM 3.1. *If there exists an integer $m \geqslant m_0$ such that B_m is not empty while*

$B_{m+1} = \varnothing$, then the set $\overline{\omega}$ of solutions of the system of equations (2.2) *on $\overline{\Omega}$ consists of not more than a finite number of isolated points not exceeding the number of elements of H_m.*

REMARK 3.1. In general, the set $\overline{\omega}_m^0$ contains in addition to the isolated points of $\overline{\omega}$ certain isolated solutions of system (2.2) belonging to $G\backslash\overline{\Omega}$ (situated "near" $\overline{\Omega}$). The iterative process (2.24) (Theorem 2.3) can be used to determine with arbitrary accuracy an isolated solution of system (2.2) contained in a closed sphere $\overline{S}(x^{\mu}_{(q)}, r)$, where $(\mu, q, r, l) \in H_m$. If, in this connection, the solution in question is not situated on the boundary $\partial\Omega$ of Ω, then estimates (2.26) and (2.27) permit one to establish in a finite number of iterations whether it belongs to Ω or to $G\backslash\overline{\Omega}$.

Let us prove the following basic theorem on the isolated solutions of system (2.2).

THEOREM 3.2. *Suppose $x^* \in \overline{\Omega}$ is an isolated solution of the system of equations* (2.2) *satisfying the condition*

$$\left.\frac{\partial(f_1, \ldots, f_n)}{\partial(x_1, \ldots, x_n)}\right|_{x=x^*} \neq 0,$$

this condition not necessarily being satisfied by the other solutions of system (2.2) *on G. Then an integer $m^{**} \geqslant m_0$ can be found such that for any integer $m \geqslant m^{**}$ there exists an element $(\mu, q, r, l) \in H_m$ such that $x^* \in \overline{S}(x^{\mu}_{(q)}, r)$.*

PROOF. Let m^* be the integer indicated in Theorem 2.2. Then either the set H_{m^*} contains an element (μ, q, r, l) such that $x^* \in \overline{S}(x^{\mu}_{(q)}, r)$, or it does not contain such an element. In the first case the assertion of Theorem 3.2 is realized for $m^{**} = m^*$, since $H_m \supset H_{m^*}$ for any $m \geqslant m^*$.

Consider the second case. We denote by r^* the distance from x^* to the set $g\backslash x^*$, where g is the set of all solutions of the system of equations (2.2) on G. By assumption, $r^* > 0$. On the basis of (2.22) there exists an integer $m^{***} > m^*$ such that the inequality

$$l_p^{m^{***}0} \leqslant r^*/4 \tag{3.3}$$

is satisfied for any $p \in \gamma_*^{m^{***}}$, where $\gamma_*^m = \{p\colon x^* \in \overline{\sigma}_p^m,\ p \in M(m)\}$. Here, in turn, either a separation of the solution x^* occurs for some $m' \in (m^*, m^{***})$, i.e. $x^* \in \overline{S}(x^{\mu'}_{(q')}, r')$, $(\mu', q', r', l') \in H_{m'}$, and the assertion of Theorem 3.2 is thereby realized for $m^{**} = m'$, or else an element possessing the indicated property does not exist in $H_{m'}$ for $m' = m^{***} - 1$. Then all of the conditions of Case 2 for the procedure described above for forming the sets H_m^k, in which a separation (determination) of an isolated solution x^* occurs, are satisfied for at least one element $p \in B_{m^{***}}$, for example, the first element $p \in \gamma_*^{m^{***}} \subset B_{m^{***}}$ in the sense of the indexing adopted in $B_{m^{***}}$ (none of the elementary cubes $\overline{\sigma}_p^m$ containing the solution x^* can be discarded before a separation of this solution occurs). In fact, Theorem 2.2 guarantees the fulfilment of all of the indicated conditions except (3.2), at least for $p \in \gamma_*^{m^{***}}$, while (3.2) follows for $p \in \gamma_+^{m^{***}}$ from (3.3), from the validity for any element $(\mu, q, r, l) \in H_m^k$, where $m_0 \leqslant m \leqslant m^{***}$ and $0 \leqslant k \leqslant N_m$, of the inequality

$$2r < l, \tag{3.4}$$

which is implied by condition (2.18) of Case 2, and from the fact that for any $m \in [m_0, m^{***}]$, any $k \in [0, N_m]$ and any element $(\mu, q, r, l) \in H_m^k$ the solution x^* cannot belong to the sphere $S(x_{(q)}^\mu, l)$ if $x^* \notin \overline{S}(x_{(q)}^\mu, r)$ (by virtue of Theorem 2.1, the sphere $S(x_{(q)}^\mu, l)$ is a sphere of uniqueness of an isolated solution of system (2.2) situated in $\overline{S}(x_{(q)}^\mu, r)$).

Thus, in the second case, the assertion of Theorem 3.2 is realized for at least $m^{**} = m^{***}$. Theorem 3.2 is proved.

REMARK 3.2. As follows from the proof of Theorem 3.2, the number m^{**} indicated in Theorem 3.2 can be regarded as independent of the order of indexing of the elements of B_m, $m \geqslant m_0$, adopted in forming the sets H_m^k, $0 \leqslant k \leqslant N_m$.

REMARK 3.3. If $\overline{\omega}$ contains a finite or infinite number of isolated solutions of system (2.2) satisfying the condition of Theorem 3.2, then each of these solutions is separated (determined in some H_m by the method described above) for a finite m generally depending on the "index" of the solution.

THEOREM 3.3. *If $\overline{\omega}$ consists entirely of a finite number of isolated solutions of system* (2.2) *satisfying the condition of Theorem* 3.2, *in particular, if $\overline{\omega}$ is the empty set (system* (2.2) *has no solutions on* $\overline{\Omega}$), *then there exists an integer $\widetilde{m} > m_0$ such that $B_{\widetilde{m}-1}$ is not empty while $B_{\widetilde{m}} = \varnothing$, i.e. the procedure considered above for separating the solutions of system* (2.2) *is finite.*

PROOF. Under the condition of Theorem 3.3 there exists a subdomain G^0 of the domain G such that $\overline{\Omega} \subset G^0 \subset \overline{G}^0 \subset G$ and the set of all solutions of system (2.2) on $\overline{G}^0$ coincides with $\overline{\omega}$. On the basis of Theorems 2.2 and 3.2 an integer $m' \geqslant m_0$ can be found such that for any point $x^* \in \overline{\omega}$ there exists a unique element $(\mu, q, r, l) \in H_m$, such that $2r < l$ and $x^* \in \overline{S}(x_{(q)}^\mu, r) \subset S(x_{(q)}^\mu, l) \subset G$.

Let

$$E' = \overline{G}^0 \setminus \bigcup_{(\mu,q,r,l)\in H_{m'}} S\left(x_{(q)}^\mu, \frac{r+l}{2}\right),$$

where the set $H_{m'}$, by construction, contains a finite number of elements. By virtue of the closedness and boundedness of the set E' and the absence on E' of any solutions of system (2.2), we have

$$\min_{E'} \sum_{i=1}^n f_i^2(x) = a_0 > 0. \tag{3.5}$$

Suppose, on the contrary, that B_m is not empty for any $m \geqslant m_0$. Then for sufficiently large m we have $\overline{\sigma}_p^m \subset E'$ for all $p \in B_m$. This result together with (2.4) and (3.5) implies the existence of an integer $m^* \geqslant m'$ such that condition (3.1), i.e. the inequality

$$\max_{1\leqslant i\leqslant n} \overline{F}_{ip}^{m^*0} \overline{F}_{ip}^{m^*1} > 1$$

is satisfied for any $p \in B_{m^*}$. But this means that B_{m^*+1} is empty. We have obtained a contradiction. Theorem 3.3 is proved.

The following two theorems give some global characteristics of our approximate method of solving a system of nonlinear equations.

THEOREM 3.4. *If B_m is not empty for any $m \geqslant m_0$, then the following limit exists:*

$$\lim_{m\to\infty} \sup_{x\in\bar\omega_m} \inf_{y\in\bar\omega} \rho(x, y) = 0, \tag{3.6}$$

where $\bar\omega_m = \bigcup_{p\in B_m} \bar\sigma_p^m$, and $\bar\omega$ is the set of all solutions of the system of equations (2.2) *on $\bar\Omega$.*

PROOF. The set $\bar\omega$ is not empty by virtue of Theorem 3.3. Let r_0 be the distance between $\bar\Omega$ and the boundary ∂G of G, let ω^δ, $\delta > 0$, be the set of points whose distance from $\bar\omega$ is less than δ, let g be the set of all solutions of system (2.2) on G, let ρ_δ be the distance between $\bar\Omega$ and $g\backslash\omega^\delta$, which is formally set equal to r_0 if $g\backslash\omega^\delta$ is empty, let $\theta_\delta = \min\{\delta, \rho_\delta\}/2$, and let $\bar\Omega_\delta$ be the set of points whose distance from the set $\bar\Omega\backslash\omega^\delta$ is not greater than θ_δ. Clearly, $\theta_\delta > 0$, $0 < \delta \leqslant r_0$, and $\bar\Omega_\delta$ is a bounded and closed set on which system (2.2) has no solutions. Consequently

$$\min_{\bar\Omega_\delta} \sum_{i=1}^{n} f_i^2(x) = b_\delta > 0. \tag{3.7}$$

We fix an arbitrary $\delta \in (0, r_0]$. There clearly exists an integer $m_\delta \geqslant m_0$ such that for any $m \geqslant m_\delta$

$$\bar\omega_m \subset \omega^\delta \cup \bar\Omega_\delta \tag{3.8}$$

and, in addition, every elementary cube $\bar\sigma_p^m \subset \bar\omega_m$ not having any points in common with ω^δ is entirely contained in $\bar\Omega_\delta$. Inequalities (2.4) and (3.7) imply the existence of an integer $m'_\delta \geqslant m_\delta$ such that condition (3.1) is satisfied for any $m \geqslant m'_\delta$ and any $\bar\sigma_p^m \subset \bar\Omega_\delta$. From this result and (3.8) it follows for at least $m \geqslant m'_\delta + 1$ that

$$\bar\omega_m \subset \omega^{\delta_m}, \tag{3.9}$$

where $\delta_m = \delta + 2^{-m}\sqrt{n}$, i.e. (3.6) is valid.

THEOREM 3.5. *Suppose $B_m \neq \varnothing$ for $m = m_0, m_0 + 1, \ldots$, and let $\bar\omega^0$ be any finite subset of the set of isolated solutions of system* (2.2) *satisfying the condition of Theorem* 3.2. *Then*

$$\lim_{m\to\infty} \sup_{x\in\bar\omega_m} \inf_{y\in\bar\omega\backslash\bar\omega^0} \rho(x, y) = 0. \tag{3.10}$$

Theorem 3.5 follows from Theorem 3.2 (see Remark 3.3) and Theorem 3.4 upon taking into account the fact that for every point $x^* \in \bar\omega^0$ there exists an integer $m' \geqslant m_0$ for which the set $H_{m'}$ contains an element (μ, q, r, l) such that $l > 2r$ and $x^* \in \bar S(x^\mu_{(q)}, r)$, with $\bar\omega_m \cap S(x^\mu_{(q)}, (r + l)/2) = \varnothing$ for all $m > m'$.

REMARK 3.4. If the system of equations (2.2) has only a finite number of isolated solutions on $\bar\Omega$ that can be separated by the considered method (in particular, all of the solutions satisfying the condition of Theorem 2.2 are separated), then, as follows from Theorems 3.3 and 3.5, the set $\bar\omega_m$ contracts for $m \to \infty$ to the set of the remaining solutions of system (2.2) on $\bar\Omega$ that are not separable by the method considered (including, possibly, isolated solutions but, in any event, none satisfying the condition of Theorem 2.2).

REMARK 3.5. If one replaces the quantities (2.5) by

$$a^m_{pij} = \frac{\partial f_i(x^m_{(p)})}{\partial x_i}, \qquad i, j = 1, 2, \ldots, n,$$

and sets $b^m_p = 0$ everywhere, then all of the assertions made in §§2 and 3 will remain in force.

BIBLIOGRAPHY

1. Ju. G. Evtušenko, *A numerical method of search for the global extremum of functions (scan on a nonuniform net)*, Ž. Vyčisl. Mat. i Mat. Fiz. **11** (1971), 1390–1403 = USSR Comput. Math. and Math. Phys. **11** (1971), no. 6, 38–54. MR **45** #7944.

2. E. A. Volkov, *Bilateral difference method for solving a boundary value problem for an ordinary differential equation,* Mat. Zametki **11** (1972), 421–430 = Math. Nates **11** (1972), 257–263. MR **45** #8013.

3. I. S. Berezin and N. P. Židkov, *Computing methods.* Vol. 2, Fizmatgiz, Moscow, 1960; English transl., Addison-Wesley, Reading, Mass.; Pergamon Press, New York, 1965. MR **22** #12685; **31** #1756.

4. L. D. Kudrjavcev, *Mathematical analysis.* Vol. 2. *Textbook for physics and physics-engineering students in higher educational institutions,* "Vysš. Škola", Moscow, 1970. (Russian) RŽMat. **1970** #7Б18.

Translated by S. SMITH

Trudy Mat. Inst. Steklov.
131 (1974)

Proc. Steklov Inst. Math.
131 (1974)

THEOREMS ON THE EXTENSION OF FUNCTIONS IN THE SPACES $S_p^{\mathbf{r}} W$ AND $S_{p,\theta}^{\mathbf{r}} B$ BEYOND THE BOUNDARY OF A DOMAIN Ω

UDC 517.518.22

A. D. DŽABRAILOV

ABSTRACT. The question is studied of extending the functions belonging to the well-known spaces $S_p^{\mathbf{r}} W$, $S_{p,\theta}^{\mathbf{r}} B$ from a domain Ω with preservation of class; i.e. for each function f belonging to $S_p^{\mathbf{r}} W(\Omega)$ $(S_{p,\theta}^{\mathbf{r}} B(\Omega))$ there is constructed a function $\widetilde{f}$ that is defined on E_n, coincides with f on Ω, belongs to $S_p^{\mathbf{r}} W(E_n)$ $(S_{p,\theta}^{\mathbf{r}} B(E_n))$ and satisfies the inequality $\|\widetilde{f}\|_{S_p^{\mathbf{r}} W(E_n)} \leqslant c\|f\|_{S_p^{\mathbf{r}} W(\Omega)}$ $(\|\widetilde{f}\|_{S_{p,\theta}^{\mathbf{r}} B(E_n)} \leqslant c\|f\|_{S_{p,\theta}^{\mathbf{r}} B(\Omega)})$, where c is a constant not depending on f.

Bibliography: 13 items.

In the present paper we study the question of extending the functions belonging to the spaces $S_p^{\mathbf{r}} W(\Omega)$ and $S_{p,\theta}^{\mathbf{r}} B(\Omega)$ from a domain $\Omega \subset E_n$ to E_n with preservation of class. Many works (see [1]–[5]) have been devoted to the investigation of the spaces $S_p^{\mathbf{r}} W$ and $S_{p,\theta}^{\mathbf{r}} B$.

The apparatus for our investigation is an integral representation of smooth functions obtained by the author in [3] and [6].

Works [7]–[9] are devoted to a theorem on the extension of functions belonging to the well-known spaces $W_p^{\mathbf{r}}$ and $B_{p,\theta}^{\mathbf{r}}$ from a domain $\Omega \subset E_n$ to E_n with preservation of class. In [9] and [10] this question is studied with the use of a well-known integral representation [8] for functions belonging to the spaces $W_p^{\mathbf{r}}$ and $B_{p,\theta}^{\mathbf{r}}$.

The present paper consists of four sections.

In §1 we give the basic definitions and introduce the necessary notation. §2 is devoted to estimates of singular integral operators, which arc individually presented in separate lemmas. In §3 we study the question of extending the functions belonging to the spaces $S_p^{\mathbf{r}} W$. In §4 this question is studied for the spaces $S_{p,\theta}^{\mathbf{r}} B$.

§1. Basic definitions and notation

Let E_n be an n-dimensional Euclidean space of points $\mathbf{x} = (x_1, \ldots, x_n)$ and let e be any fixed subset of the set $e_n = \{1, \ldots, n\}$ (including the empty set $\varnothing$ and the set e_n itself). The number of possible subsets e of e_n is equal to 2^n. For any vector $\mathbf{k} = (k_1, \ldots, k_n)$ we let $\mathbf{k}^e = (k_1^e, \ldots, k_n^e)$ denote the vector with components $k_j^e = k_j$ for $j \in e$ and $k_j^e = 0$ for $j \in e_n \setminus e$.

Let $f(\mathbf{x})$ be a function defined on $\Omega \subset E_n$. We put

$$\Delta^{\mathbf{k}}(\mathbf{t}) f(\mathbf{x}) = \Delta_1^{k_1}(t_1) \ldots \Delta_n^{k_n}(t_n) f(\mathbf{x}),$$

where $\Delta_i^{k_i}(t_i) f$ is the finite difference of f of order k_i with step t_i in the direction x_i, with $\Delta_i^0(t_i) f = f$. Consequently

AMS (MOS) subject classifications (1970). Primary 46E15; Secondary 46E35.

$$\Delta^{\mathbf{k}^e}(\mathbf{t})\, f(\mathbf{x}) = \Big(\prod_{j\in e} \Delta_j^{k_j}(t_j)\Big) f(\mathbf{x}).$$

Further,

$$D^{\mathbf{k}} f(\mathbf{x}) = D_1^{k_1} \dots D_n^{k_n} f(\mathbf{x}) \equiv \frac{\partial^{k_1}}{\partial x_1^{k_1}} \dots \frac{\partial^{k_n}}{\partial x_n^{k_n}} f,$$

$$\int_{\mathbf{a}}^{\mathbf{b}} f(\mathbf{x})\, d\mathbf{x}^e = \int_{a_1}^{b_1} dx_1^e \dots \int_{a_n}^{b_n} f(\mathbf{x})\, dx_n^e \equiv \Big(\prod_{j\in e} \int_{a_j}^{b_j} dx_j\Big) f(\mathbf{x}) \equiv \int_{\mathbf{a}}^{\mathbf{b}} f(\mathbf{x}) \prod_{j\in e} dx_j,$$

i.e. the integration is taken only with respect to those variables x_j whose index $j \in e$.

We put

$$\Delta^{\mathbf{k}}(\mathbf{t};\ \Omega) f(\mathbf{x}) = \Delta^{\mathbf{k}}(\mathbf{t}) f(\mathbf{x}),$$

if the difference is constructed with respect to points that are vertices of parallelepipeds and belong to the domain Ω; otherwise, we put

$$\Delta^{\mathbf{k}}(\mathbf{t};\ \Omega) f(\mathbf{x}) = 0.$$

Suppose $\mathbf{r} = (r_1, \dots, r_n) \geqslant 0$, i.e. $r_j \geqslant 0$ for $j = 1, \dots, n$. We denote by $e_{\mathbf{r}}$ the support of the vector $\mathbf{r}$ (see [1] or [2]), i.e. the smallest of the subsets $e \subseteq e_n$ for which $\mathbf{r}^e = \mathbf{r}$. To each vector $\mathbf{r}$ there correspond the vectors $[\mathbf{r}] = ([r_1], \dots, [r_n])$ and $\bar{\mathbf{r}} = (\bar{r}_1, \dots, \bar{r}_n)$, where $[r_j]$ is the integral part of r_j and $\bar{r}_j$ is the greatest integer that is less than r_j for $r_j > 0$ and vanishes for $r_j = 0$. Hence $0 \leqslant r_j - [r_j] < 1$, while for $r_j > 0$ we have $0 < r_j - \bar{r}_j \leqslant 1$.

The n-dimensional unit vector will always be denoted in the sequel by $\omega = (1, \dots, 1)$ and, if $\mathbf{h} = (h_1, \dots, h_n)$ and $\mathbf{y} = (y_1, \dots, y_n)$, then $\mathbf{y}/\mathbf{h}$ will denote the vector $\mathbf{y}/\mathbf{h} = (y_1/h_1, \dots, y_n/h_n)$, where $y_j/h_j = 0$ if $y_j = h_j = 0$.

DEFINITION 1. We will say that a function f belongs to $\mathscr{L}_{p,\theta}^{\langle\mathbf{r}\rangle}(\Omega)$ if it has on the domain Ω a generalized derivative $D^{\bar{\mathbf{r}}} f$ (in the sense of S. L. Sobolev [7]) satisfying the condition

$$\|f\|_{\mathscr{L}_{p,\theta}^{\langle\mathbf{r}\rangle}(\Omega)} = \left(\int \dots \int \frac{\|\Delta^{\omega_{\mathbf{r}}+[\mathbf{r}-\bar{\mathbf{r}}]}(\mathbf{t};\Omega)\, D^{\bar{\mathbf{r}}} f\|_{L_p(\Omega)}^{\theta}}{\prod_{j\in e_{\mathbf{r}}} |t_j|^{1+\theta(r_j-\bar{r}_j)}}\, d\mathbf{t}^{e_{\mathbf{r}}}\right)^{1/\theta} < \infty,$$

where $\omega_{\mathbf{r}} \equiv \omega^{e_{\mathbf{r}}}$,

$$\|\cdot\|_{L_p(\Omega)} = \Big(\int_\Omega |\cdot|^p\, d\mathbf{x}\Big)^{1/p} \qquad (1 \leqslant p < \infty),$$

$$\|f\|_{L_\infty(\Omega)} = \operatorname*{ess\,sup}_{x\in\Omega} |f(\mathbf{x})|,$$

$$\|f\|_{\mathscr{L}_{p,\infty}^{\langle\mathbf{r}\rangle}(\Omega)} = \operatorname*{ess\,sup}_{t_j\neq 0\ (j\in e_{\mathbf{r}})} \frac{\|\Delta^{\omega_{\mathbf{r}}+[\mathbf{r}-\bar{\mathbf{r}}]}(\mathbf{t};\Omega)\, D^{\bar{\mathbf{r}}} f\|_{L_p(\Omega)}}{\prod_{j\in e_{\mathbf{r}}} |t_j|^{r_j-\bar{r}_j}}.$$

DEFINITION 2. We will say that a function f belongs to $S_{p,\theta}^{\mathbf{r}} B(\Omega)$ if it simultaneously belongs to each $\mathscr{L}_{p,\theta}^{\langle\mathbf{r}^e\rangle}(\Omega)$ for $e \subseteq e_{\mathbf{r}}$; i.e.

$$S_{p,\theta}^{\mathbf{r}}B(\Omega) \equiv \bigcap_{e \subseteq e_{\mathbf{r}}} \mathscr{L}_{p,\theta}^{\langle \mathbf{r}^e \rangle}(\Omega).$$

The norm in the space $S_{p,\theta}^{\mathbf{r}}B(\Omega)$ is defined as follows:

$$\|f\|_{S_{p,\theta}^{\mathbf{r}}B(\Omega)} = \sum_{e \subseteq e_{\mathbf{r}}} \|f\|_{\mathscr{L}_{p,\theta}^{\langle \mathbf{r}^e \rangle}(\Omega)}, \tag{1}$$

where the sum is taken over all of the subsets e of $e_{\mathbf{r}}$ (including the empty set $\varnothing$ and $e_{\mathbf{r}}$ itself).

Clearly,

$$S_{p,\infty}^{\mathbf{r}}B(\Omega) = S_p^{\mathbf{r}}H(\Omega).$$

The spaces $S_p^{\mathbf{r}}H(\Omega)$ were first defined and studied by S. M. Nikol'skiĭ [2].

Now let $e_{\mathbf{r}}^* \equiv e_{\mathbf{r}-[\mathbf{r}]}$, i.e. let $e_{\mathbf{r}}^*$ be the set of indices of the nonintegral components of the vector $\mathbf{r}$.

DEFINITION. We will say that a function f belongs to $L_p^{\langle \mathbf{r} \rangle}(\Omega)$ if it has on the domain Ω a generalized derivative $D^{[\mathbf{r}]}f$ satisfying the inequality(1)

$$\|f\|_{L_p^{\langle \mathbf{r} \rangle}(\Omega)} = \left(\int \cdots \int \frac{\|\Delta^{\overset{*}{\omega}_{\mathbf{r}}}(\mathbf{t};\Omega) D^{[\mathbf{r}]}f\|_{L_p(\Omega)}^p}{\prod_{j \in \overset{*}{e}_{\mathbf{r}}} |t_j|^{1+p(r_j-[r_j])}} \, dt^{\overset{*}{e}_{\mathbf{r}}} \right)^{1/p} < \infty,$$

where $\omega_{\mathbf{r}}^* \equiv \omega^{e_{\mathbf{r}}^*}$, $\omega = (1, \ldots, 1)$, and

$$\|f\|_{L_\infty^{\langle \mathbf{r} \rangle}(\Omega)} = \operatorname*{ess\,sup}_{t_j \neq 0\ (j \in \overset{*}{e}_{\mathbf{r}})} \frac{\operatorname{ess\,sup} |\Delta^{\overset{*}{\omega}_{\mathbf{r}}}(\mathbf{t};\Omega) D^{[\mathbf{r}]}f(\mathbf{x})|}{\prod_{j \in \overset{*}{e}_{\mathbf{r}}} |t_j|^{r_j-[r_j]}}.$$

DEFINITION 3. We will say that a function f belongs to $S_p^{\mathbf{r}}W(\Omega)$ if it belongs to $L_p^{\langle \mathbf{r}^e \rangle}(\Omega)$ for all $e \subseteq e_{\mathbf{r}}$; i.e.

$$S_p^{\mathbf{r}}W(\Omega) = \bigcap_{e \subseteq e_{\mathbf{r}}} L^{\langle \mathbf{r}^e \rangle}(\Omega).$$

The norm in this space is defined as follows:

$$\|f\|_{S_p^{\mathbf{r}}W(\Omega)} = \sum_{e \subseteq e_{\mathbf{r}}} \|f\|_{L_p^{\langle \mathbf{r}^e \rangle}(\Omega)}. \tag{2}$$

This space was studied in the case of integral vectors $\mathbf{r}$ by S. M. Nikol'skiĭ [1].

We will introduce some classes of domains. Let $\delta = (\delta_1, \ldots, \delta_n)$ be a vector with components $\delta_j = \pm 1$ $(j = 1, \ldots, n)$. There are 2^n distinct vectors δ.

Further, let $\mathbf{H} = (H_1, \ldots, H_n) > 0$, i.e. $H_j > 0$ $(j = 1, \ldots, n)$, and let

$$\square_\delta \equiv \square(\delta; \mathbf{H}) = \{\mathbf{y};\ 0 \leqslant \delta_j y_j \leqslant H_j\ (j = 1, \ldots, n)\}. \tag{3}$$

The number of $\square_\delta$ for each fixed $\mathbf{H} = (H_1, \ldots, H_n) > 0$ will be 2^n.

DEFINITION 4. We will say that an n-dimensional domain Ω belongs to the class

(1)When $e_{\mathbf{r}}^* = \varnothing$, i.e. when $\mathbf{r}$ is an integral vector, it is assumed that the product in the denominator is equal to unity and hence that $\|f\|_{L_p^{\langle \mathbf{r} \rangle}(\Omega)} = \|D^{\mathbf{r}}f\|_{L_p(\Omega)}$.

$C(H_1, \ldots, H_n)$ if there exist a finite covering by open sets U_k, $k = 1, \ldots, N$, and a collection of corresponding parallelepipeds $\square_k \equiv \square_{\delta k}$ satisfying the conditions

1) $\bigcup_{k=1}^N U_k = \Omega$;

2) $U_k + \square_k \subset \Omega$ ($k = 1, \ldots, N$), where the sign + indicates the vector sum of the sets.

Suppose the covering $\{U_k\}$ also satisfies the condition

3) $\operatorname{meas}(U_k \cap U_j) \geqslant \alpha > 0$ ($k \neq j$).

The above-indicated class of domains is then denoted by $C_\alpha(H_1, \ldots, H_n)$. Consequently,

$$C(H_1, \ldots, H_n) = C_0(H_1, \ldots, H_n).$$

We will say that a domain Ω belongs to $C_{\alpha,\epsilon}(H_1, \ldots, H_n)$, where $\epsilon > 0$, if it belongs to $C_\alpha(H_1, \ldots, H_n)$ and the covering $\{U_k\}$ satisfies the additional condition

$$\Omega \subset \bigcup_{k=1}^{N} U_k^{(\varepsilon)},$$

where

$$U_k^{(\varepsilon)} = \{\mathbf{x};\ \mathbf{x} \in U_k,\ \ \rho(\mathbf{x}, \Omega \setminus U_k) > \varepsilon\}.$$

The following integral representation was obtained by the author in [6] for sufficiently smooth functions $f(x)$ defined on $\Omega \subset E_n$ (we indicate a special form of this representation):

$$D^{\mathbf{v}} f(\mathbf{x}) = \sum_{e \subseteq \mathscr{E}} (-1)^{|\mathbf{v}|+|\omega^e|+|\mathbf{m}^e|}\, 2^{|\omega^e|} \prod_{j \in \mathscr{E} \setminus e} h_j^{-1-\nu_j}$$
$$\times \prod_{j \in e^*} (k_j)^{k_j} \int_0^{\mathbf{h}} \frac{d\mathbf{v}^e}{\prod_{j \in e} v_j^{2-m_j+\nu_j+\omega_j^{e^*}}} \iint \Phi_{e,\delta}\left(\frac{\mathbf{y}^{\mathscr{E}}}{\mathbf{v}^e + \mathbf{h}^{\mathscr{E} \setminus e}}; \frac{\mathbf{z}^{e^*}}{\mathbf{v}}\right) \tag{4}$$
$$\times \Delta^{\mathbf{k}^{e^*}}\left(\frac{\mathbf{z}}{\mathbf{k}}\right) D^{\mathbf{m}^e} f(\mathbf{x} + \mathbf{y}^{\mathscr{E}})\, d\mathbf{y}^{\mathscr{E}}\, d\mathbf{z}^{e^*},$$

where $\mathscr{E}^* \subseteq \mathscr{E} \subseteq e_n$, $e^* = e\mathscr{E}^*$, $\omega = (1, \ldots, 1)$, $\omega^{e^*} = (\omega_1^{e^*}, \ldots, \omega_n^{e^*})$ and $|\mathbf{a}| = a_1 + \cdots + a_n$ for any vector $\mathbf{a} > 0$.

The kernel $\Phi_{e,\delta}(\mathbf{y}^{\mathscr{E}}; \mathbf{z}^{e^*})$ in (4) belongs to $\mathring{C}^\infty(E_{\mathbf{y}^{\mathscr{E}}} \times E_{\mathbf{z}^{e^*}})$, and its support is

$$\{(\mathbf{y}^{\mathscr{E}}, \mathbf{z}^{e^*});\ 0 \leqslant \delta \mathbf{y}^{\mathscr{E}} \leqslant \delta\omega^{\mathscr{E}},\ 0 \leqslant \delta \mathbf{z}^{e^*} \leqslant \delta\omega^{e^*}\},$$

where $\delta = (\delta_1, \ldots, \delta_n)$, $\delta_i = \pm 1$ ($i = 1, \ldots, n$), and $\delta\mathbf{y} = (\delta_1 y_1, \ldots, \delta_n y_n)$; also

$$\Phi_{e,\delta}(\mathbf{y}^{\mathscr{E}}, \mathbf{z}^{e^*}) = N(\mathbf{y}^{\mathscr{E}_1}; \mathbf{z}^{e^*}) \prod_{j \in e \setminus \mathscr{E}^*} M_{\delta_j}(y_j),$$

where $\mathscr{E}_1 = \mathscr{E} \setminus (e \setminus \mathscr{E}^*)$, and

$$\int M_{\delta_j}(y_j)\, dy_j = 0. \tag{$*$}$$

§2. Auxiliary inequalities

Let

$$K_{j, \varepsilon_j, h_j}(y_j) = \int_{\varepsilon_j}^{h_j} M_{\delta_j}\left(\frac{y_j}{v_j}\right) \frac{dv_j}{v_j^2}, \tag{5}$$

$$A_{j, \varepsilon_j, h_j} u(\mathbf{x}) = \int_{E_1} K_{j, \varepsilon_j, h_j}(y_j)\, u(x_1, \ldots, x_j + y_j, \ldots, x_n)\, dy_j. \tag{6}$$

LEMMA 1. *Suppose $u(\mathbf{x}) \in S$, i.e. $u(\mathbf{x})$ is infinitely differentiable and such that for any α and β*

$$\sup_{\mathbf{x}\in E_n} |x_1^{\alpha_1}\ldots x_n^{\alpha_n} D^{\beta} u(\mathbf{x})| < \infty.$$

Then the following inequalities are valid for any $0 < \epsilon_j < h_j \leqslant H_j < \infty$ $(j = 1, \ldots, n)$:

$$\| A_{j,\epsilon_j,h_j} u \|_{L_2(E_n)} \leqslant c \| u \|_{L_2(E_n)}, \tag{7}$$

$$| \widehat{K}_{j,\epsilon_j,h_j}(\xi) | \leqslant c, \tag{8}$$

where $\widehat{K}_{j,\epsilon_j,h_j}(\xi)$ *is the Fourier transform of* $\widehat{K}_{j,\epsilon_j,h_j}(y_j)$ *and* c *is a constant not depending on* $u(\mathbf{x})$, ϵ_j, h_j *or* ξ $(\xi \in E_1)$.

PROOF. The lemma is proved with the use of arguments presented in [9] and [11]–[13]. Since

$$A_{j,\epsilon_j,h_j} u(\mathbf{x}) = (K_{j,\epsilon_j,h_j} * \check{u}(\mathbf{x})),$$

$$\check{u}(\mathbf{x}) = u(x_1, \ldots, -x_j, \ldots, x_n)$$

and the Fourier transform of a convolution is equal to the product of the Fourier transforms, it follows that assertions (7) and (8) are equivalent, and we will therefore prove (8).

Clearly,

$$\begin{aligned} \widehat{K}_{j,\epsilon_j,h_j}(\xi) &= \iint_{E_1} e^{-2\pi i x\xi} M_{\delta_j}\left(\frac{x}{v}\right)\frac{dv}{v^2}\,dx \\ &= \int_{E_1}\int_{\frac{\epsilon_j}{\alpha}}^{\frac{h_1}{\alpha}} e^{-2\pi i x\xi\alpha} M_{\delta_j}\left(\frac{x}{t}\right)\frac{dt}{t^2}\,dt = \widehat{K}_{j,\frac{\epsilon_j}{\alpha},\frac{h_j}{\alpha}}(\alpha\xi); \end{aligned} \tag{9}$$

hence it suffices to prove (8) for $|\xi| = 1$.

Taking into account the equality

$$\widehat{K}_{j,\epsilon_j,h_j}(\xi) = \widehat{K}_{j,\epsilon_j,\lambda_j}(\xi) + \widehat{K}_{j,\lambda_j,h_j}(\xi),$$

we first consider the case $0 < \epsilon_j < h_j < 1$. Bearing in mind properties $(*)$, we get

$$| \widehat{K}_{j,\epsilon_j,h_j}(\xi) | = \left| \iint_{E_1} (e^{-2\pi i x\xi} - 1) M_{\delta_j}\left(\frac{x}{v}\right)\frac{dv}{v^2}\,dx \right| \leqslant c_1 \int_{\epsilon_j}^{h_j}\int_0^v | x | \frac{dv}{v^2}\,dx \leqslant c_2.$$

Suppose now $1 \leqslant \epsilon_j < h_j \leqslant H_j < \infty$. Clearly,

$$\widehat{K}_{j,\epsilon_j,h_j}(\xi) - e^{-2\pi i y\xi}\widehat{K}_{j,\epsilon_j,h_j}(\xi) = \int_{E_1}\int_{\epsilon_j}^{h_j} e^{-2\pi i x\xi}\left[M_{\delta_j}\left(\frac{x}{v}\right) - M_{\delta_j}\left(\frac{x-y}{v}\right)\right]\frac{dv}{v^2}\,dx,$$

from which, taking into account the finiteness of $M_{\delta_j}(y)$, using a theorem on differences and setting $y = \theta\xi$ $(0 < \theta < 1)$, we get

$$| \widehat{K}_{j,\epsilon_j,h_j}(\xi) | \leqslant \int_1^{\infty}\int_0^{h_j} \frac{dv}{v^3}\,dx \leqslant c.$$

LEMMA 2. *Suppose* $u(\mathbf{x}) \in S$, $0 < \epsilon_j < h_j \leqslant H_j < \infty$ *and* $1 < p < \infty$. *Then the following inequality is valid*:

$$\| A_{j, \varepsilon_j, h_j} u \|_{L_p(E_n)} \leqslant c_p \| u \|_{L_p(E_n)}, \tag{10}$$

where c is a constant not depending on $u(\mathbf{x})$, ϵ_j *or* h_j.

The proof of the lemma rests on the following theorem of Krée (see [9]).*

THEOREM (Krée). *Suppose*

$$U_a = \{\mathbf{x}; \quad \mathbf{x} \in E_n, \quad | x_i | \leqslant a_i \ (i = 1, \ldots, n\},$$

$$\begin{gathered} K(\mathbf{x}) \in L^{\mathrm{loc}}(E_n), \\ \int\limits_{E_n \setminus U_{2^s}} | K(\mathbf{x} - \mathbf{y}) - K(\mathbf{x}) | \, d\mathbf{x} \leqslant c \end{gathered} \tag{11}$$

for any integer s and any $\mathbf{y} \in U_{2^{s-s_0}}$, *where* s_0 *is some natural number. If*

$$\| K * u \|_{L_2(E_n)} \leqslant c \| u \|_{L_2(E_n)},$$

for any function $u(\mathbf{x}) \in S$, *then*

$$\| K * u \|_{L_p(E_n)} \leqslant c_p \| u \|_{L_p(E_n)},$$

for any $p \in (1, \infty)$, *where* c_p *is a constant not depending on* $u(\mathbf{x})$.

Let us see that the kernels $K_{j,\epsilon_j,h_j}(y_j)$ satisfy the condition of Krée's theorem (for $n = 1$). Since

$$K_{j, \varepsilon_j, h_j}(y_j) = t^{-1} K_{j, \varepsilon_j/t, h_j/t}\left(\frac{y_j}{t}\right),$$

it suffices to prove (11) in the case of our kernel for $s = 0$.

Suppose $|y_j| \leqslant \theta_j < 1$. Then by virtue of the finiteness of $M_{\delta_j}(t)$ we have

$$\begin{aligned} &\int\limits_{E_1 \setminus U_1} | K_{j, \varepsilon_j, h_j}(x_j - y_j) - K_{j, \varepsilon_j, h_j}(x_j) | \, dx_j \\ &\leqslant \int\limits_{E_1 \setminus U_1} \int\limits_{\varepsilon_j}^{h_j} \left| M_{\delta_j}\left(\frac{x_j - y_j}{v_j}\right) - M_{\delta_j}\left(\frac{x_j}{v_j}\right) \right| \frac{dv_j}{v_j^2} \, dx_j \\ &\leqslant \int\limits_{1-\theta_j}^{h_j} dx_j \int\limits_{|x_j| > 1} \left| M_{\delta_j}\left(\frac{x_j - y_j}{v_j}\right) - M_{\delta_j}\left(\frac{x_j}{v_j}\right) \right| dx_j \frac{dv_j}{v_j^2} \\ &\leqslant c \int\limits_{1-\theta_j}^{h_j} \left(\int\limits_0^{1+v_j} dx_j \right) \frac{dv_j}{v_j^3} \leqslant c. \end{aligned}$$

Assertion (11) is satisfied for the kernels $K_{j,\epsilon_j,h_j}(x_j)$ ($n = 1$). Consequently the following estimate is valid for almost all $\mathbf{x}^* = (x_1, \ldots, x_{j-1}, x_{j+1}, \ldots, x_n)$:

$$\| A_{j, \varepsilon_j, h_j} u(\mathbf{x}) \|_{L_p(E_{x_j})} \leqslant c_p \| u(\mathbf{x}) \|_{L_p(E_{x_j})},$$

Editor's note.* The original reference is Paul Krée, *Sur les multiplicateurs dans* FL^p, C. R. Acad. Sci. Paris **260 (1965), 4400–4403; MR **34** #2634.

where $E_{x_j} = \{x_j; -\infty < x_j < \infty\}$ and c_p does not depend on the point $\mathbf{x}^*$. The assertion of Lemma 2 follows from the latter inequality. Q.E.D.

On the other hand,

$$|A_{j,\epsilon_j^1,\epsilon_j^2}u(\mathbf{x})| = \left| \int_{\epsilon_j^1}^{\epsilon_j^2} \int_{E_1} M_{\delta_j}\left(\frac{y_j}{v_j}\right) v_j^{-2} [u(x_1, \dots, x_j + y_j, \dots, x_n) - u(\mathbf{x})]\, dy_j\, dv_j \right|$$

$$\leqslant c \sup_{|y_j| \leqslant h_j} \left| \frac{\partial}{\partial y_j} u(x_1, \dots, x_j + y_j, \dots, x_n) \right| \left| \int_{\epsilon_j^1}^{\epsilon_j^2} \int_0^{v_j} |y_j|\, dy_j \frac{dv_j}{v_j^2} \right| \to 0$$

as $\epsilon_j^1 < \epsilon_j^2 \to 0$ ($\epsilon_j^2 > \epsilon_j^1 \to \infty$). Consequently

$$A_{j,\epsilon_j,h_j}u(\mathbf{x}) \to A_{j,0,h_j}u(\mathbf{x}) \tag{12}$$

in the sense of $L^{\mathrm{loc}}(E_n)$ ($u \in S$). Taking (10) into account, we obtain the estimate

$$\| A_{j,0,h_j}u \|_{L_p(E_n)} \leqslant c \| u \|_{L_p(E_n)}. \tag{13}$$

§3. Extension of the functions in the spaces $S_p^{\mathbf{r}}W(\Omega)$

Let $\mathbf{r} = (r_1, \dots, r_n)$ be a vector with nonnegative components and support $e_{\mathbf{r}} \equiv \mathscr{E} \neq \varnothing$, and let $e_{\mathbf{r}}^* \equiv \mathscr{E}^*$ be the set of indices of the nonintegral components of $\mathbf{r}$, so that $\mathscr{E}^* \subseteq \mathscr{E} \subseteq e_n \equiv \{1, \dots, n\}$.

Suppose the domain Ω belongs to $C_{\epsilon,\alpha}(H_1, \dots, H_n)$, and let $f(\mathbf{x})$ be a smooth function defined in Ω. With the use of this function, we construct the function

$$\begin{aligned} \tilde{f}_k(\mathbf{x}) = & \sum_{e \subseteq \mathscr{E}} (-1)^{|\omega^e|} 2^{|\omega^e|} \prod_{j \in \mathscr{E} \setminus e} h_j^{-1} \\ & \times \int_0^{\mathbf{h}} \frac{d\mathbf{v}^e}{\prod_{j \in e} v_j^{2+\omega_j^{e^*} - [r_j]}} \iint \Phi_{e,\delta^k}\left(\frac{\mathbf{y}^{\mathscr{E}}}{\mathbf{v}^e + \mathbf{h}^{\mathscr{E} \setminus e}}; \frac{\mathbf{z}^{e^*}}{\mathbf{v}} \right) \\ & \times \Delta^{\omega^{e^*}}(\mathbf{z}; U_k + \square_k)\, D^{[\mathbf{r}]^e} f(\mathbf{x} + \mathbf{y}^{\mathscr{E}})\, d\mathbf{y}^{\mathscr{E}}\, d\mathbf{z}^{e^*}, \end{aligned} \tag{14}$$

where $[\mathbf{r}] = ([r_1], \dots, [r_n])$, $[r_j]$ being the integral part of r_j, $e^* = e\mathscr{E}^*$, $\Delta^0(\mathbf{z}; U_k + \square_k)\psi = \chi(U_k + \square_k)\psi$ and $\psi(G)$ is the characteristic function of the set G.

It follows from (4) that $\tilde{f}_k(\mathbf{x})$ is defined on E_n and coincides with $f(x)$ on $U_k + \square_k$.

LEMMA 3. *Suppose $1 < p < \infty$, $\tilde{e} \subseteq E$, f is a sufficiently smooth function belonging to $S_p^{\mathbf{r}}W(\Omega)$ and $\Omega \in C_{0,\epsilon}(H_1, \dots, H_n)$. Then the following estimate is valid*:

$$\| D^{[\mathbf{r}]^{\tilde{e}}} \tilde{f}_k \|_{L_p(E_n)} \leqslant c \| f \|_{S_p^{\mathbf{r}}W(U_k + \square_k)}, \tag{15}$$

where c is a constant not depending on f.

PROOF. From (14), integrating by parts, we have

$$D^{[\mathbf{r}]^{\tilde{e}}} \tilde{f}_k(\mathbf{x}) = \sum_{e \subseteq \mathscr{E}} (-1)^{|\omega^e + [\mathbf{r}]^{\tilde{e}}|} 2^{|\omega^e|} \Big(\prod_{j \in e \cdot (\tilde{e} \setminus \tilde{e}^*)} A_{j,0,h_j} \Big) \Psi_{\mathscr{E}_1}(\mathbf{x}; U_k + \square_k). \tag{16}$$

Here

$$\Psi_{\mathscr{E}_2}(\mathbf{x};\ U_k+\square_k)=\prod_{j\in\mathscr{E}\setminus e} h_j^{-1-[r_j]\omega_j^{\hat{e}}}\int_0^{\mathbf{h}}\frac{d\mathbf{v}^{\mathscr{E}_1}}{\prod_{j\in\mathscr{E}_1} v_j^{2+\omega_j^{e^*}-[r_j]\omega_j^{e\setminus\tilde{e}}}}$$
$$\times\iint\Phi_{e,\delta k}\left(\frac{\mathbf{y}^{\mathscr{E}_2}}{\mathbf{v}^e+\mathbf{h}^{\mathscr{E}\setminus e}};\ \frac{\mathbf{z}^{e^*}}{\mathbf{v}}\right)\Delta^{\omega^{e^*}}(\mathbf{z};\ U_k+\square_k)\,D^{[\mathbf{r}]^e}f(\mathbf{x}+\mathbf{y}^{\mathscr{E}_2})\,d\mathbf{y}^{\mathscr{E}_2}\,d\mathbf{z}^{e^*},\tag{17}$$

where the kernel Φ_{e,δ_k} is a sufficiently smooth function with compact support, $\widetilde{e}^*=\widetilde{e}\cdot\mathscr{E}^*$ and $\mathscr{E}_1=e\setminus(\widetilde{e}\setminus\widetilde{e}^*)$, $\mathscr{E}_2=\mathscr{E}\setminus e\cdot(\widetilde{e}\setminus\widetilde{e}^*)$.

With the use of the arguments presented in [6] it is proved that

$$\Psi_{\mathscr{E}_2}(\mathbf{x};\ U_k+\square_k)\in L_p(E_n),$$

with

$$\|\Psi_{\mathscr{E}_2}(\cdot;\ U_k+\square_k)\|_{L_p(E_n)}\leqslant c\,\|f\|_{L_p^{\langle \mathbf{r}^e\rangle}(U_k+\square_k)}.\tag{18}$$

Applying Lemma 2 now to the right side of (16) $|\omega^{e\cdot(\widetilde{e}\setminus\widetilde{e}^*)}|$ times, we arrive at the estimate

$$\|D^{[\mathbf{r}]^{\widetilde{e}}}f_k\|_{L_p(E_n)}\leqslant c\sum_{e\subseteq\mathscr{E}}\|\Psi_{\mathscr{E}_2}(\cdot;\ U_k+\square_k)\|_{L_p(E_n)},$$

which together with (18) proves Lemma 3.

We break up the integral expression (17) into a sum of two expressions:

$$\Psi_{\mathscr{E}_2}(\mathbf{x};\ U_k+\square_k)=c(\mathbf{h})\left(\int_0^{\mathbf{t}}+\int_{\mathbf{t}}^{\mathbf{h}}\right)\frac{d\mathbf{v}^{e^*\cdot\widetilde{e}^*}}{\prod_{j\in e^*\widetilde{e}^*} v_j^3}$$
$$\times\int_0^{\mathbf{h}}\frac{d\mathbf{v}^{e\setminus\widetilde{e}}}{\prod_{j\in e\setminus\widetilde{e}} v_j^{2+\omega_j^{e^*}-[r_j]}}\iint\Phi_{e,\delta k}\left(\frac{\mathbf{y}^{\mathscr{E}_2}}{\mathbf{v}^e+\mathbf{h}^{\mathscr{E}\setminus e}};\ \frac{\mathbf{z}^{e^*}}{\mathbf{v}}\right)\Delta^{\omega^{e^*}}(\mathbf{z};\ U_k+\square_k)\,D^{[\mathbf{r}]^e}f\tag{19}$$
$$\times(\mathbf{x}+\mathbf{y}^{\mathscr{E}_2})\,d\mathbf{y}^{\mathscr{E}_2}\,d\mathbf{z}^{e^*}\equiv\overline{\Psi}_{\mathscr{E}_2}(\mathbf{x};\ U_k+\square_k)+\overline{\Psi}^*_{\mathscr{E}_2}(\mathbf{x};\ U_k+\square_k).$$

LEMMA 4. *Under the conditions of Lemma* 3

$$D_{\mathbf{x}}^{\omega^{\widetilde{e}^*\setminus e}}\overline{\Psi}_{\mathscr{E}_2}(\mathbf{x};\ U_k+\square_k)\in L_p(E_n),$$
$$D_{\mathbf{x}}^{\omega^{e^*}}\overline{\Psi}^*_{\mathscr{E}_2}(\mathbf{x};\ U_k+\square_k)\in L_p(E_n),$$

with the following inequalities being valid:

$$\|D_{\mathbf{x}}^{\omega^{\widetilde{e}^*\setminus e}}\overline{\Psi}_{\mathscr{E}_2}(\cdot;\ U_k+\square_k)\|_{L_p(E_n)}\leqslant c(\mathbf{h})\prod_{j\in e^*\cdot\widetilde{e}^*} t_j^{r_j-[r_j]-(\gamma-\xi_j)}$$
$$\times\left(\int_0^{\mathbf{t}} d\mathbf{z}^{e^*\cdot\widetilde{e}^*}\int_0^{\mathbf{h}}\frac{\|\cdot\|^p_{L_p(E_n)}\,d\mathbf{z}^{e^*\setminus\widetilde{e}^*}}{\prod_{j\in e^*} z_j^{1+p(r_j-[r_j])-(\gamma_j-\xi_j)p}}\right)^{1/p},\tag{20}$$

$$\|D_{\mathbf{x}}^{\omega^{e^*}}\overline{\Psi}^*_{\mathscr{E}_2}(\cdot;\ U_k+\square_k)\|_{L_p(E_n)}\leqslant c(\mathbf{h})\prod_{j\in e^*\cdot\widetilde{e}^*} t_j^{-(1-r_j+[r_j]-\xi_j)}$$
$$\times\left\{\int_0^{\mathbf{h}}\frac{\prod_{j\in e^*\setminus\widetilde{e}^*} z_j^{-1-p(r_j-[r_j])}\|\cdot\|^p_{L_p(E_n)}\,d\mathbf{z}^{e^*}}{\prod_{j\in e^*\cdot\widetilde{e}^*} z_j^{1+p(r_j-[r_j])-\gamma_jp}\left(\sqrt{z_j^2+t_j^2}\right)^{(\gamma_j+\xi_j)p}}\right\}^{1/p},\tag{21}$$

where $\gamma_j > \xi_j > 0$ *are arbitrary small numbers and*

$$\|\cdot\| \equiv \|\Delta^{\omega^{e^*}}(\mathbf{z};\ U_k + \square_k)\, D^{[\mathbf{r}]^e} f\|.$$

The proof of (20) and (21) is conducted with the use of arguments presented in [6].

LEMMA 5. *The following inequality is valid under the conditions of Lemma* 3:

$$\left(\int_0^{\mathbf{h}} \|\Delta^{\omega^{\tilde{e}^*}}(\mathbf{t})\, D^{[\mathbf{r}]^{\tilde{e}}} \tilde{f}_k\|^p_{L_p(E_n)} \prod_{j\in\tilde{e}^*} \frac{dt_j}{t_j^{1+p(r_j-[r_j])}}\right)^{1/p} \leqslant c\|f\|_{S_p^{\mathbf{r}}W(U_k+\square_k)}, \tag{22}$$

where c is a constant not depending on f.

PROOF. From (16) and (19) it follows that

$$\begin{aligned} &\|\Delta^{\omega^{\tilde{e}^*}}(\mathbf{t})\, D^{[\mathbf{r}]^{\tilde{e}}} \tilde{f}_k\|_{L_p(E_n)} \\ &\quad\leqslant c \sum_{e\subseteq\mathscr{E}} \Big\{ \prod_{j\in\tilde{e}^*\setminus e} t_j \Big\| \Big(\prod_{j\in e\cdot(\tilde{e}\setminus\tilde{e}^*)} A_{j,0,h_j}\Big) D^{\omega^{\tilde{e}^*\setminus e}} \overline{\Psi}_{\mathscr{E}_2}\Big\|_{L_p(E_n)} \\ &\qquad + \prod_{j\in\tilde{e}^*} t_j \Big\| \Big(\prod_{j\in e\cdot(\tilde{e}\setminus\tilde{e}^*)} A_{j,0,h_j}\Big) D^{\omega^{e^*}} \overline{\Psi}^*_{\mathscr{E}_2}\Big\|_{L_p(E_n)}\Big\}. \end{aligned} \tag{23}$$

On the other hand, since

$$D_{\mathbf{x}}^{\omega^{\tilde{e}^*\setminus e}} \overline{\Psi}_{\mathscr{E}_2} \in L_p(E_n), \qquad D_{\mathbf{x}}^{\omega^{\tilde{e}^*}} \overline{\Psi}^*_{\mathscr{E}_2} \in L_p(E_n),$$

by successively applying (13) to the integral expressions in the right side of (23), we arrive at the estimate

$$\begin{aligned} \|\Delta^{\omega^{\tilde{e}^*}}(\mathbf{t})\, D^{[\mathbf{r}]^{\tilde{e}}} \tilde{f}_k\|_{L_p(E_n)} \leqslant c \prod_{j\in\tilde{e}^*\setminus e} t_j \|D^{\omega^{\tilde{e}^*\setminus e}} \overline{\Psi}_{\mathscr{E}_2}\|_{L_p(E_n)} \\ + c \prod_{j\in\tilde{e}^*} t_j \|D^{\omega^{\tilde{e}^*}} \overline{\Psi}^*_{\mathscr{E}_2}\|_{L_p(E_n)}. \end{aligned} \tag{24}$$

Applying (24), (20) and (21), we have

$$\begin{aligned} &\left(\int_0^{\mathbf{h}} \|\Delta^{\omega^{\tilde{e}^*}}(\mathbf{t})\, D^{[\mathbf{r}]^{\tilde{e}}} \tilde{f}_k\|^p_{L_p(E_n)} \prod_{j\in\tilde{e}^*} \frac{dt_j}{t_j^{1+p(r_j-[r_j])}}\right)^{1/p} \leqslant \sum_{e\subseteq\mathscr{E}} c_e(\mathbf{h}) \left(\int_0^{\mathbf{h}} \prod_{j\in\tilde{e}^*\setminus e} \frac{dt_j}{t_j^{1-(1-r_j+[r_j])p}}\right)^{1/p} \\ &\quad\times \left\{ \left[\int_0^{\mathbf{h}} \prod_{j\in e^*\cdot\tilde{e}^*} \frac{dt_j}{t_j^{1+(\gamma_j-\xi_j)p}} \times \int_0^{\mathbf{t}} dz^{e^*\cdot\tilde{e}^*} \int_0^{\mathbf{h}} \frac{\|\cdot\|^p_{L_p(E_n)}\, dz^{e^*\setminus\tilde{e}^*}}{\prod_{j\in e^*} z_j^{1+p(r_j-[r_j]-\gamma_j+\xi_j)}}\right]^{1/p} \right. \\ &\quad\left. + \left[\int_0^{\mathbf{h}} \prod_{j\in e^*\cdot\tilde{e}^*} \frac{dt_j}{t_j^{1-\xi_j p}} \times \int_0^{\mathbf{h}} \frac{\prod_{j\in e^*\setminus\tilde{e}^*} z_j^{-1-p(r_j-[r_j])} \|\cdot\|^p_{L_p(E_n)}\, dz^{e^*}}{\prod_{j\in e^*\cdot\tilde{e}^*} z_j^{1+p(r_j-[r_j]-\gamma_j)p} (\sqrt{z_j^2+t_j^2})^{(\gamma_j+\xi_j)p}}\right]^{1/p} \right\} \\ &\equiv \sum_{e\subseteq\mathscr{E}} c_e(\mathbf{h}) \{\mathfrak{M}_{e,k}(f) + \mathfrak{M}^*_{e,k}(f)\}, \end{aligned} \tag{25}$$

where $\gamma > \xi > 0$ are arbitrary small numbers,

$$\|\cdot\| \equiv \| \Delta^{\omega^{e^*}}(\mathbf{z};\, U_k + \square_k)\, D^{[\mathbf{r}]^e} f \|,$$

and $c_e(\mathbf{h})$ is a constant not depending on f.

Further, it can be proved by simple arguments that

$$\mathfrak{M}_{e,k}(f) \leqslant c \| f \|_{S_p^{\mathbf{r}} W(U_k + \square_k)}, \tag{26}$$

$$\mathfrak{M}^*_{e,k}(f) \leqslant c \| f \|_{S_p^{\mathbf{r}} W(U_k + \square_k)}. \tag{27}$$

Assertion (22) of Lemma 5 follows from (25)–(27). Q.E.D.

LEMMA 6. *Suppose* $1 < p < \infty$ *and* $f(\mathbf{x})$ *is a sufficiently smooth function belonging to the space* $S_p^{\mathbf{r}} W(\Omega)$. *Then the following estimate is valid*:

$$\| \tilde{f}_k \|_{S_p^{\mathbf{r}} W(E_n)} \leqslant c \| f \|_{S_p^{\mathbf{r}} W(U_k + \square_k)}, \tag{28}$$

where $\tilde{f}_k(\mathbf{x})$ *is defined by* (14) *and* c *is a constant not depending on* f.

PROOF. In fact, since

$$\begin{aligned} \| \tilde{f}_k \|_{L_p^{\langle \mathbf{r}^{\tilde{e}} \rangle}(E_n)} &\leqslant c \left(\int_{\mathbf{h}}^{\infty} \prod_{j \in \tilde{e}^*} \frac{dt_j}{t_j^{1+(r_j - [r_j])p}} \right)^{1/p} \| D^{[\mathbf{r}]^{\tilde{e}}} \tilde{f}_k \|_{L_p(E_n)} \\ &+ \left\{ \int_0^{\mathbf{h}} \| \Delta^{\omega^{\tilde{e}^*}}(\mathbf{t})\, D^{[\mathbf{r}]^{\tilde{e}}} \tilde{f}_k \|^p_{L_p(E_n)} \prod_{j \in \tilde{e}^*} \frac{dt_j}{t_j^{1+p(r_j - [r_j])}} \right\}^{1/p} \\ + \left\{ \int_{-\mathbf{h}}^{0} \| \Delta^{\omega^{\tilde{e}^*}}(\mathbf{t})\, D^{[\mathbf{r}]^{\tilde{e}}} \tilde{f}_k \|^p_{L_p(E_n)} \prod_{j \in \tilde{e}^*} \frac{dt_j}{|t_j|^{1+p(r_j - [r_j])}} \right\}^{1/p} &\equiv Q_1(\tilde{f}_k) + Q_2(\tilde{f}_k) + Q_3(\tilde{f}_k), \end{aligned}$$

estimate (28) is a consequence of the inequality

$$Q_i(\tilde{f}_k) \leqslant c \| f \|_{S_p^{\mathbf{r}} W(U_k + \square_k)}, \tag{29}$$

which follows from Lemma 3 when $i = 1$ and from Lemma 5 when $i = 2$, and is proved in the same way as Lemma 5 when $i = 3$. Lemma 6 is proved.

THEOREM 1. *Suppose* $f \in S_p^{\mathbf{r}} W(\Omega)$, *where* $\mathbf{r} = (r_1, \ldots, r_n) \geqslant 0$, $e_{\mathbf{r}} \equiv \mathscr{E} \supseteq e^*_{\mathbf{r}} \equiv \mathscr{E}^*$, $1 < p < \infty$, *and* $\Omega \in C_{\alpha,\epsilon}(H_1, \ldots, H_n)$. *Then* f *can be extended from the domain* $\Omega \subset E_n$ *to* E_n *with preservation of class*; *i.e. a function* $\tilde{f}$ *can be constructed that is defined on* E_n, *coincides with* f *on* Ω, *belongs to the space* $S_p^{\mathbf{r}} W(E_n)$ *and satisfies the inequality*

$$\| \tilde{f} \|_{S_p^{\mathbf{r}} W(E_n)} \leqslant c \| f \|_{S_p^{\mathbf{r}} W(\Omega)}, \tag{30}$$

where c *is a constant not depending on* f.

REMARK. In the case of an integral vector $\mathbf{r}$ (when the r_i are nonnegative integers) the theorem is also true for $\Omega \in C_{0,\epsilon}(H_1, \ldots, H_n)$ (see the definition of the class $C_{\alpha,\epsilon}(\mathbf{H})$ of domains Ω).

PROOF OF THEOREM 1. It can be assumed without diminishing the generality of the assertion that $f(\mathbf{x})$ is a sufficiently smooth function belonging to $S_p^{\mathbf{r}}W(\Omega)$.

We define the function $\tilde{f}(\mathbf{x})$ as follows:

$$\tilde{f}(\mathbf{x}) = \sum_{k=1}^{N} \eta_k(\mathbf{x}) \tilde{f}_k(\mathbf{x}), \tag{31}$$

where the $f_k(\mathbf{x})$ ($k = 1, \ldots, N$) are defined by (14) and the $\eta_k(\mathbf{x})$ ($k = 1, \ldots, N$) are infinitely differentiable functions forming a partition of unity subordinate to the covering $\{U_k\}_1^N$ of Ω, i.e.

$$0 \leqslant \eta_k(\mathbf{x}) \leqslant 1 \quad \text{for} \quad \mathbf{x} \in E_n,$$

$$\sum_{k=1}^{N} \eta_k(\mathbf{x}) = 1 \quad \text{for} \quad x \in \Omega,$$

$$\eta_k(\mathbf{x}) = 0 \text{ outside an } \epsilon\text{-contradiction } U_k^{(\epsilon)}.$$

also, the $\eta_k(\mathbf{x})$ satisfy the conditions

$$\sup_{\mathbf{x} \in E_n} |D^{\mathbf{v}} \eta_k(\mathbf{x})| \leqslant c \quad \text{for} \quad |\mathbf{v}| \leqslant |[\mathbf{r}] + \boldsymbol{\omega}|. \tag{32}$$

From (31), (14), (28) and (32) it follows that

$$\|\tilde{f}\|_{S_p^{\mathbf{r}}W(E_n)} \leqslant c \sum_{k=1}^{N} \|\tilde{f}_k\|_{S_p^{\mathbf{r}}W(E_n)} \leqslant c \sum_{k=1}^{N} \|f\|_{S_p^{\mathbf{r}}W(U_k + \square_k)} \leqslant c \|f\|_{S_p^{\mathbf{r}}W(\Omega)}.$$

Q.E.D.

§4. On the extension of functions belonging to the spaces $S_{p,\theta}^{\mathbf{r}}B(\Omega)$

THEOREM 2. *Suppose $f \in S_{p,\theta}^{\mathbf{r}}B(\Omega)$, where $1 < p \leqslant \theta < \infty$, $\mathbf{r} = (r_1, \ldots, r_n) \geqslant 0$, $e_{\mathbf{r}} \equiv \mathscr{E} \neq \varnothing$ and $\Omega \in C_{\alpha,\epsilon}(H_1, \ldots, H_n)$. Then a function $\tilde{f}(\mathbf{x})$ can be constructed that is defined on E_n, coincides with $f(\mathbf{x})$ on Ω and satisfies the inequality*

$$\|\tilde{f}\|_{S_{p,\theta}^{\mathbf{r}}B(E_n)} \leqslant c \|f\|_{S_{p,\theta}^{\mathbf{r}}B(\Omega)}, \tag{33}$$

where c is a constant not depending on f.

Suppose $f(\mathbf{x})$ is a sufficiently smooth function belonging to $S_{p,\theta}^{\mathbf{r}}B(\Omega)$ and consider the functions

$$\begin{aligned} \tilde{f}_k(\mathbf{x}) = {} & \sum_{e \subseteq \mathscr{E}} (-1)^{|\omega^e|} 2^{3|\omega^e|} \prod_{j \in \mathscr{E} \setminus e} h_j^{-1} \\ & \times \int_0^{\mathbf{h}} \frac{d\mathbf{v}^e}{\prod_{j \in e} v_j^{3-\bar{r}_j}} \iint \Phi_{e,\delta^k}\left(\frac{\mathbf{y}^{\mathscr{E}}}{\mathbf{v}^e + \mathbf{h}^{\mathscr{E} \setminus e}}; \frac{\mathbf{z}^e}{\mathbf{v}}\right) \\ & \times \Delta^{2\omega^e}\left(\frac{\mathbf{z}}{2\omega}; U_k + \square_k\right) D^{\bar{\mathbf{r}}^e} f(\mathbf{x} + \mathbf{y}^{\mathscr{E}})\, d\mathbf{y}^{\mathscr{E}}\, d\mathbf{z}^e, \end{aligned} \tag{34}$$

where $k = 1, \ldots, N$.

Putting $\mathscr{E} \equiv \mathscr{E}^*$ in (4) and comparing (34) with (14), we arrive at the assertion that the functions $\tilde{f}_k$ and f coincide on the set $U_k + \square_k$ (we put $\nu = 0$ in (4)).

LEMMA 7. *Suppose f is a sufficiently smooth function belonging to $S^{\mathbf{r}}_{p,\theta}B(\Omega)$, where $1 \leqslant p \leqslant \theta \leqslant \infty$, $\mathbf{r} = (r_1, \ldots, r_n) \geqslant 0$, $e_{\mathbf{r}} \equiv \mathscr{E} \neq \varnothing$ and $\Omega \in C_{0,\epsilon}(H_1, \ldots, H_n)$. Then*

$$\| D^{\nu} \tilde{f}_k \|_{L_p(E_n)} \leqslant c \| f \|_{S^{\mathbf{r}}_{p,\theta}B(U_k + \square_k)}, \tag{35}$$

where $\widetilde{f}_k$ is defined by (34), $\mathbf{r} - \nu \geqslant \mathbf{0}$, $e_{\mathbf{r}-\nu} = \mathscr{E}$ *and c is a constant not depending on f.*

LEMMA 8. *Suppose $f(\mathbf{x})$ is a sufficiently smooth function belonging to $S^{\mathbf{r}}_{p,\theta}B(\Omega)$, where $1 \leqslant p \leqslant \theta \leqslant \infty$, $\Omega \in C_{\alpha,\epsilon}(H_1, \ldots, H_n)$, $\mathbf{r} = (r_1, \ldots, r_n) \geqslant \mathbf{0}$ and $e_{\mathbf{r}} \equiv \mathscr{E} \neq \varnothing$. Suppose, further, $\mathbf{0} \leqslant \rho = (\rho_1, \ldots, \rho_n) \leqslant \mathbf{r} - \nu$, where $e_{\mathbf{r}-\nu} = \mathscr{E}$. Then*

$$\| D^{\nu} \tilde{f}_k \|_{\mathscr{L}^{\langle\rho\rangle}_{p,\theta}(E_n)} \leqslant c \| f \|_{S^{\mathbf{r}}_{p,\theta}B(U_k + \square_k)}, \qquad k = 1, \ldots, N, \tag{36}$$

where $\widetilde{f}_k$ is defined by (34) *and c is a constant not depending on f.*

Lemmas 7 and 8 are proved by arguments that, as usual, were developed in the proofs of imbedding theorems (for the spaces $S^{\mathbf{r}}_{p,\theta}B(\Omega)$) with the use of integral representations of type (4) (see [6]).

PROOF OF THEOREM 2. It suffices to carry out the proof for sufficiently smooth functions f belonging to the spaces $S^{\mathbf{r}}_{p,\theta}B(\Omega)$.

Let

$$\tilde{f}(\mathbf{x}) = \sum_{k=1}^{N} \eta_k(\mathbf{x}) \tilde{f}_k(\mathbf{x}), \tag{37}$$

where the $\widetilde{f}_k(\mathbf{x})$ $(k = 1, \ldots, N)$ are defined by (34) and the $\eta_k(\mathbf{x})$ $(k = 1, \ldots, N)$ are infinitely differentiable functions forming a partition of unity subordinate to the covering $\{U_k\}_1^N$ of Ω.

From (35) and (36) it follows that

$$\| \tilde{f} \|_{S^{\mathbf{r}}_{p,\theta}B(E_n)} \leqslant c \sum_{k=1}^{N} \| f \|_{S^{\mathbf{r}}_{p,\theta}B(U_k + \square_k)},$$

which proves Theorem 2.

BIBLIOGRAPHY

1. S. M. Nikol'skiĭ, *A variational problem*, Mat. Sb. **62** (**104**) (1963), 53–75; English transl., Amer. Math. Soc. Transl. (2) **51** (1966), 132–154. MR **28** #1391.

2. ———, *Functions with dominant mixed derivative satisfying a multiple Hölder condition*, Sibirsk. Mat. Ž. **4** (1963), 1342–1364; English transl., Amer. Math. Soc. Transl. (2) **102** (1973), 27–51. MR **28** #3322.

3. A. D. Džabrailov, *On the theory of "imbedding theorems"*, Trudy Mat. Inst. Steklov. **89** (1967), 80–118 = Proc. Steklov Inst. Math. **89** (1967), 92–135. MR **36** #676.

4. T. I. Amanov, *Study of properties of classes of functions with dominant mixed derivatives. Representation, embedding, extension and interpolation theorems*, Doctoral Dissertation, Novosibirsk, 1967. (Russian)

5. P. I. Lizorkin and S. M. Nikol'skiĭ, *A classification of differentiable functions in some fundamental spaces with dominant mixed derivative*, Trudy Mat. Inst. Steklov. **77** (1965), 143–167 = Proc. Steklov Inst. Math. **77** (1965), 160–187. MR **33** #450.

6. A. D. Džabrailov, *Families of spaces of functions whose mixed derivatives satisfy a multiple-integral Hölder condition*, Trudy Mat. Inst. Steklov. **117** (1972), 139–158 = Proc. Steklov Inst. Math. **117** (1972), 165–188.

7. S. L. Sobolev, *Applications of functional analysis in mathematical physics*, Izdat. Leningrad. Gos. Univ., Leningrad, 1950; English transl., Transl. Math. Monographs, vol. 7, Amer. Math. Soc., Providence, R. I., 1963. MR **14**, 565; **29** #2624.

8. V. P. Il'in, *The properties of some classes of differentiable functions of several varaibles defined in an n-dimensional region,* Trudy Mat. Inst. Steklov. **66** (1962), 227–363; English transl., Amer. Math. Soc. Transl. (2) **81** (1969), 91–256. MR **27** #3750.

9. O. V. Besov, *The continuation of functions in* L_p^l *and* W_p^l, Trudy Mat. Inst. Steklov. **89** (1967), 5–17 = Proc. Steklov Inst. Math. **89** (1967), 1–15. MR **35** #5920.

10. O. V. Besov and V. P. Il'in, *Natural extension of the class of regions in imbedding theorems,* Mat. Sb. **75 (117)** (1968), 483–495 = Math. USSR Sb. **4** (1968), 445–456. MR **37** #1987.

11. O. V. Besov and P. I. Lizorkin, *Singular integral operators and sequences of convolutions in* L_p*-spaces,* Mat. Sb. **73 (115)** (1967), 65–88 = Math. USSR Sb. **2** (1967), 57–76. MR **36** #1936.

12. O. V. Besov, V. P. Il'in and V. I. Lizorkin, L_p*-estimates of a certain class of nonisotropic singular integrals,* Dokl. Akad. Nauk SSSR **169** (1966), 1250–1253 = Soviet Math. Dokl. **7** (1966), 1065–1069. MR **35** #2103.

13. V. P. Il'in, *Conditions of validity of inequalities between* L_p*-norms of partial derivatives of functions of several variables,* Trudy Mat. Inst. Steklov. **96** (1968), 205–242 = Proc. Steklov Inst. Math. **96** (1968), 259–306. MR **42** #6605.

Translated by S. SMITH

Trudy Mat. Inst. Steklov.
131 (1974)

Proc. Steklov Inst. Math.
131 (1974)

ON THE COMPARISON OF DIFFERENTIAL OPERATORS AND ON DIFFERENTIAL OPERATORS OF CONSTANT STRENGTH

UDC 517.518.22

G. G. KAZARJAN

ABSTRACT. Conditions are studied for the existence of the estimate

$$|P(\xi)| \leqslant c(1+|Q(\xi)|) \quad \forall \xi \in R_n$$

for two differential operators $P(D)$ and $Q(D)$ of the form $\Sigma\, \gamma_\alpha D^\alpha$.

By developing the method of V. P. Mihaĭlov, the author finds necessary and sufficient conditions for the existence of such an estimate in the case when not all of the parts of the polynomial $Q(\xi)$ that correspond to the faces of a characteristic polyhedron satisfy "regularity" conditions. The results are then used to solve problems on the comparison of differential operators and on the constancy of the strength of an operator $P(x, D)$.

Bibliography: 7 items.

Statement of the problem

Let $\xi = (\xi_1, \ldots, \xi_n)$ denote a point of an n-dimensional Euclidean space R_n, let $x = (x_1, \ldots, x_n)$ denote a point of an n-dimensional Euclidean space E_n and let A_n denote the space of points (vectors) $\alpha = (\alpha_1, \ldots, \alpha_n)$ with nonnegative integral coordinates. The points $\alpha \in A_n$ will also be called multi-indices. We put, in addition, $\xi^\alpha = \xi_1^{\alpha_1} \cdots \xi_n^{\alpha_n}$ and $D^\alpha = D_1^{\alpha_1} \cdots D_n^{\alpha_n}$, where $D_k = \partial/i\partial x_k$ $(k = 1, \ldots, n)$, and

$$R_n^{(0)} = \Big\{\xi : \xi \in R_n,\ \prod_{k=1}^{n} \xi_k \neq 0\Big\}.$$

By the *characteristic polynomial* (in $\xi \in R_n$) or *full symbol* of a given differential operator $P(x, D) = \Sigma_\alpha\, \gamma_\alpha(x) D^\alpha$ is meant the polynomial $P(x, \xi) = \Sigma_\alpha\, \gamma_\alpha(x)\xi^\alpha$, $x \in \Omega$, Ω being a domain in E_n. Here, for each $x \in \Omega$ the sum is taken over the finite set $\mathfrak{A}(x) = \{\alpha\colon \alpha \in A_n, \gamma_\alpha(x) \neq 0\}$.

We will study both differential operators with constant (generally complex) coefficients and differential operators with variable coefficients $\gamma_\alpha(x)$, $x \in \Omega$.

In connection with a study of the regularity of a solution of the problem $P(D)U = f$ in the space of generalized functions, L. Hörmander (see, for example, [1]) introduced the following definitions.

DEFINITION 0.1. We say that an operator $Q(D)$ (polynomial $Q(\xi)$) with constant coefficients is *weaker* than an operator $P(D)$ (polynomial $P(\xi)$), and write $Q \prec P$, if

$$\frac{\tilde{Q}(\xi)}{\tilde{P}(\xi)} \leqslant C \qquad \forall \xi \in R_n, \tag{0.1}$$

AMS (MOS) subject classifications (1970). Primary 47F05, 35B45.

where, for a given polynomial $R(\xi)$, the polynomial $\widetilde{R}(\xi)$ (called a Hörmander function) is defined as follows:

$$\widetilde{R}(\xi)^2 = \sum_{|\nu|\geq 0} |R^{(\nu)}(\xi)|^2, \qquad R^{(\nu)}(\xi) = D^\nu R(\xi).$$

DEFINITION 0.2. We say that a differential operator $P(x, D)$ (polynomial $P(x, \xi)$) has *constant strength* in Ω if for any fixed points $x, y \in \Omega$ the differential operators $P(x, D)$ and $P(y, D)$ with constant coefficients are equally strong, i.e. if there exist constants C' and C'', generally depending on x and y, such that

$$C'_{xy} \leq \frac{\widetilde{P}(x, \xi)}{\widetilde{P}(y, \xi)} \leq C''_{x,y} \qquad \forall \xi \in R_n, \tag{0.2}$$

where $\widetilde{P}(x, \xi)^2 = \Sigma_{|\nu|\geq 0}|D_\xi^\nu P(x, \xi)|^2$ is a Hörmander function.

To Hörmander belongs the following result: a homogeneous polynomial $P(\xi)$ of degree m is stronger than any polynomial $Q(\xi)$ of degree not greater than m if and only if $P(\xi)$ is elliptic, i.e. $P(\xi) = 0$ implies $\xi = 0$.

DEFINITION 0.3. By the *characteristic polyhedron* (c.p.) $\mathfrak{N}$ for a given collection $\mathfrak{A} = \{\alpha^k\}_1^N$ of multi-indices is meant the least convex polyhedron in A_n containing all of the points α^k $(k = 1, \ldots, N)$; the c.p. for a given polynomial $P(x, \xi) = \Sigma_\alpha \gamma_\alpha(x)\xi^\alpha$ is defined as the c.p. for the collection $\mathfrak{A}(x) = \{\alpha\colon \alpha \in A_n, \gamma_\alpha(x) \neq 0, x \in \Omega\}$.

DEFINITION 0.4. A c.p. $\mathfrak{N}$ is said to be *full* if it has vertices at the origin and on all of the coordinate axes in A_n. A full c.p. $\mathfrak{N}$ is said to be *proper* (*completely proper*) if all of the coordinates of the outward normals of its noncoordinate $(n-1)$-dimensional faces are nonnegative (positive).

L. R. Volevič and S. G. Gindikin [2] gave a necessary and sufficient condition for a polynomial $P(\xi)$ with a proper c.p. $\mathfrak{N}(P)$ to be stronger than any polynomial $Q(\xi)$ with a c.p. $\mathfrak{N}(Q) \subseteq \mathfrak{N}(P)$. They also posed and solved the following problem: given in A_n a collection of points $\{\alpha^k\}_1^N$ with a proper c.p. $\mathfrak{N}$, find algebraic conditions on the coefficients of a polynomial $P(\xi) = \Sigma_1^N \gamma_k \xi^{\alpha^k}$ for it to be stronger than any polynomial of the form $Q(\xi) = \Sigma_1^N q_k \xi^{\alpha^k}$.

Here, guided by the results of V. P. Mihaĭlov [3] and our own results [4], [5], we solve the following problems.

I. For a given collection of multi-indices $\mathfrak{A} = \{\alpha^k\}_1^N$ with an arbitrary c.p. $\mathfrak{N}$, find necessary and sufficient conditions on the coefficients $\{\gamma_k\}_1^N$ for the polynomial $P(\xi) = \Sigma_1^N \gamma_k \xi^{\alpha^k}$ to be stronger than any polynomial $Q(\xi) = \Sigma_1^N q_k \xi^{\alpha^k}$.(1)

II. For a given polynomial $P(\xi)$ with an arbitrary c.p. $\mathfrak{N}$, describe the set $\{Q(\xi)\}$ of polynomials that are weaker than $P(\xi)$.

III. Find algebraic conditions on the characteristic polynomial $P(x, \xi)$ for an operator $P(x, D)$ with an arbitrary c.p. $\mathfrak{N}(P)$ to have constant strength in Ω.

Problem I is solved in general. Problems II and III are solved in general in regard to the c.p. but under certain restrictions (less rigid than in [2]) on the character of the polynomials $P(\xi)$ and $P(x, \xi)$.

---

(1) In another note we intend to give explicit conditions on the collection $\mathfrak{A} = \{\alpha^k\}_1^N$ for the existence of such a polynomial $P(\xi)$. In a special case the existence of such a polynomial follows from a result of Barozzi [6].

§1. Auxiliary problem

In the solution of the problems, an important role is played by the following problem, which in our opinion is of independent interest. Suppose given two polynomials $R(\xi)$ and $Q(\xi)$ with constant coefficients. It is required to find algebraic conditions on the polynomials $R(\xi)$ and $Q(\xi)$ for the validity of the inequality

$$\theta C_1 \leqslant \frac{1+|R(\xi)|}{1+|Q(\xi)|} \leqslant C_2 \qquad \forall \xi \in R_n, \tag{1.1}$$

where $\theta = 0$ or $\theta = 1$, $C_1 > 0$ and $C_2 > 0$.

Let $\mathfrak{N}$ be a full c.p. in A_n and let $\mathfrak{N}_i^k$ ($i = 1, \ldots, M_k'$ and $k = 0, 1, \ldots, n-1$) denote its k-dimensional faces. The zero-dimensional faces $\mathfrak{N}_i^0$ ($i = 1, \ldots, M_0'$) will sometimes be called the vertices of $\mathfrak{N}$ and denoted by e^i ($i = 1, \ldots, M_0'$).

DEFINITION 1.1. A face $\mathfrak{N}_i^k$ ($i = 1, \ldots, M_k$ and $k = 0, 1, \ldots, n-1$) of a c.p. $\mathfrak{N}$ is called a *principal face* if among the $(n-1)$-dimensional hyperplanes of support of $\mathfrak{N}$ containing the face $\mathfrak{N}_i^k$ and not containing any face of dimension greater than k there exists a hyperplane whose outward normal has at least one positive coordinate. A principal face $\mathfrak{N}_i^k$ of a c.p. $\mathfrak{N}$ is said to be *proper* (*completely proper*) if among these hyperplanes there exists a hyperplane all of the coordinates of the outward normal of which are nonnegative (positive).

REMARK 1.1. Clearly, all of the subfaces of a principal (completely proper) face are principal (completely proper) faces, whereas a proper face can have an improper (but principal) subface.

DEFINITION 1.2. A polynomial $P(\xi)$ is said to be λ-*homogeneous* (of degree d) if for any $\xi \in R_n$ and $t > 0$

$$P(t^\lambda \xi) = t^d P(\xi),$$

where $\lambda = (\lambda_1, \ldots, \lambda_n)$, $t^\lambda \xi = (t^{\lambda}\xi_1, \ldots, t^{\lambda}\xi_n)$ and d is a real number.

For a given polynomial $P(\xi) = \Sigma_\alpha \gamma_\alpha \xi^\alpha$ with a c.p. $\mathfrak{N}$ we denote by $P^{i,k}(\xi)$ the polynomial

$$P^{i,k}(\xi) = \sum_{\alpha \in \mathfrak{N}_i^k} \gamma_\alpha \xi^\alpha \qquad (i = 1, \ldots, M_k', \; k = 0, \ldots, n-1).$$

V. P. Mihaĭlov in [3] proved

LEMMA 1.1. *For a given polynomial $P(\xi)$ with a c.p. $\mathfrak{N}(P)$, a polynomial $P^{i,k}(\xi)$ ($i = 1, \ldots, M_k^i$ and $k = 0, \ldots, n-1$) is λ-homogeneous, where λ is the outward normal to any hyperplane of support of $\mathfrak{N}$ containing the face $\mathfrak{N}_i^k$ and not containing any face of dimension greater than k.*

In the sequel, the outward normal to a hyperplane of support of a c.p. $\mathfrak{N}$ containing some face $\mathfrak{N}_i^k$ and not containing any other face of dimension greater than k will simply be called *an outward normal to the face* $\mathfrak{N}_i^k$. Thus a given vector λ can serve as an outward normal to one and only one face of a convex c. p. $\mathfrak{N}$.

DEFINITION 1.3. A face $\mathfrak{N}_i^k$ ($i = 1, \ldots, M_k'$ and $k = 0, \ldots, n-1$) of the c.p. $\mathfrak{N}(P)$ for a polynomial $P(\xi) = \Sigma_\alpha \gamma_\alpha \xi^\alpha$ is said to be *P-regular* if the polynomial $P^{i,k}(\xi)$ does not have any zeros in $R_n^{(0)}$.

It clearly follows from this definition that the zero-dimensional faces (vertices) of a c.p. $\mathfrak{N}(P)$ are always P-regular.

V. P. Mihaĭlov [3] proved(2)

LEMMA 1.2. *Suppose given a polynomial $P(\xi)$ with a full c.p. $\mathfrak{N}(P)$ all of whose principal faces are P-regular. Then*

$$\sum_{\nu\in\mathfrak{N}} |\xi^\nu| \leqslant C(|P(\xi)|+1) \qquad \forall \xi \in R_n.$$

This lemma directly implies

THEOREM 1.1. *Suppose given a polynomial $Q(\xi)$ with a full c.p. $\mathfrak{N}(Q)$ all of whose principal faces are Q-regular. Then inequality* (1.1) *holds for $\theta = 0$ if the c.p. $\mathfrak{N}(R) \subseteq \mathfrak{N}(Q)$; it also holds for $\theta = 1$ if $\mathfrak{N}(R) \equiv \mathfrak{N}(Q)$ and all of the principal faces of $\mathfrak{N}(Q) \equiv \mathfrak{N}(R)$ are R-regular.*

The proof of the theorem follows from the obvious inequality

$$1+|R(\xi)| \leqslant C_1\Big(1+\sum_{\nu\in\mathfrak{N}(R)} |\xi^\nu|\Big) \leqslant C \sum_{\nu\in\mathfrak{N}(Q)} |\xi^\nu|$$

and from Lemma 1.2.

A proof of the theorem not relying on Lemma 1.2 is contained in the proof of Theorem 1.2 below.

In the sequel, the faces of the c.p. $\mathfrak{N}(P)$ for a given polynomial $P(\xi)$ that are not P-regular will be said to be *P-irregular.*

Let $P(\xi) = \Sigma_\alpha \gamma_\alpha \xi^\alpha$, let $\mathfrak{N}_i^k$ be a P-irregular principal face of the c.p. $\mathfrak{N}(P)$ and let λ be an outward normal to the face $\mathfrak{N}_i^k$. We put

$$P_{\lambda,d}^{i,k}(\xi) = \sum_{(\lambda,\alpha)=d} \gamma_\alpha \xi^\alpha, \qquad d_\lambda(P) = \max_{\substack{\alpha\in\mathfrak{N}\setminus\mathfrak{N}_i^k \\ \gamma_\alpha\neq 0}} (\lambda,\alpha);$$
$$P_\lambda^{i,k}(\xi) = P_{\lambda, d_\lambda(P)}^{i,k}(\xi). \tag{1.2}$$

Let $P(\xi) = \Sigma_\alpha \gamma_\alpha \xi^\alpha$, and let λ be a unit vector. We denote by $\delta(P, \lambda, \xi)$ the maximum of zero and the numbers δ for which $\Sigma_{(\lambda,\alpha)=\delta} \gamma_\alpha \xi^\alpha \neq 0$.

The following proposition gives a necessary condition for the validity of inequality (1.1). It is similar to the sufficient condition, as will be seen below.

LEMMA 1.3. *For the validity of inequality* (1.1) *for two polynomials $R(\xi)$ and $Q(\xi)$ with the same c.p. $\mathfrak{N}$ it is necessary that the following condition be satisfied for any outward unit normals λ to the principal faces of $\mathfrak{N}$: $\delta(R, \lambda, \xi) \leqslant \delta(Q, \lambda, \xi)$ when $\theta = 0$ and $\delta(R,\lambda,\xi) = \delta(Q, \lambda, \xi)$ when $\theta = 1$.*

PROOF. Suppose $\delta^0 \equiv \delta(R, \lambda^0, \xi^0) > \delta(Q, \lambda^0, \xi^0) \equiv \delta^1 \geqslant 0$ for some $\lambda^0 = (\lambda_1^0, \ldots, \lambda_n^0)$ and $\xi^0 \in R^n$.

We first note that if λ^0 is a unit vector, then, clearly, any polynomial $P(\xi)$ with a full c.p. $\mathfrak{N}$ can be represented in the form of a sum of λ^0-homogeneous polynomials:

(2) An analogous result for a proper c.p. $\mathfrak{N}(P)$ has been obtained in other terms in [2].

$$P(\xi)=\sum_{k=1}^{N} P_k(\xi)=\sum_{k=1}^{N}\sum_{(\lambda^0,\alpha)=\sigma_k}\gamma_\alpha\xi^\alpha \qquad (\sigma_1>\sigma_2>\ldots>\sigma_N),$$

where the set of numbers $\{\sigma_k\}$ coincides with the set of values of (λ^0,α) for $\alpha\in\mathfrak{N}$ and is consequently finite.

We represent in accordance with this remark the polynomials $R(\xi)$ and $Q(\xi)$ in the form

$$R(\xi)=\sum_{k=1}^{N_1} R_k(\xi)+R_{\lambda^0,\delta^0}(\xi)+\sum_{k=N_1+2}^{N} R_k(\xi), \tag{1.3a}$$

$$Q(\xi)=\sum_{k=1}^{N_1} Q_k(\xi)+Q_{\lambda^0,\delta^0}(\xi)+\sum_{k=N_1+2}^{N} Q_k(\xi), \tag{1.3b}$$

where $R_k(\xi)$ and $Q_k(\xi)$ $(k=1,\ldots,N)$ are λ^0-homogeneous polynomials of degree of homogeneity σ_k, with $\sigma_k>\delta^0$ for $k=1,\ldots,N_1$ and $\sigma_k<\delta^0$ for $k=N_1+2,\ldots,N$, while $R_{\lambda^0,\delta^0}(\xi)$ and $Q_{\lambda^0,\delta^0}(\xi)$ are λ^0-homogeneous polynomials of degree of homogeneity δ^0.

We put $\xi=t^{\lambda^0}\xi^0=(t^{\lambda_1^0}\xi_1^0,\ldots,t^{\lambda_n^0}\xi_n^0)$. From (1.3a) and (1.3b) we get by virtue of the λ^0-homogeneity of the polynomials $R_k(\xi)$ and $Q_k(\xi)$ $(k=1,\ldots,N)$ and the definition of δ^0 that

$$R(\xi)=t^{\delta^0}R_{\lambda^0,\delta^0}(\xi^0)+\sum_{k=N_1+2}^{N} t^{\sigma_k}R_k(\xi^0)=t^{\delta^0}R_{\lambda^0,\delta^0}(\xi^0)+o(t^{\delta^0}) \tag{1.4a}$$

for $t\to\infty$ and

$$Q(\xi)=t^{\delta^0}Q_{\lambda^0,\delta^0}(\xi^0)+\sum_{k=N_1+2}^{N} t^{\sigma_k}Q_k(\xi^0). \tag{1.4b}$$

Since, by assumption, $\delta^1<\delta^0$, it follows that $Q_{\lambda^0,\delta^0}(\xi^0)=0$ and $R_{\lambda^0,\delta^0}(\xi^0)\neq 0$, and we finally get from (1.4a) and (1.4b) that for $t\to\infty$

$$R(\xi)=O(t^{\delta^0}),\qquad Q(\xi)=o(t^{\delta^0}).$$

But this contradicts (1.1) for $\theta=0$ and $\theta=1$ since for $t\to\infty$, $\xi=t^{\lambda^0}\xi^0$, we have

$$\frac{R(\xi)}{1+|Q(\xi)|}\to\infty.$$

In the case $\theta=1$ we analogously obtain a contradiction with (1.1) by assuming that $\delta(R,\lambda^0,\xi^0)<\delta(Q,\lambda^0,\xi^0)$. The lemma is proved.

REMARK 1.2. It immediately follows from the lemma that inequality (1.1) does not hold for $\theta=1$ in the following cases, for example:

1) a principal face of the c.p. $\mathfrak{N}\equiv\mathfrak{N}(Q)\equiv\mathfrak{N}(R)$ for the polynomials $R(\xi)$ and $Q(\xi)$ is R-regular but Q-irregular (or conversely);

2) a principal face $\mathfrak{N}_i^k$ of the c.p. $\mathfrak{N}$ is R-irregular and Q-irregular with $d_\lambda(P)=d_\lambda(Q)>0$ (λ is a normal to the face $\mathfrak{N}_i^k$) and there exists a point $\xi^0\in R_n$ such that $R^{i,k}(\xi^0)=Q^{i,k}(\xi^0)=R_\lambda^{i,k}(\xi^0)=0\neq Q_\lambda^{i,k}(\xi^0)$;

3) the same as case 2) except that the condition $Q_\lambda^{i,k}(\xi^0)\neq 0$ is replaced by the condition $Q_\lambda^{i,k}(\xi^0)=R_{\lambda\lambda}^{i,k}(\xi^0)=0\neq Q_{\lambda\lambda}^{i,k}(\xi^0)$, where the polynomials $R_{\lambda\lambda}^{i,k}(\xi)$ and $Q_{\lambda\lambda}^{i,k}(\xi)$, corresponding to the next "higher" level, are defined as follows:

$$d_{\lambda\lambda}(R) = \max_{\substack{\alpha \in \mathfrak{N} \setminus \mathfrak{N}_i^k \\ (\lambda, \alpha) < d_\lambda, \gamma_\alpha \neq 0}} (\lambda, \alpha), \qquad R_{\lambda\lambda}^{i,k}(\xi) = R_{\lambda, d_{\lambda\lambda}(R)}^{i,k}(\xi)$$

and analogously for $Q(\xi)$.

In fact, since a principal face does not pass through the origin, we have, taking a suitable normal to this face as λ^0,

$$\delta^0 \equiv \delta(R, \lambda^0, \xi^0) > \delta(Q, \lambda^0, \xi^0) \geqslant 0$$

for some $\xi^0 \in R_n$.

The following proposition gives a sufficient condition for the fulfilment of inequality (1.1) in a case not covered by Theorem 1.1.

THEOREM 1.2. *Inequality* (1.1) *holds if the polynomial* $Q(\xi)$ *has a full c.p.* $\mathfrak{N}(Q)$ *and the following conditions are satisfied*:

1) *All of the principal faces of* $\mathfrak{N}(Q)$ *with the exception of possibly one* $(n-1)$*-dimensional face* $\mathfrak{N}_{i_0}^{n-1}$ *are* Q*-regular for* $\theta = 0$ *and* R*- and* Q*-regular for* $\theta = 1$.

2) *In the event that* $\mathfrak{N}_{i_0}^{n-1}$ *with outward normal* $\lambda = (\lambda_1, \ldots, \lambda_n)$ *is* Q*-irregular*,

a) $Q^{i_0, n-1}(\xi) \geqslant 0$, $\theta R^{i_0, n-1}(\xi) \geqslant 0 \, \forall \xi \in R_n$,

b) *the polynomials* $R^{i_0, n-1}(\xi)$ *and* $Q^{i_0 n-1}(\xi)$ *have the form*

$$Q^{i_0, n-1}(\xi) = \overline{Q}^{i_0, n-1}(\xi)\, p(\xi), \tag{1.5a}$$

$$R^{i_0, n-1}(\xi) = \overline{R}^{i_0, n-1}(\xi)\, p(\xi), \tag{1.5b}$$

where $\overline{Q}^{i_0, n-1}(\xi)$ *does not vanish in* $R_n^{(0)}$ *for* $\theta = 0$ *and* $\overline{Q}^{i_0, n-1}(\xi)$ *and* $\overline{R}^{i_0, n-1}(\xi)$ *do not vanish in* $R_n^{(0)}$ *for* $\theta = 1$,

c) $Q_\lambda^{i_0, n-1}(\xi) > 0$ *for* $\theta = 0$ *and* $Q_\lambda^{i_0, n-1}(\xi) > 0$, $R_\lambda^{i_0, n-1}(\xi) > 0$ *for* $\theta = 1$ *at those points* $\xi \in R_n^{(0)}$ *at which* $R^{i_0, n-1}(\xi) = Q^{i_0, n-1}(\xi) = 0$,

d) $\mathfrak{N}(R) \subseteq \mathfrak{N}(Q)$, $d_\lambda(R) \leqslant d_\lambda(Q)$ *for* $\theta = 0$, *and* $\mathfrak{N}(R) \equiv \mathfrak{N}(Q)$, $d_\lambda(R) = d_\lambda(Q)$ *for* $\theta = 1$.

REMARK 1.3. In the two-dimensional case two homogeneous polynomials of degree of homogeneity m have the same real zeros $\xi \in R_n^{(0)}$ if and only if they have representations of type (1.5a), (1.5b). This follows from the fact that in the two-dimensional case homogeneous polynomials can be reduced by a change of the independent variables to polynomials in a single variable, whereas in the one-dimensional case the existence of representations of type (1.5a), (1.5b) for two polynomials having the same real zeros is well known.

Condition 2b), at least when $n = 2$ and for the homogeneous polynomial $Q^{i_0, n-1}(\xi)$, is necessary for the validity of inequality (1.1) by virtue of Remark 1.2.

REMARK 1.4. The following examples show that conditions 2a) and 2c) are essential for the validity of inequality (1.1).

EXAMPLE 1. Suppose $n = 2$, $Q(\xi) = -(\xi_1^3 - \xi_2^3)^2 + \xi_1^4 + 1$ and $R(\xi) = \theta(\xi_1^3 - \xi_2^3)^2 + \xi_1^4 + 1$, where $\theta = 0$ or $\theta = 1$. Clearly, all of the conditions of Theorem 1.2 except condition 2a) are satisfied for the polynomials $Q(\xi)$ and $R(\xi)$. Here the only Q- and R-irregular face is the one-dimensional face $\mathfrak{N}_1^1$ with vertices (6, 0) and (0, 6). The vector $\lambda = (1/\sqrt{2}, 1/\sqrt{2})$

is an outward (unit) normal to this face, $R^{1,1}(\xi) \geqslant 0$, $Q^{1,1}(\xi) \leqslant 0$, and $Q^{1,1}_{(\lambda)}(\xi) = R^{1,1}_{(\lambda)}(\xi) = \xi_1^4 \geqslant 0$.

Let $\xi_1 = t\sqrt[3]{t}$ and $\xi_2 = t\sqrt[3]{t - t^\sigma}$, where $t > 0$ and $\sigma = 1/3$. Then

$$\frac{|R(\xi)|}{1 + (Q(\xi)|} = \frac{|\theta t^{6-2\sigma} + t^{16/3} + 1|}{1 + |-t^{6-2\sigma} + t^{16/3} + 1|} = \frac{(1+\theta)\, t^{16/3} + 1}{2} \to \infty \quad \text{as} \quad t \to \infty.$$

EXAMPLE 2. It can be shown analogously that inequality (1.1) does not hold for the polynomials

$$Q(\xi) = (\xi_1^3 - \xi_2^3)^2 - \xi_1^4 + 1, \qquad R(\xi) = \theta(\xi_1^3 - \xi_2^3) + \xi_1^4 + 1.$$

Here condition 2c) of Theorem 1.2 is not satisfied.

PROOF OF THEOREM 1.2. Suppose, on the contrary, that there exists a sequence $\{\xi^s\}$, $\xi^s \to \infty$, such that, for example,

$$\frac{|R(\xi^s)|}{1 + |Q(\xi^s)|} \to \infty \quad \text{as} \quad \xi^s \to \infty. \tag{1.6}$$

It can be assumed without loss of generality that all of the coordinates of the vectors ξ^s are positive. Let

$$\rho_s = \exp\sqrt{\sum_{k=1}^{n} (\ln \xi_k^s)^2}, \quad \lambda_i^s = \frac{\ln \xi_i^s}{\ln \rho_s} \quad (i = 1, \dots, n),$$

$$\xi^s = \rho_s^{\lambda^s} \qquad (\xi_i^s = \rho_s^{\lambda_i^s},\ i = 1, \dots, n),$$

so that $\lambda^s = (\lambda_1^s, \dots, \lambda_n^s)$ is a unit vector. Clearly, $\rho_s \to \infty$ if $\xi^s \to \infty$ or if a coordinate of ξ^s tends to zero.

Since the vectors λ^s are on the unit sphere, the sequence $\{\lambda^s\}$ has a limit point λ^∞. It can be assumed that $\lambda^s \to \lambda^\infty$. From the convexity of the c.p. $\mathfrak{R}(Q)$ it follows that λ^∞ is an outward normal to one and only one face of $\mathfrak{R}(Q)$.

We take in A_n a basis $(e^{1,1}, e^{1,2}, \dots, e^{1,n})$ in which $e^{1,1} \equiv \lambda^\infty$. Then $\lambda^s = \Sigma_{i=1}^n \lambda_{1,i}^s e^{1,i}$, and $\lambda_{1,i}^s = o(\lambda_{1,1}^s) = o(1)$ for $i = 2, 3, \dots, n$ since $\lambda^s \to \lambda^\infty \equiv e^{1,1}$.

It can be assumed (perhaps for some subsequence) that as $s \to \infty$

$$\frac{\sum_{i=2}^{n} \lambda_{1,i}^s e^{1,i}}{\left|\sum_{i=2}^{n} \lambda_{1,i}^s e^{1,i}\right|} \to e^{2,2}.$$

We go over in the subspace spanned by $(e^{1,2}, e^{1,3}, \dots, e^{1,n})$ to a new basis $(e^{2,2}, \dots, e^{2,n})$. Then

$$\lambda^s = \lambda_{1,1}^s e^{1,1} + \sum_{i=2}^{n} \lambda_{2,i}^s e^{2,i},$$

in which clearly

$$\lambda_{2,2}^s = o(\lambda_{1,1}^s), \ \lambda_{2,i}^s = o(\lambda_{2,2}^s), \ i = 3, \dots, n.$$

Proceeding analogously in the subspace with basis $(e^{2,3}, \dots, e^{2,n})$ and so on, we get (after modifying the notation)

$$\lambda^s = \sum_{i=1}^{n} \lambda_i^s e^i, \qquad \lambda_1^s \to 1, \qquad \lambda_{i+1}^s = o(\lambda_i^s) \qquad i = 1, \ldots, n-1.$$

In this connection, there exists an s_0 such that for all $s \geqslant s_0$ we have $\lambda_i^s > 0$ $(i = 1, \ldots, m)$ and $\lambda_i^s = 0$ $(i = m+1, \ldots, n)$, $m \leqslant n$.

Let $\mathfrak{N}_{i_2}^{k_1}, \mathfrak{N}_{i_2}^{k_2}, \ldots, \mathfrak{N}_{i_m}^{k_m}$ denote the faces of $\mathfrak{N}(Q)$ satisfying the condition that $\mathfrak{N}_{i_1}^{k_1}$ lies in the hyperplane of support of $\mathfrak{N}(Q)$ with outward normal e^1 while each face $\mathfrak{N}_{i_j}^{k_j}$ $(j = 2, \ldots, m)$ lies in the hyperplane of support of the preceding face (considered as an isolated object) with outward normal e^j. If there is more than one subface of $\mathfrak{N}_{i_j}^{k_j}$ with normal e^{j+1}, we take as $\mathfrak{N}_{i_{j+1}}^{k_{j+1}}$ the subface containing the point α at which the value of (e^{j+1}, α) is maximal.

From the construction of the faces $\mathfrak{N}_{i_1}^{k_1}, \ldots, \mathfrak{N}_{i_m}^{k_m}$ we see that their dimensions are nonincreasing: $k_1 \geqslant k_2 \geqslant \cdots \geqslant k_m$.

Let, as above, $Q^{i_j k_j}(\xi)$ denote the part of the polynomial $Q(\xi)$ whose multi-indices belong to $\mathfrak{N}_{i_j}^{k_j}$, and let α denote an arbitrary point (vector) belonging to all of the $\mathfrak{N}_{i_j}^{k_j}$ $(j = 1, \ldots, m)$; i.e. $\alpha \in \mathfrak{N}_{i_m}^{k_m}$. We will study the behavior of the polynomials $Q(\xi)$ and $R(\xi)$ for $\rho_s \to \infty$, $\xi^s = \rho_s^{\Sigma_1^n \lambda_i^s e^i}$. The index s will be omitted for the sake of simplicity in notation. By choosing a subsequence, if necessary, it can be assumed that $\rho^{\lambda_r} \to \infty$, $\rho^{\lambda_{r+1}} \to b \geqslant 1$ for some $r \in [1, m]$ (when $r = n$ we put, by definition, $\lambda_{n+1} = 0$). Then, from the e^j-homogeneity of the polynomials $Q^{i_j, k_j}(\xi)$ (see Lemma 1.1) and from the convexity of $\mathfrak{N}(Q)$ and its faces we get for certain positive $\sigma_1, \ldots, \sigma_r$ and $1 \leqslant r \leqslant m \leqslant n$ (e^{n+1} is a unit vector)

$$\begin{aligned} Q(\xi) &= \rho^{(\alpha, \lambda_1 e^1)} [Q^{i_1, k_1}(\rho^{\sum_{j=2}^{n+1} \lambda_j e^j}) + o(\rho^{-\sigma_1 \lambda_1})] \\ &= \rho^{(\alpha, \lambda_1 e^1 + \lambda_2 e^2)} [Q^{i_2 k_2}(\rho^{\sum_{j=3}^{n+1} \lambda_j e^j}) + o(\rho^{-\sigma_2 \lambda_2})] \\ &\cdots\cdots\cdots\cdots\cdots\cdots \\ &= \rho^{(\alpha, \sum_{j=1}^{r} \lambda_j e^j)} [Q^{i_r, k_r}(\rho^{\sum_{j=r+1}^{n+1} \lambda_j e^j}) + o(\rho^{-\sigma_r \lambda_r})]. \end{aligned} \tag{1.7}$$

An analogous representation is obtained for $R(\xi)$ if $R^{i_r, k_r}(\xi) \neq 0$. But if $R^{i_r, k_r}(\xi) = 0$ (then $\theta = 0$ in Theorem 1.2), we have

$$R(\xi) = o(\rho^{(\alpha, \sum_{j=1}^{r} \lambda_j e^j)}), \qquad \alpha \in \mathfrak{N}_{i_r}^{k_r}(Q). \tag{1.8}$$

Since $\rho^{\lambda_{r+1}} \to b \geqslant 1$, it follows that $\rho^{\Sigma_{r+1}^{n+1} \lambda_j e^j} \to b^{e^{r+1}} \equiv \eta$. Clearly, $0 < \eta_i < \infty$ for all $i = 1, \ldots, n$ (in accordance with the definition of the η_i as finite powers of a positive number).

We consider two cases: a) $(e^1, \alpha) > 0$ and b) $(e^1, \alpha) = 0$. The case $(e^1, \alpha) < 0$ is impossible, as can be seen from the fact that the equation for the hyperplane of support with outward unit normal λ of a c.p. $\mathfrak{N}$ can be written in the form $(\lambda, \alpha) = d$, where $d \geqslant 0$ is the distance from the origin to the given hyperplane and α is a roving point of the hyperplane (see, for example, [7]).

Case a). Suppose first $Q^{i_r, k_r}(\eta) \neq 0$. Since $(e^1, \alpha) > 0$ and $\lambda_i = o(\lambda_1)$ for $i = 2, \ldots, n$, we eventually have, beginning at some point, $(\alpha, \Sigma_1^r \lambda_j e^j) > 0$. Therefore (1.7) implies

$$Q(\xi) = \rho^{(\alpha, \sum_{j=r}^{r} \lambda_j e^j)} [Q^{i_r, k_r}(\eta) + o(1)]. \tag{1.9a}$$

We analogously get for $\theta = 1$ or for $\theta = 0, R^{i_r, k_r}(\eta) \neq 0$ that

$$R(\xi) = \rho^{(\alpha, \sum_{j=1}^{r} \lambda_j e^j)} [R^{i_r, k_r}(\eta) + o(1)]. \tag{1.9b}$$

But representations (1.9a), (1.9b) for $\theta = 1$ or for $\theta = 0, R^{i_r, k_r}(\eta) \neq 0$ and representations (1.9a), (1.8) for $R^{i_r, k_r}(\eta) = 0$ (and hence for $\theta = 0$) together contradict (1.6).

Suppose now $Q^{i_r, k_r}(\eta) = 0$. This means that $P(\eta) = 0$ (see (1.5a)) and that the face $\mathfrak{N}_{i_r}^{k_r}$ coincides with the Q- and R-irregular face $\mathfrak{N}_{i_0}^{n-1}$, $r = m = 1, k_r = n - 1$. From (1.5a), (1.5b) it follows that $R^{i_0, n-1}(\eta) = 0$.

In this case we respectively represent $Q(\xi)$ and $R(\xi)$ in the form

$$Q(\xi) = \bar{Q}^{i_0, n-1}(\xi)\, p(\xi) + Q_{e1}^{i_0, n-1}(\xi) + q(\xi), \tag{1.10a}$$

$$R(\xi) = \bar{R}^{i_0, n-1}(\xi)\, p(\xi) + R_{e1}^{i_0, n-1}(\xi) + r(\xi). \tag{1.10b}$$

Substituting the value

$$\xi = \rho^{\sum_{i=1}^{n+1} \lambda_i e^i}$$

in (1.10a), (1.10b) and using the e^1-homogeneity of the polynomials $Q^{i_0, n-1}(\xi)$, $R^{i_0, n-1}(\xi)$, $R_{e1}^{i_0, n-1}(\xi)$ and $Q_{e1}^{i_0, n-1}(\xi)$, we get

$$Q(\xi) = \rho^{(\alpha, \lambda_1 e^1)} \bar{Q}^{i_0, n-1}(\rho^{\sum_{j=2}^{n+1} \lambda_j e^j}) p(\rho^{\sum_{j=2}^{n+1} \lambda_j e^j}) + \rho^{(\beta, \lambda_1 e^1)} Q_{e1}^{i_0, n-1}(\rho^{\sum_{j=2}^{n+1} \lambda_j e^j}) + q(\xi), \tag{1.11a}$$

$$R(\xi) = \rho^{(\alpha, \lambda_1 e^1)} \bar{R}^{i_0, n-1}(\rho^{\sum_{j=2}^{n+1} \lambda_j e^j}) p(\rho^{\sum_{j=2}^{n+1} \lambda_i e^j}) + \rho^{(\beta, \lambda_1 e^1)} R_{e1}^{i_0, n-1}(\rho^{\sum_{j=2}^{n+1} \lambda_j e^i}) + r(\xi). \tag{1.11b}$$

Let $\xi \to \infty$. Then, by condition 2b) of the theorem, $Q_{e1}^{i_0, n-1}(\eta) > 0$; consequently for sufficiently large $|\xi|$ we have $Q_{e1}^{i_0, n-1}(\rho^{\Sigma_2^{n+1} \lambda_j e^j}) > 0$, and hence, since $q(\xi) = o(\rho^{(\beta, \lambda_1 e^1)})$,

$$\rho^{(\beta, \lambda_1 e^1)} Q_{e1}^{i_0, n-1}(\rho^{\sum_{j=2}^{n+1} \lambda_j e^j}) + q(\xi) > 0. \tag{1.12}$$

On the other hand, by condition 2a) of the theorem, $Q^{i_0, n-1}(\xi) > 0$. It therefore follows from (1.11a) and (1.12) that for sufficiently large $|\xi|$

$$\frac{|R(\xi)|}{1+|Q(\xi)|}\leqslant\frac{\rho^{(\alpha,\lambda_1e^1)}\,|R^{i_0,n-1}(\rho^{\sum_{j=2}^{n+1}\lambda_je^j})|}{1+|Q(\xi)|}+\frac{|\rho^{(\beta,\lambda_1e^1)}R^{i_0,n-1}_{e^1}(\rho^{\sum_{j=2}^{n+1}\lambda_je^j})+q(\xi)|}{1+|Q(\xi)|}$$

$$\leqslant\frac{\rho^{(\alpha,\lambda_1e^1)}\,|\overline{R}^{i_0,n-1}(\rho^{\sum_{j=2}^{n+1}\lambda_je^j})|\cdot|p(\rho^{\sum_{j=2}^{n+1}\lambda_je^j})|}{1+\rho^{(\alpha,\lambda_1e^1)}\,|\overline{Q}^{i_0,n-1}(\rho^{\sum_{j=2}^{n+1}\lambda_je^j})|\cdot|p(\rho^{\sum_{j=2}^{n+1}\lambda_je^j})|}+\frac{\rho^{(\beta,\lambda_1e^1)}\,|R^{i_0,n-1}_{e^1}(\eta)+o(1)|}{1+\rho^{(\beta,\lambda_1e^1)}\,|Q^{i_0,n-1}_{e^1}(\eta)+o(1)|}\tag{1.13}$$

$$\leqslant C_1\frac{|\overline{R}^{i_0,n-1}(\eta)+o(1)|}{|\overline{Q}^{i_0,n-1}(\eta)+o(1)|}+C_2\frac{|R^{i_0,n-1}_{e^1}(\eta)+o(1)|}{|Q^{i_0,n-1}_{e^1}(\eta)+o(1)|}\leqslant C.$$

Relation (1.13) contradicts (1.6) and completes the consideration of case a).

Case b). $(e^1, \alpha)=0$.

REMARK 1.5. Clearly, if a polynomial $R(\xi)$ has a full c.p. $\mathfrak{R}$, then the c.p. for a polynomial $R(\xi)|_{\xi_j=0}$ for $j\in[1,n]$ is also full in the corresponding $(n-1)$-dimensional subspace. Moreover, if two polynomials $R(\xi)$ and $Q(\xi)$ have the same full c.p. $\mathfrak{R}$, then the polynomials $R(\xi)|_{\xi_j=0}$ and $Q(\xi)|_{\xi_j=0}$ also have the same full c.p. in the corresponding $(n-1)$-dimensional subspace.

In the case b) under consideration, the face whose outward normal is e^1 clearly passes through the origin and hence is not a principal face of $\mathfrak{R}$; consequently $e^1_i\leqslant 0$ $(i=1,\dots,n)$. In this connection, if the nonprincipal face with outward normal e^1 has dimension k, then k of the numbers e^1_i $(i=1,\dots,n)$ are equal to zero with the remaining numbers being negative. It can clearly be assumed without loss of generality that $e^1_1=\cdots=e^1_k=0$, $e^1_{k+1}<0,\dots,e^1_n<0$. Since $e^1_j<0$ $(j=k+1,\dots,n)$ and

$$e^1_j=\lim_{\xi\to\infty}\ln\xi_j\Big/\Big(\sum_{k=1}^n(\ln\xi_k)^2\Big)^{1/2},$$

we have, beginning at some point, $\xi_j<1$ $(j=k+1,\dots,n)$; on the other hand, since $\xi\to\infty$, we have $\xi_i\to\infty$ for certain $i\in[1,k]$. But since $e^1_i=0$ for those i, it follows (at least for a subsequence of the points ξ) that $\xi_j\to 0$ for at least one $j\in(k,n]$.

Suppose (after modifying the notation) $\xi_1\to\infty,\dots,\xi_l\to\infty$ $(l\leqslant k)$ and $\xi_{l+1}\to 0,\dots,\xi_{l+m}\to 0$ $(l+m\leqslant n)$. Let $\psi(\xi)=\max_{1\leqslant j\leqslant l}\xi_j$. Then, clearly,

$$\ln\psi(\xi)\Big/\Big(\sum_{k=1}^n(\ln\xi_k)^2\Big)^{1/2}\to 0.\tag{1.14}$$

On the other hand, there clearly exist positive constants m and M such that

$$m\leqslant\sum_{k=1}^l(\ln\xi_k)^2/(\ln\psi(\xi))^2\leqslant M.\tag{1.15}$$

It follows from (1.14) and (1.15) that

$$\sum_{k=l+1}^{n} (\ln \xi_k)^2/(\ln \psi(\xi))^2 \to \infty. \tag{1.16}$$

From this result, by going over to a subsequence, if necessary, we get that for some $j \in [l+1, n]$

$$|\ln \xi_j| / \ln \psi(\xi) \to \infty, \tag{1.17}$$

i.e. $|\ln \xi_j| \to \infty$ "faster" than $\ln \psi(\xi) \to \infty$. Hence $\xi_j = o([\psi(\xi)]^{-\sigma})$ for $\sigma > 0$ or, equivalently,

$$\xi_j^{\alpha_1} \cdot [\psi(\xi)]^{\alpha_2} \to 0. \tag{1.18}$$

for $\alpha_1 > 0$ and $\alpha_2 \geqslant 0$. Let $\breve{\xi} = (\breve{\xi}_1, \ldots, \breve{\xi}_n)$, where $\breve{\xi}_j = 0$ if j satisfies condition (1.17) and $\breve{\xi}_j = \xi_j$ otherwise.

In view of (1.18) it follows from (1.6) that

$$|R(\breve{\xi})|/(1 + |Q(\breve{\xi})|) \to \infty \quad \text{as} \quad \xi \to \infty \tag{1.19}$$

(under our limit process, i.e. with the possibility of repeatedly going over to subsequences of the sequence $\{\xi^s\}$ of (1.6)).

The resulting polynomials $R(\breve{\xi})$ and $Q(\breve{\xi})$ have the same full c.p. $\breve{\mathfrak{R}}$ for $\theta = 1$, and $\breve{\mathfrak{R}}(R) \subseteq \breve{\mathfrak{R}}(Q)$ for $\theta = 0$ in Theorem 1.2. Also the dimension of $\breve{\mathfrak{R}}(Q)$ is less than the dimension of $\mathfrak{R}(Q)$. In accordance with Remark 1.5, the principal faces of $\breve{\mathfrak{R}}$ are those and only those faces of $\breve{\mathfrak{R}}(Q)$ that are principal faces of $\mathfrak{R}$.

Since the conditions of $R(\breve{\xi})$- and $Q(\breve{\xi})$-regularity of a face of the polyhedron $\breve{\mathfrak{R}}(Q)$ coincide with the conditions of $R(\xi)$- and $Q(\xi)$-regularity of the same face of $\mathfrak{R}(Q)$ and the only Q-irregular face of $\mathfrak{R}(Q)$ is an $(n-1)$-dimensional face, it follows that all of the principal faces of $\breve{\mathfrak{R}}(Q)$ are Q- and R-regular for $\theta = 1$ and Q-regular for $\theta = 0$, inasmuch as their dimension is less than $n - 1$.

Thus, in the course of the proof of the theorem relation (1.6) leads either to a contradiction or to relation (1.19), which is analogous to (1.6) but corresponds to a space of dimension less than n (and, furthermore, only to the case when all of the principal faces are $Q(\breve{\xi})$-regular for $\theta = 0$ and $Q(\breve{\xi})$- and $R(\breve{\xi})$-regular for $\theta = 1$).

Repeating the arguments presented above in the proof of this theorem now with respect to the c.p. $\breve{\mathfrak{R}}(Q)$ and the polynomials $Q(\breve{\xi})$, $R(\breve{\xi})$, and so on, we clearly arrive after a finite number of steps at either a contradiction or relation (1.19) for a one-dimensional polyhedron $\breve{\mathfrak{R}}(Q)$. But in the one-dimensional case the fullness of $\breve{\mathfrak{R}}(Q)$ means that the polynomials $Q(\breve{\xi})$ and $R(\breve{\xi})$ have nonzero constant terms and are of the same positive degree for $\theta = 1$, while the degree of $R(\breve{\xi})$ does not exceed the degree of $Q(\breve{\xi})$ for $\theta = 0$. In both cases we obtain a contradiction with (1.19). By the same token Theorem 1.2 is proved.

We give an example of polynomials for which the conditions of Theorem 1.2 are satisfied and inequality (1.1) is consequently valid.

Let $(n = 2)$

$$Q(\xi) = a_1 + a_2\xi_1^8 + (a_3\xi_1^3\xi_2^2 + a_4\xi_1^2\xi_2^3)^2 + a_5\xi_2^4,$$
$$R(\xi) = b_1 + \theta b_2\xi_1^8 + c(a_3\xi_1^3\xi_2^2 + a_4\xi_1^2\xi_2^3)^2 + \theta b_5\xi_2^4,$$

where a_2, b_2, a_5, b_5 and c are positive numbers, a_3 and a_4 are real numbers, $\theta = 0$ and $\theta = 1$.

We will actually need for our purposes a generalization of Theorem 1.2.

Suppose given a polynomial $Q(\xi) = \Sigma_\alpha \gamma_\alpha \xi^\alpha$ with a full c.p. $\mathfrak{N}$ having only one $(n-1)$-dimensional face $\mathfrak{N}_{i_0}^{n-1}$, and suppose $Q^{i_0,n-1}(\xi^0) = Q_\lambda^{i_0,n-1}(\xi^0) = 0$ at some point $\xi^0 \in R_n^{(0)}$. Here λ is the outward normal to the face $\mathfrak{N}_{i_0}^{n-1}$, and the polynomial $Q_\lambda^{i_0,n-1}(\xi)$ is defined as in (1.2). Suppose there exists a number $\delta = \delta(Q, \xi^0) > 0$ such that $Q_{\lambda,\delta}^{i_0,n-1}(\xi^0) \neq 0$ and $Q_{\lambda,\sigma}^{i_0,n-1}(\xi^0) = 0$ for $\sigma > \delta$ (see (1.2) for the notation). Let $(\lambda, \alpha) = d$ be the equation of the face $\mathfrak{N}_{i_0}^{n-1}$. We put

$$K_{\lambda,\delta(Q)}(Q,\xi) = \sum_{\delta(Q)<\sigma<d} Q_{\lambda,\sigma}^{i_0,(n-1}(\xi) = \sum_{\delta(Q)<\sigma<d} \gamma_\alpha\xi^\alpha$$

and analogously for $R(\xi)$. Clearly, by definition, $K_{\lambda,\delta(Q)}^{i_0,n-1}(Q, \xi^0) = 0$.

THEOREM 1.2$'$. *Inequality* (1.1) *holds if the polynomials* $Q(\xi)$ *and* $R(\xi)$ *satisfy conditions* 1), 2a) *and* 2b) *of Theorem* 1.2 *and the conditions*

c$'$) $K_{\lambda,\delta(Q)}^{i_0,n-1}(Q,\xi) \geqslant 0$, $Q_{\lambda,\delta(Q)}^{i_0,n-1}(\xi) \geqslant 0$ *for* $\theta = 0$ *and* $K_{\lambda,\delta(R)}^{i_0,n-1}(R,\xi) \geqslant 0$, $R_{\lambda,\delta(R)}^{i_0,n-1}(\xi) \geqslant 0$ *for* $\theta = 1$ *in a neighborhood of those points at which*

$$R^{i_0,n-1}(\xi) = Q^{i_0,n-1}(\xi) = K_{\lambda,\delta(Q)}^{i_0,n-1}(Q,\xi) = K_{\lambda,\delta(R)}^{i_0,n-1}(R,\xi) = 0,$$

d$'$) $\mathfrak{N}(R) \subseteq \mathfrak{N}(Q)$, $\delta(R,\xi) \leqslant \delta(Q,\xi)$ *for* $\theta = 0$ *and* $\mathfrak{N}(R) \equiv \mathfrak{N}(Q)$, $\delta(R,\xi) = \delta(Q,\xi)$ *for* $\theta = 1$.

The proof follows the same plan as the proof of Theorem 1.2.

§2. Comparison of differential operators

DEFINITION 2.1. By the *characteristic polyhedron* (c.p.) $\mathfrak{N}$ for a given collection of polynomials $P_j(\xi) = \Sigma_\alpha \gamma_\alpha \xi^\alpha$ $(j = 1, \ldots, N)$ with constant coefficients is meant the minimal convex polyhedron in A_n containing all of the sets $\mathfrak{A}_j = \{\alpha\colon \alpha \in A_n, \gamma_\alpha^j \neq 0\}$ $(j = 1, \ldots, N)$.

DEFINITION 2.2. A face $\mathfrak{N}_i^k$ of the c.p. $\mathfrak{N}$ for a collection of polynomials $\{P_j(\xi)\}_1^N$ is said to be $\{P_j\}$*-regular* if the polynomials $P_j^{i,k}(\xi) = \Sigma_{\alpha\in\mathfrak{N}_i^k} \gamma_\alpha^j\xi^\alpha$ $(j = 1, \ldots, N)$ do not have a common zero in $R_n^{(0)}$.

For $N = 1$, V. P. Mihailov [3], and for $N \geqslant 1$, the present author [5] proved

LEMMA 2.1. *Suppose given a collection of polynomials* $\{P_j(\xi)\}_1^N$ *with a full c.p.* $\mathfrak{N}$, *and suppose* $\mathfrak{M}$ *is the c.p. for the polynomial* $P(\xi) = \Sigma_1^N |P_j(\xi)|^2$. *Then* $\mathfrak{M}$ *is similar to* $\mathfrak{N}$ *with ratio of similitude* 2 *and center of similitude at the origin. Further, if similar faces are denoted by the same indices* (i, k), *then*

$$P^{i,k}(\xi) = \sum_{j=1}^N |P_j^{i,k}(\xi)|^2; \tag{2.1}$$

this means, in particular, that if a face $\mathfrak{N}_i^k$ of $\mathfrak{N}$ is $\{P_j\}$-regular, then the face $\mathfrak{M}_i^k$ of $\mathfrak{M}$ is P-regular, and conversely.

From this lemma and Theorem 1.1, applied to the polynomials $P(\xi) = \Sigma_1^N |P_j(\xi)|^2$ and $Q(\xi) = \Sigma_{\mu \in \mathfrak{m}} \xi^\mu$, we immediately obtain

LEMMA 2.2. *Suppose given a collection of polynomials $\{P_j(\xi)\}_1^N$ with a full c.p. $\mathfrak{N}$. Then*

$$\sum_{\nu \in \mathfrak{n}} \xi^{2\nu} \leqslant \sum_{\mu \in \mathfrak{m}} \xi^\mu \leqslant C\left(1 + \sum_{j=1}^{N} |P_j(\xi)|^2\right) \quad \forall \xi \in R_n \tag{2.2}$$

if and only if all the principal faces of $\mathfrak{N}$ are $\{P_j\}$-regular.

We introduce some more notation, which we will use throughout the remainder of the paper. For a given point $\alpha \in A_n$, let

$$D\alpha = \{\beta : \beta \in A_n,\ 0 \leqslant \beta_i \leqslant \alpha_i,\ i = 1, \ldots, n\}.$$

For a collection $\mathfrak{A} = \{\alpha^k\}_1^N$ with c.p. $\mathfrak{N}$, let

$$D\mathfrak{A} = \bigcup_{\alpha^k \in \mathfrak{A}} D\alpha^k,$$

and let $\widetilde{\mathfrak{N}}$ denote the c.p. for the collection $D\mathfrak{A}$.

It is obvious that: a) $\mathfrak{N} \subseteq \widetilde{\mathfrak{N}}$; b) if $\mathfrak{N}$ is a proper polyhedron (see Definition 0.4), then $\widetilde{\mathfrak{N}} = \mathfrak{N}$; and c) the completely proper faces (see Definition 1.1) of $\mathfrak{N}$ and $\widetilde{\mathfrak{N}}$ coincide.

A point $\alpha \in A_n$ will be regarded as belonging to a face $\mathfrak{N}_i^k$ of a c.p. $\mathfrak{N}$ if $\alpha \in \mathfrak{N}_i^k$ and α does not lie on any other face of $\mathfrak{N}$ of dimension $k_1 < k$.

DEFINITION 2.3. A point $\alpha \in A_n$ will be said to be *proper* (*completely proper*) if it lies on a proper (completely proper) face of a c.p. $\mathfrak{N}$; the remaining points will be said to be *improper* (*not completely proper*). The c.p. spanned by only the completely proper points of a (not necessarily full) convex polyhedron $\mathfrak{N}$ in A_n will be denoted by $\mathfrak{N}'$.

Clearly, $\mathfrak{N}' \subseteq \mathfrak{N}$. We will prove that $\widetilde{\mathfrak{N}}' \equiv \widetilde{\mathfrak{N}}$ (see above for the notation).

LEMMA 2.3. *For any convex polyhedron $\mathfrak{N} \subset A_n$, $\widetilde{\mathfrak{N}}' \equiv \widetilde{\mathfrak{N}}$.*

PROOF. The inclusion $\widetilde{\mathfrak{N}}' \subseteq \widetilde{\mathfrak{N}}$ is obvious, so that it suffices to prove the reverse inclusion. Since a convex polynedron is uniquely determined by its vertices and $\widetilde{\mathfrak{N}}$ is a convex polyhedron, it suffices to prove that all of the vertices of $\widetilde{\mathfrak{N}}$ either are completely proper points, and hence belong to $\widetilde{\mathfrak{N}}'$, or are obtained from completely proper vertices, and hence again belong to $\widetilde{\mathfrak{N}}'$.

Thus we will prove that each vertex e of $\widetilde{\mathfrak{N}}$ has the form

$$e = e^* - \mu, \tag{2.3}$$

where e^* is a completely proper vertex of $\mathfrak{N}$ (and hence $e^* \in \mathfrak{N}'$) and $\mu = (\mu_1, \ldots, \mu_n)$, in which each μ_i is a nonnegative integer not greater than e_i^*. If e is completely proper, then (2.3) is obvious ($\mu \equiv 0$); it will therefore be assumed below that e is not a completely proper vertex.

Since in a proper polyhedron all of the vertices not lying on the coordinate hyperplanes of A_n are completely proper, it suffices to establish (2.3) for an arbitrary vertex e lying in some coordinate hyperplane.

Suppose for the sake of definiteness that the equation of the hyperplane on which the point e lies is $\alpha_n = 0$, so that $e = (e_1, \ldots, e_{n-1}, 0)$, and consider the projection $\mathfrak{N}|_{\alpha_n=0}$ of the polyhedron $\mathfrak{N}$ on the hyperplane $\alpha_n = 0$. Then $\mathfrak{N}|_{\alpha_n=0}$ is an $(n-1)$-dimensional proper polyhedron (in an $(n-1)$-dimensional space). We will assume that $e' = (e_1, \ldots, e_{n-1})$ is a completely proper vertex of $\mathfrak{N}|_{\alpha_n=0}$ and show that it is the orthogonal projection on the hyperplane $\alpha_n = 0$ of a vertex of $\mathfrak{N}'$. If our assumption concerning the complete properness of e' is not satisfied, then the question clearly reduces to a solution of the same problem but for a dimension smaller by one. Thus, after a finite number of steps we will either obtain a completely proper vertex e' or reach a one-dimensional polyhedron. In the one-dimensional case the assertion of the lemma is obvious.

Thus suppose $e' = (e_1, \ldots, e_{n-1})$ is a completely proper vertex of $\mathfrak{N}|_{\alpha_n=0}$. We represent e' in the form of an intersection of $n-1$ $(n-2)$-dimensional faces of the polyhedron $\mathfrak{N}|_{\alpha_n=0}$. We then construct, in the n-dimensional space A_n, $n-1$ $(n-1)$-dimensional hyperplanes whose orthogonal projections on the hyperplane $\alpha_n = 0$ coincide with the chosen faces of the polyhedron $\mathfrak{N}|_{\alpha_n=0}$.

The intersection of the latter hyperplanes (in A_n) is a one-dimensional straight line l projecting into the point e.

Let us show that there are points of the polyhedron $\widetilde{\mathfrak{N}}$ on this straight line l that are different from e. Assuming the contrary, we construct an $(n-2)$-dimensional hyperplane of support to the point e' of $\mathfrak{N}|_{\alpha_n=0}$ with a normal $\lambda' = (\lambda_1, \ldots, \lambda_{n-1})$ all of whose coordinates are positive (the existence of such a hyperplane follows from the fact that e' is a completely proper point of $\mathfrak{N}|_{\alpha_n=0}$). We then construct an $(n-1)$-dimensional hyperplane containing the point e with normal $\lambda_\sigma = (\lambda_1, \ldots, \lambda_{n-1}, \sigma)$. When $\sigma = 0$ our hyperplane does not contain any other points of $\mathfrak{N}$; the same statement holds, by continuity, for some $\sigma > 0$. But this contradicts the fact that e is not a completely proper vertex.

Thus we have established the existence of points of $\widetilde{\mathfrak{N}}$ on the straight line l that are different from e. It follows that l contains a one-dimensional face of $\mathfrak{N}$ and hence a vertex $e^{(1)} \neq e$.

If $e^{(1)}$ is a completely proper vertex, the lemma is proved; if $e^{(1)}$ is not completely proper, then it lies on an $(n-1)$-dimensional coordinate face of $\mathfrak{N}$ (that is different from the face $\alpha_n = 0$).

By repeating the arguments, we will arrive after a finite number $(\leqslant n)$ of steps at either a completely proper vertex or a vertex not lying on a coordinate hyperplane (and hence a vertex that is also completely proper). This proves Lemma 2.3.

Let us prove the following proposition, which solves problem I.

THEOREM 2.1. *Suppose given a collection of multi-indices* $\mathfrak{A} = \{\alpha^k\}_1^N$ *with c.p.* $\mathfrak{N}$. *A polynomial* $P(\xi) = \Sigma_1^N \gamma_k \xi^{\alpha^k}$ *is stronger than any polynomial* $Q(\xi) = \Sigma_1^N q_k \xi^{\alpha^k}$ *if and only if all of the completely proper faces of* $\mathfrak{N}$ *are P-regular.*

PROOF. *Sufficiency.* We must prove the inequality (see Definition 0.1)

$$\tilde{Q}(\xi)/\tilde{P}(\xi) \leqslant C \qquad \forall \xi \in R_n.$$

Since, by the condition of the theorem, all of the completely proper faces of $\mathfrak{N}$ are P-regular, it follows, in particular, that all of the completely proper vertices (zero-dimensional faces) of $\mathfrak{N}$ are P-regular. This means that the corresponding coefficients γ_k are different from zero. We form, as above, the c.p. $\mathfrak{N}'$ consisting of the completely proper points of $\mathfrak{N}$. According to Lemma 2.3, $\widetilde{\mathfrak{N}}' \equiv \widetilde{\mathfrak{N}}$. Since, on the other hand, the inclusion $\widetilde{\mathfrak{N}}(Q) \subseteq \widetilde{\mathfrak{N}}$ clearly holds for any polynomial $Q(\xi)$ with c.p. $\mathfrak{N}(Q)$, we have $\widetilde{\mathfrak{N}}(Q) \subseteq \widetilde{\mathfrak{N}}(P)$.

Let $\mathfrak{M}(\widetilde{P}^2)$ and $\mathfrak{M}(\widetilde{Q}^2)$ respectively denote the c.p.'s for the polynomials $\widetilde{P}^2(\xi)$ and $\widetilde{Q}^2(\xi)$. According to Lemma 2.1, the fact that $\widetilde{\mathfrak{N}}(Q) \subseteq \widetilde{\mathfrak{N}}(P)$ implies $\mathfrak{M}(\widetilde{Q}^2) \subseteq \mathfrak{M}(\widetilde{P}^2)$. We will now prove (2.4) by proving the stronger assertion

$$\sum_{\nu \in \mathfrak{m}(\widetilde{P}^2)} \xi^\nu \leqslant C\widetilde{P}(\xi^2) = C \sum_{|\mu| \geqslant 0} |P^{(\mu)}(\xi)|^2 \qquad \forall \xi \in R_n. \tag{2.4}$$

Clearly, if not all of the points of the collection $\mathfrak{A}$ lie on a coordinate hyperlane of A_n, the polyhedron $\widetilde{\mathfrak{N}}$ is n-dimensional and consequently proper. According to Lemmas 2.3 and 2.1, the c.p. $\mathfrak{M}(\widetilde{P}^2)$ is also proper. Therefore (see Lemma 2.2 and representation (2.1)), in order to prove inequality (2.5), we must prove that all of the principal faces of $\mathfrak{M}(\widetilde{P}^2)$ are $\widetilde{P}^2$-regular.

The $\widetilde{P}^2$-regularity of a face $\mathfrak{M}_i^k$ of $\mathfrak{M}(\widetilde{P}^2)$ that is similar to a completely proper face $\widetilde{\mathfrak{N}}_i^k$ of $\widetilde{\mathfrak{N}}(P)$ (and hence, according to Lemma 2.1, is itself completely proper) follows from the P-regularity of the face $\mathfrak{N}_i^k$ and from representation (2.1), which in this case takes the form

$$[P(\xi)^2]^{i,k} = \sum_{|\mu| \geqslant 0} |[P^{(\mu)}(\xi)]^{i,k}|^2 = |P^{i,k}(\xi)|^2, \tag{2.5}$$

since, for a completely proper face $\widetilde{\mathfrak{N}}_i^k$ of $\mathfrak{N}$, $[P^{(\mu)}(\xi)]^{i,k} \equiv 0$ for $|\mu| > 0$.

Now suppose a principal face $\mathfrak{M}_i^k$ of $\mathfrak{M}(\widetilde{P}^2)$ does not correspond to a completely proper face of $\widetilde{\mathfrak{N}}$ and lies on a coordinate hyperplane.

According to Lemmas 2.3 and 2.1, the face $\widetilde{\mathfrak{N}}_i^k$ of $\widetilde{\mathfrak{N}}$ to which $\mathfrak{M}_i^k$ corresponds can be obtained by the displacement (parallel translation) of a completely proper face of $\widetilde{\mathfrak{N}}'$. But this means that representation (2.1), which in our case has the form

$$[\widetilde{P}(\xi)^2]^{i,k} = \sum_{|\mu| \geqslant 0} |[P^{(\mu)}(\xi)]^{i,k}|^2, \tag{2.6}$$

contains a term of the form

$$[P^{(\mu)}(\xi)]^{i,k} = \frac{1}{\mu!} \xi^{-\mu} P^{i_1,k_1}(\xi), \tag{2.7}$$

where $P^{i_1,k_1}(\xi)$ corresponds to the part of the polynomial $P(\xi)$ on a completely proper face $\mathfrak{N}_{i_1}^{k_1}$. The $\widetilde{P}^2$-regularity of the face $\mathfrak{M}_i^k$ of $\mathfrak{M}(\widetilde{P}^2)$ now follows from representations (2.6), (2.7) and the P-regularity of the completely proper face $\widetilde{\mathfrak{N}}_{i_1}^{k_1}$ of $\widetilde{\mathfrak{N}}(P)$.

Finally, suppose a principal face $\mathfrak{M}_i^k$ of $\mathfrak{M}(\widetilde{P}^2)$ does not lie entirely on a coordinate hyperplane and does not correspond to a completely proper face of $\widetilde{\mathfrak{N}}(P)$. Then it clearly has points in common with the coordinate hyperplanes of A_n. Let $\mathring{\mathfrak{M}}_i^k$ denote the part of $\mathfrak{M}_i^k$ lying on the coordinate hyperplanes. This part is a union of principal faces of dimensions less than k. The right side of (2.6) can then be represented in the form

$$\sum_{|\mu| \geqslant 0} \sum_m |[P^{(\mu)}(\xi)]^{i_m k_m}|^2 + \ldots, \tag{2.8}$$

where the nonnegative terms have not been written out and $[P^{(\mu)}(\xi)]^{i_m,k_m}$ denotes the polynomial corresponding to the coordinate face $\overset{\circ}{\mathfrak{M}}{}^{k_m}_{i_m}$ $(m = 1, \ldots, M_i^k)$ of the face $\mathfrak{M}_i^k$. But the $\widetilde{P}^2$-regularity of these faces has already been proved. Hence (2.8) does not vanish in $R_n^{(0)}$, which means that the face $\mathfrak{M}_i^k$ is $\widetilde{P}^2$-regular.

Necessity. Suppose a completely proper face $\mathfrak{N}^{k_0}_{i_0}$ of $\mathfrak{N}(P)$ is P-irregular. Since the completely proper faces of $\mathfrak{N}(P)$ and $\widetilde{\mathfrak{N}}(P)$ coincide, the c.p. $\widetilde{\mathfrak{N}}(P)$ has a P-irregular face $\widetilde{\mathfrak{N}}^{k_0}_{i_0}$ (which we denote with the same indices for the sake of simplicity). We write the equation of an $(n-1)$-dimensional hyperplane of support of $\widetilde{\mathfrak{N}}$ that passes through this face in the form

$$(\lambda, \alpha) = d, \tag{2.9}$$

where $\lambda = (\lambda_1, \ldots, \lambda_n)$ is the outward unit normal to this hyperplane and α is a generic point of it.

By definition of the c.p. $\mathfrak{N}$ for the collection $\mathfrak{A}$, there exist points of this collection belonging to the face $\mathfrak{N}^{k_0}_{i_0} \equiv \widetilde{\mathfrak{N}}^{k_0}_{i_0}$. Suppose α^1 is one of these points. We take the polynomial $Q(\xi) = \xi^{\alpha^1}$ and assume that inequality (2.4) is satisfied for the polynomials $Q(\xi)$ and $P(\xi)$. We represent $\widetilde{P}^2(\xi)$ in the form

$$\widetilde{P}^2(\xi) = [\widetilde{P}(\xi)^2]^{i_0,k_0} + r(\xi) = |P^{i_0,k_0}(\xi)|^2 + r(\xi). \tag{2.10}$$

The satisfaction of (2.4) means, in particular, that

$$\frac{\xi^{2\alpha^1}}{\widetilde{P}(\xi)^2} \leqslant \frac{\widetilde{Q}(\xi)^2}{\widetilde{P}(\xi)^2} \leqslant C \quad \forall \xi \in R_n. \tag{2.11}$$

But the P-irregularity of the face $\mathfrak{N}^{k_0}_{i_0}$ of $\mathfrak{N}(P)$ means that there exists a point $\xi^0 \in R_n^{(0)}$ such that

$$P^{i_0,k_0}(\xi^0) = 0. \tag{2.12}$$

Replacing $\widetilde{P}^2(\xi)$ and ξ in (2.11) by formula (2.10) and the expression $t^\lambda \xi^0 = (t^{\lambda_1}\xi_1^0, \ldots, t^{\lambda_n}\xi_n^0)$ respectively and using the λ-homogeneity of the polynomial $P^{i_0,k_0}(\xi)$ (Lemma 1.1), we get for any $t > 0$ that

$$\frac{t^{2(\lambda,\alpha^1)}\cdot(\xi^0)^{2\alpha^1}}{t^{2d}P^{i_0,k_0}(\xi^0) + r(t^\lambda\xi^0)} = \frac{t^{2(\lambda,\alpha^1)}\cdot(\xi^0)^{2\alpha^1}}{r(t^\lambda\xi^0)} \leqslant C. \tag{2.13}$$

The convexity of the c.p. $\mathfrak{M}(\widetilde{P}^2)$ implies on the basis of obvious geometrical arguments that $r(t^\lambda\xi^0) = o(t^{2d})$ for $t \to \infty$.

Therefore, as $t \to \infty$ in (2.13), the left side tends to infinity, whereas the right side is finite. This contradiction proves Theorem 2.1.

Let us now turn to problem II. We first prove the following simple proposition ($\widetilde{\mathfrak{N}}(R)$ is the c.p. of the collection of polynomials $R^{(\mu)}(\xi)$, $|\mu| \geqslant 0$, for a given polynomial $R(\xi)$).

LEMMA 2.4. *Suppose given polynomials $P(\xi)$ and $Q(\xi)$ with c.p.'s $\widetilde{\mathfrak{N}}(P)$ and $\widetilde{\mathfrak{N}}(Q)$. Then the polynomials $P(\xi)$ and $Q(\xi)$ are comparable only if either $\widetilde{\mathfrak{N}}(P) \subseteq \widetilde{\mathfrak{N}}(Q)$ or $\widetilde{\mathfrak{N}}(Q) \subseteq \widetilde{\mathfrak{N}}(P)$.*

PROOF. Since a convex polyhedron is determined by its vertices, it suffices to prove that $P(\xi)$ and $Q(\xi)$ are not comparable if $\widetilde{\mathfrak{N}}(P)$ has a vertex e^1 not belonging to $\widetilde{\mathfrak{N}}(Q)$ and $\mathfrak{N}(Q)$ has a vertex e^2 not belonging to $\widetilde{\mathfrak{N}}(P)$.

Suppose, on the contrary, that $P(\xi)$ and $Q(\xi)$ are comparable and, say $Q \prec P$, i.e.

$$\widetilde{Q}(\xi)^2/\widetilde{P}(\xi)^2 \leqslant C \qquad \forall \xi \in R_n. \tag{2.13}$$

Since the only nonprincipal vertex of a full polyhedron is the origin $(0, \ldots, 0)$ of A_n, and since both of the polyhedrons $\widetilde{\mathfrak{N}}(Q)$ and $\widetilde{\mathfrak{N}}(P)$ are proper and hence contain the origin as a vertex, it follows that e^1 and e^2 are principal vertices of $\widetilde{\mathfrak{N}}(P)$ and $\widetilde{\mathfrak{N}}(Q)$ respectively. There therefore clearly exists a hyperplane of support $(\lambda, \alpha) = d > 0$ of $\widetilde{\mathfrak{N}}(Q)$ that passes through the vertex e^2 and does not contain any points of $\widetilde{\mathfrak{N}}(P)$, so that for all points $\alpha \in \widetilde{\mathfrak{N}}(P)$ and $\beta \in \widetilde{\mathfrak{N}}(Q)$, $\beta \neq e^2$,

$$(\lambda, \alpha) < (\lambda, e^2) = d, \qquad (\lambda, \beta) < d. \tag{2.14}$$

We replace ξ in (2.13) by $t^\lambda \xi^0$ for some $\xi^0 \in R_n^{(0)}$ and $t > 0$. Relations (2.14) show that $\widetilde{P}(\xi)^2 = o(t^d)$ and $\widetilde{Q}(\xi)^2 = O(t^d)$ for $t \to \infty$. This contradicts (2.13) and proves the lemma.

Lemma 2.4 and inequality (2.5), established in the proof of Theorem 2.1, directly imply

THEOREM 2.2. *Suppose given a polynomial $P(\xi)$ with c.p. $\mathfrak{N}(P)$. The polynomial $P(\xi)$ is stronger than a polynomial $Q(\xi)$ with c.p. $\mathfrak{N}(Q)$ if and only if $\widetilde{\mathfrak{N}}(Q) \subseteq \widetilde{\mathfrak{N}}(P)$ and all of the completely proper faces of $\mathfrak{N}(P)$ are P-regular.*

We now consider the case when the c.p. $\mathfrak{N}$ for a given polynomial $P(\xi)$ has a P-irregular $(n-1)$-dimensional face $\mathfrak{N}_{i_0}^{n-1}$. From representation (2.1) it follows that the face $\mathfrak{M}_{i_0}^{n-1}$ of $\mathfrak{M}(\widetilde{P}^2)$ corresponding according to Lemma 2.1 to the face $\widetilde{\mathfrak{N}}_{i_0}^{n-1} \equiv \mathfrak{N}_{i_0}^{n-1}$ of $\widetilde{\mathfrak{N}}(P)$ will then be $\widetilde{P}^2$-irregular.

Let $\lambda = (\lambda_1, \ldots, \lambda_n)$ be the outward unit normal to the face $\widetilde{\mathfrak{N}}_{i_0}^{n-1}$. Then $(\lambda, \alpha) = d > 0$ for $\alpha \in \widetilde{\mathfrak{N}}_{i_0}^{n-1}$, and $(\lambda, \alpha) = 2d$ for $\alpha \in \mathfrak{M}_{i_0}^{n-1}$.

We introduce the notation (cf. (1.2))

$$P^{(\mu)}(\xi) = \sum_\alpha \gamma_{\alpha,\mu}\xi^\alpha = \sum_\alpha \frac{\alpha!}{(\alpha-\mu)!}\gamma_{\alpha-\mu}\xi^{\alpha-\mu}, \quad \alpha-\mu \in A_n,$$

$$d_\lambda = d_\lambda\{P^{(\mu)}(\xi)\} = \max_{\substack{\alpha \in \widetilde{\mathfrak{N}} \setminus \widetilde{\mathfrak{N}}_{i_0}^{n-1} \\ \sum_\mu |\gamma_{\alpha,\mu}| > 0}} (\lambda, \alpha), \tag{2.15}$$

$$[P_{\lambda,\delta}^{(\mu)}(\xi)]^{i_0,n-1} = \sum_{(\lambda,\alpha)=\delta} \gamma_{\alpha,\mu}\xi^\alpha,$$

$$[P_\lambda^{(\mu)}(\xi)]^{i_0,n-1} = [P_{\lambda,d_\lambda}^{(\mu)}(\xi)]^{i_0,n-1}. \tag{2.16}$$

Let $P_{\lambda\lambda}^{i_0,n-1}(\xi)$ denote the part of the polynomial $P(\xi)$ with multi-indices α such that $(\lambda, \alpha) \leqslant d_\lambda$, $(\lambda, \alpha) + d \geqslant 2d_\lambda$, and let

$$\widetilde{\mathfrak{N}}^* = \{\nu \colon \nu \in \widetilde{\mathfrak{N}},\ (\lambda, \nu) \leqslant d_\lambda\}. \tag{2.17}$$

The following proposition is a partial solution of problem II.

THEOREM 2.3. *Suppose given a polynomial $q(\xi) = \Sigma_\alpha\, \gamma_\alpha \xi^\alpha$ with c.p. $\mathfrak{N}(q)$. Suppose all of the completely proper faces of $\mathfrak{N}(q)$ with the exception of an $(n-1)$-dimensional face $\mathfrak{N}_{i_0}^{n-1}$ are q-regular. Further, if, in this connection, $q^{i_0,n-1}(\xi^0) = 0$ for some point $\xi^0 \in R_n^{(0)}$, suppose*

a) $\sum_{|\mu|\geqslant 0}|[q_\lambda^{(\mu)}(\xi^0)]^{i_0,n-1}|^2 \neq 0$; *and*

b) ([3])*there exists a neighborhood of* ξ^0 *in which* ($\bar{P}(\xi)$ *is the complex conjugate of* $P(\xi)$)

$$K_\lambda^{i_0,\, n-1}(\xi) = \bar{q}^{i_0,\, n-1}(\xi)\, q_{\lambda\lambda}^{i_0,\, n-1}(\xi) + q^{i_0,\, n-1}(\xi)\, \bar{q}_{\lambda\lambda}^{i_0,\, n-1}(\xi) \geqslant 0.$$

Then any polynomial $r(\xi)$ *with c.p.* $\mathfrak{N}(r) \subseteq \mathfrak{N}^*(q)$ *is weaker than the polynomial* $q(\xi)$.

PROOF. It suffices to show that the polynomials $Q(\xi) = \tilde{q}(\xi)^2$ and $R(\xi) = \tilde{r}(\xi)^2$ satisfy all of the conditions of Theorem 1.2′ for $\theta = 0$.

In regard to condition 1), the Q-regularity of all of the principal faces of $\mathfrak{M}(Q)$ (except the face $\mathfrak{M}_{i_0}^{n-1}$) is proved in the same way as in the proof of Theorem 2.1.

Condition 2a) is obvious, since $Q^{i_0,n-1}(\xi) = |q^{i_0,n-1}(\xi)|^2$.

Condition 2b) is obvious, since $\mathfrak{N}(r) \subseteq \mathfrak{N}^*(q)$ implies $R^{i_0,n-1}(\xi) \equiv 0$.

For condition 2c′) we consider the polynomial

$$|Q(\xi)| = \sum_{|\mu|\geqslant 0} |q^{(\mu)}(\xi)|^2 = |q(\xi)|^2 + \sum_{|\mu|>0} |q^{(\mu)}(\xi)|^2.$$

and represent $q(\xi)$ in the form

$$q(\xi) = q^{i_0,\, n-1}(\xi) + q_{\lambda\lambda}^{i_0,\, n-1}(\xi) + \dots \tag{2.18}$$

Then (see the notation in condition b) of Theorem 2.3)

$$|Q(\xi)|^2 = |q^{i_0,\, n-1}(\xi)|^2 + K_\lambda^{i_0,\, n-1}(\xi) + \sum_{|\mu|\geqslant 0} |[q_\lambda^{(\mu)}(\xi)]^{i_0,\, n-1}|^2 + \dots \tag{2.19}$$

where the terms for the multi-indices α such that $(\lambda, \alpha) < 2d_\lambda$, i.e. the points lying in $\tilde{\mathfrak{N}}^*(q)$, have not been written out. Condition 2c′) of Theorem 1.2′ now follows from representation (2.19) and conditions a) and b) of Theorem 2.3.

Condition 2d′) follows from the condition $\mathfrak{N}(r) \subseteq \tilde{\mathfrak{N}}^*(q)$.

Theorem 2.3 is proved.

Let us make some remarks before proceeding to the general case.

REMARK 2.1. Suppose $\mathfrak{N}(P)$ is the c.p. for a polynomial $P(\xi)$, $\mathfrak{N}_{i_0}^{n-1}$ is a P-irregular $(n-1)$-dimensional face of $\mathfrak{N}(P)$, λ is the outward unit normal to this face, $(\lambda, \alpha) = d > 0$ is the equation of the hyperplane of support of $\mathfrak{N}$ containing the face $\mathfrak{N}_{i_0}^{n-1}$, and d_λ and $\tilde{\mathfrak{N}}^*(P)$ are defined as in (2.15) and (2.17). Then any polynomial $Q(\xi)$ with c.p. $\mathfrak{N}(Q) \subseteq \tilde{\mathfrak{N}}(P)$ (or, equivalently, $\tilde{\mathfrak{N}}(Q) \subseteq \tilde{\mathfrak{N}}(P)$) can be represented in the form

$$Q(\xi) = Q_0(\xi) + \sum_{k=1}^{M} Q_k(\xi) = Q_0(\xi) + \sum_{k=1}^{M} \sum_{(\lambda,\alpha)=\sigma_k} \gamma_\alpha^k \xi^\alpha, \tag{2.20}$$

where the set $\{\sigma_k\}$ coincides with the values of (λ, α) for $\alpha \in \tilde{\mathfrak{N}}(Q)$ and hence, of course, $\tilde{\mathfrak{N}}(Q_0) \subseteq \tilde{\mathfrak{N}}^*(P)$ and $d_\lambda < \sigma_1 < \sigma_2 < \cdots < \sigma_M \leqslant d$.

REMARK 2.2. It is easily shown that if the c.p. $\tilde{\mathfrak{N}}(P)$ of a given polynomial $P(\xi)$ has a P-irregular completely proper face $\tilde{\mathfrak{N}}_{i_0}^{n-1}$ and a polynomial $Q(\xi)$ can be represented in the form (2.20), then in order for $Q \prec P$ it is necessary that all of the polynomials $Q_k(\xi)$ $(k = 1, \dots, M)$ vanish at all of the points $\xi \in R_n^{(0)}$ at which the polynomial $P^{i_0,n-1}(\xi)$ vanishes.

([3]) Condition b) is automatically satisfied if $q_{\lambda\lambda}^{i_0,n-1}(\xi) \equiv 0$. This latter condition is satisfied in turn if, for example, the polynomial $q(\xi) \equiv q^{i_0,n-1}(\xi)$.

In turn, in order for two λ-homogeneous polynomials $R_1(\xi)$ and $R_2(\xi)$ to have the same zeros in $R_n^{(0)}$ it is sufficient (and necessary in the case of $n = 2$ and homogeneous polynomials) that they can be represented in the form $R_i(\xi) = \bar{R}_i(\xi)R(\xi)$ $(i = 1, 2)$, where the polynomials $\bar{R}_i(\xi)$ $(i = 1, 2)$ do not vanish at any points $\xi \in R_n^{(0)}$.

We prove a proposition solving problem II in a quite general form.

THEOREM 2.4. *Suppose $q(\xi)$ is a polynomial satisfying all of the conditions of Theorem 2.3, suppose $r(\xi)$ is a polynomial satisfying the condition $\widetilde{\mathfrak{N}}(r) \subseteq \widetilde{\mathfrak{N}}(q)$ and suppose the polynomials $q^{i_0,n-1}(\xi)$ and $r(\xi)$ can be represented in the form*

$$q^{i_0, n-1}(\xi) = \bar{q}^{i_0, n-1}(\xi)\cdot p(\xi), \tag{2.21}$$

$$r(\xi) = r_0(\xi) + \sum_{k=1}^{M} \bar{r}_k(\xi)\cdot p(\xi), \tag{2.22}$$

where the polynomials $\bar{q}^{i_0,n-1}(\xi)$ do not vanish in $R_n^{(0)}$, the polynomials $r_k(\xi) = \bar{r}_k(\xi)P(\xi)$ are defined as in Remark 2.2 and $\widetilde{\mathfrak{N}}(r_0) \subseteq \widetilde{\mathfrak{N}}^*(q)$. Then $r \prec q$.

PROOF. It suffices to prove the inequality (see [1], Theorem 3.3.2)

$$|r(\xi)|^2 / \tilde{q}(\xi)^2 \leqslant c \qquad \forall \xi \in R_n.$$

But since $|r(\xi)|^2 \leqslant \Sigma_0^M |r_j(\xi)|^2$ and the inequality $|r_0(\xi)|^2/\tilde{q}(\xi)^2 \leqslant C$ has been proved in Theorem 2.3 ($\mathfrak{N}(r_0) \subseteq \widetilde{\mathfrak{N}}^*(q)$), it remains to prove the inequality

$$\frac{|r_k(\xi)|^2}{\tilde{q}(\xi)^2} \leqslant C \qquad \forall \xi \in R_n, \qquad k = 1, \ldots, M.$$

Suppose, on the contrary, that this inequality is not satisfied for some k (say, $k = 1$). This means that there exists a sequence $\{\xi^s\}$, $\xi^s \to \infty$, such that

$$\frac{|r_1(\xi^s)|^2}{\tilde{q}(\xi^s)^2} \to \infty, \tag{2.23}$$

and the polynomial $Q(\xi) = \tilde{q}(\xi)^2$ satisfies all of the conditions of Theorem 1.2. Repeating the arguments presented in the proof of Theorem 1.2, we obtain a contradiction with (2.23) in all of the cases in which $\mathfrak{N}_{i_r}^{k_r}$ does not coincide with the Q-irregular face $\mathfrak{M}_{i_0}^{n-1}$. It is therefore only necessary to consider the case in which $\lambda \equiv e^1$ is the outward unit normal to the face $\mathfrak{N}_{i_0}^{n-1}$. We represent the polynomial $Q(\xi)$ in this case in the form (2.19).

Using representation (2.19) and the condition concerning the nonnegativeness of $K_\lambda^{i_0,n-1}(\xi)$ in a neighborhood of a point η at which $q^{i_0,n-1}(\eta) = 0$, we obtain the inequality

$$|Q(\xi)| \geqslant |q^{i_0, n-1}(\xi)|^2 + \sum_{|\mu| \geqslant 0} |[q_\lambda^{(\mu)}(\xi)]^{i_0, n-1}|^2 + \ldots \tag{2.24}$$

for points $\xi \in R_n$ close to η. From the λ-homogeneity of the polynomials $q^{i_0,n-1}(\xi)$ and $\{[q_\lambda^{(\mu)}(\xi)]^{i_0,n-1}\}$ and from condition (2.21) of Theorem 2.4 we get that for $\xi^s \to \infty$

$$|Q(\xi)| \geqslant \rho^{2d}\left|\bar{q}^{i_0,n-1}\left(\rho^{\sum_{j=2}^{n+1}\lambda_j e^j}\right)\right|^2\cdot\left|p\left(\rho^{\sum_{j=2}^{n+1}\lambda_j e^j}\right)\right|^2 + \rho^{2d_\lambda}\sum_{|\mu|\geqslant 0}\left|\left[q_\lambda^{(\mu)}\left(\rho^{\sum_{j=2}^{n+1}\lambda_j e^j}\right)\right]^{i_0,n-1}\right|^2 - o(\rho^{2d_\lambda})$$
$$\geqslant \rho^{2d}\left|\bar{q}^{i_0,n-1}\left(\rho^{\sum_{j=2}^{n+1}\lambda_j e^j}\right)\right|^2\cdot\left|p\left(\rho^{\sum_{j=2}^{n+1}\lambda_j e^j}\right)\right|^2 + M\cdot\rho^{2d_\lambda}, \tag{2.25}$$

where $M > 0$, since, by assumption,

$$\sum_{|\mu|\geqslant 0}|[q_\lambda^{(\mu)}(\eta)]^{i_0,n-1}|^2 \neq 0.$$

We write the analogous representation for the polynomial $R(\xi) = |r_1(\xi)|^2$:

$$R(\xi) = \rho^{2\sigma_1}\left|\bar{r}_1\left(\rho^{\sum_{j=2}^{n+1}\lambda_j e^j}\right)\right|^2\left|p\left(\rho^{\sum_{j=2}^{n+1}\lambda_j e^j}\right)\right|^2. \tag{2.26}$$

Substituting the right sides of (2.25) and (2.26) for $Q(\xi)$ and $R(\xi)$ in (2.23), we get that for $\xi \to \infty$

$$\frac{R(\xi)}{Q(\xi)} \leqslant \frac{\rho^{2\sigma_1}\left|\bar{r}_1\left(\rho^{\sum_{j=2}^{n+1}\lambda_j e^j}\right)\right|^2\left|p\left(\rho^{\sum_{j=2}^{n+1}\lambda_j e^j}\right)\right|^2}{\rho^{2d}\left|\bar{q}^{i_0,n-1}\left(\rho^{\sum_{j=2}^{n+1}\lambda_j e^j}\right)\right|^2\left|p\left(\rho^{\sum_{j=2}^{n+1}\lambda_j e^j}\right)\right|^2 + M\rho^{2d_\lambda}}$$
$$\leqslant \rho^{2(\sigma_1-d)}\left|\frac{\bar{r}_1\left(\rho^{\sum_{j=2}^{n+1}\lambda_j e^j}\right)}{\bar{q}^{i_0,n-1}\left(\rho^{\sum_{j=2}^{n+1}\lambda_j e^j}\right)}\right| \leqslant C,$$

since $\sigma_1 - d \leqslant 0$, $q^{i_0,n-1}(\eta) \neq 0$ and $\rho^{\sum_2^{n+1}\lambda_j e^j} \to \eta$. This contradicts (2.23) and proves Theorem 2.4.

Let us give an example illustrating Theorem 2.4. Suppose $n = 2$, $q(\xi) = \xi_1^4 - \xi_2^4$, $r(\xi) = (\xi_1^2 - \xi_2^2)\xi_1\xi_2$ and the c.p. $\widetilde{\mathfrak{R}}(q)$ is the triangle in A_2 with vertices (4, 0), (0, 4) and (0, 0). The only completely proper face of $\widetilde{\mathfrak{R}}(q)$ is the q-irregular (one-dimensional) face $\widetilde{\mathfrak{R}}_1^1$ with vertices (4, 0) and (0, 4).

It is easily seen that

a) the outward unit normal to $\mathfrak{R}_1^1$ has the form $\lambda = (1/\sqrt{2}, 1/\sqrt{2})$;

b) $q^{1,1}(\xi) \equiv 0$;

c) $\Sigma_{|\mu|\geqslant 0}|[q_\lambda^{(\mu)}(\xi)]^{1,1}|^2 = 16(\xi_1^6 + \xi_2^6) \neq 0$ for $\xi \in R_2^{(0)}$;

d) the polynomials $q^{1,1}(\xi) \equiv q(\xi)$ and $r(\xi)$ have the form

$$q^{1,1}(\xi) = (\xi_1^2 + \xi_2^2)(\xi_1^2 - \xi_2^2) = \bar{q}^{1,1}(\xi)\,p(\xi),$$
$$r(\xi) = \xi_1\xi_2(\xi_1^2 - \xi_2^2) = \bar{r}(\xi)\,p(\xi),$$

in which the polynomials $\bar{q}^{1,1}(\xi)$ and $\bar{r}(\xi)$ do not vanish in $R_2^{(0)}$;

e) $\widetilde{\mathfrak{N}}(r) \subset \widetilde{\mathfrak{N}}(q)$.

Thus all of the conditions of Theorem 2.3 are satisfied, and hence $r \prec q$. We note in addition that $\widetilde{\mathfrak{N}}(r) \subseteq \widetilde{\mathfrak{N}}^*(q)$.

§3. Differential operators of constant strength

We now consider the problem of finding algebraic conditions on the characteristic polynomial $P(x, \xi)$ of a given differential operator $P(x, D)$ with variable coefficients $\gamma_\alpha(x)$ defined in Ω for it to have constant strength in Ω, i.e. for inequality (0.2) to hold for any two points $x, y \in \Omega$.

Let $\mathfrak{A}(x)$, $x \in \Omega$, denote the set of those multi-indices α for which the coefficients $\gamma_\alpha(x)$ of the polynomial $P(x, \xi) = \Sigma_\alpha \gamma_\alpha(x)\xi^\alpha$ are different from zero.

As in the case of constant coefficeints, we let

$$D\mathfrak{A}(x) = \{\beta: \beta \in A_n, \quad 0 \leqslant \beta_i \leqslant \alpha_i, \quad i = 1, \ldots, n, \quad \forall \alpha \in \mathfrak{A}(x)\},$$

and let $\widetilde{\mathfrak{N}}(x)$ denote the c.p. for the set $D\mathfrak{A}(x)$. Clearly, $\widetilde{\mathfrak{N}}(x)$ coincides for $x \in \Omega$ with the c.p. for the collection of polynomials $\{P^{(\mu)}(x, \xi) = D_\xi^\mu P(x, \xi), |\mu| \geqslant 0\}$.

It will be assumed everywhere below that $\widetilde{\mathfrak{N}}(x)$ does not depend on the point $x \in \Omega$. The necessity of this condition in order for the polynomial $P(x, \xi)$ to have constant strength in Ω is obvious.

REMARK 3.1. For a given point $x \in \Omega$, let $\mathfrak{N}(x)$ denote the c.p. for the polynomial $P(x, \xi)$ and let $\mathfrak{N}'(x)$ denote the c.p. for all of the completely proper points of the polyhedron $\mathfrak{N}(x)$. It then follows from Lemma 2.3 that $|\widetilde{\mathfrak{N}}|(x)$ does not depend on $x \in \Omega$ if and only if $|\mathfrak{N}'|(x)$ does not depend on $x \in \Omega$.

We also let $\widetilde{\mathfrak{M}}(x) \equiv \widetilde{\mathfrak{M}}$ denote the c.p. for the collection of polynomials

$$\{|P^{(\mu)}(x, \xi)|^2, |\mu| \geqslant 0\},$$

or, equivalently, for the polynomial $\widetilde{P}(x, \xi)^2$ (see Definition 0.2).

For a given $(n-1)$-dimensional face $\widetilde{\mathfrak{N}}_{i_0}^{n-1}$ (of $\widetilde{\mathfrak{N}}$) with outward unit normal λ and equation $(\lambda, \alpha) = d$, we introduce the following notation (cf. (2.15) and (2.16)):

$$[P_{\lambda,\delta}^{(\mu)}(x, \xi)]^{i_0, n-1} = \sum_{(\lambda, \alpha) = \delta} \gamma_{\alpha,\mu}(x)\, \xi^\alpha, \tag{3.1}$$

$$[P_\lambda^{(\mu)}(x, \xi)]^{i_0, n-1} = [P_{\lambda, d_\lambda(x)}(x, \xi)]^{i_0, n-1},$$

where

$$d_\lambda(x) = \max_{\alpha \in \widetilde{\mathfrak{N}} \setminus \widetilde{\mathfrak{N}}_{i_0}^{n-1}} (\lambda, \alpha). \tag{3.2}$$

We denote by $\delta(P, \lambda, \xi, x)$ the maximum among zero and the numbers δ for which

$$P_{\lambda,\delta}^{(\mu)}(x, \xi) \neq 0$$

for at least one μ, $|\mu| \geqslant 0$.

Lemma 1.3 directly implies

LEMMA 3.1. *A polynomial $P(x, \xi)$ has constant strength only if*

$$\delta(P, \lambda, \xi, x) = \delta(P, \lambda, \xi, y)$$

for any pair of points $x, y \in \Omega$.

We will henceforth assume that this condition is satisfied.

Suppose now that the polynomial $P(x, \xi)$ satisfies the following conditions:

a) The c.p. $\widetilde{\mathfrak{N}}(x)$ does not depend on the point $x \in \Omega$, which according to Remark 3.1 is equivalent to the assertion $\mathfrak{N}'(x) \equiv \mathfrak{N}'$.

b) If a completely proper $(n-1)$-dimensional face $\widetilde{\mathfrak{N}}_{i_0}^{n-1}$ (with outward unit normal λ) of $\widetilde{\mathfrak{N}}$ is P-irregular, then $d_\lambda(x) \equiv d_\lambda \equiv \max_{x\in\Omega} d_\lambda(x)$ and the polynomial $P^{i_0,n-1}(x, \xi)$ has the form[4]

$$P^{i_0, n-1}(x, \xi) = \mathcal{P}^{i_0, n-1}(x, \xi) r(\xi),$$

where $\mathcal{P}^{i_0,n-1}(x, \xi)$ does not vanish at any point $(x, \xi) \in \Omega \times R_n^{(0)}$.

We denote by $P_{\lambda\lambda}^{i_0,n-1}(x, \xi)$ the part of the polynomial $P(x, \xi)$ with multi-indices α such that $(\lambda, \alpha) \leqslant d_\lambda$, $(\lambda, \alpha) + d \geqslant 2d_\lambda$ and impose on the polynomial $P(x, \xi)$ the additional condition

c) if $P^{i_0,n-1}(x, \xi^0) = 0$ for some point $\xi^0 \in R_n^{(0)}$, then (see (3.1) and (3.2) for the notation)

1) $\sum_{|\mu|\geqslant 0} |[P_\lambda^{(\mu)}(x, \xi^0)]^{i_0,n-1}|^2 \neq 0$;

2) there exists a neighborhood of ξ^0 in which

$$K(x, \xi) = \bar{P}^{i_0, n-1}(x, \xi) P_{\lambda\lambda}^{i_0, n-1}(x, \xi) + P^{i_0, n-1}(x, \xi) \bar{P}_{\lambda\lambda}^{i_0, n-1}(x, \xi) \geqslant 0.$$[5]

THEOREM 3.1. *Suppose a polynomial $P(x, \xi)$ satisfies condition* a), *if all of the completely proper faces of the c.p. $\mathfrak{N}(P)$ are P-regular, and conditions* a) – c), *if all of the completely proper faces of $\mathfrak{N}(P)$ are P-regular with the exception of a P-irregular completely proper $(n-1)$-dimensional face $\mathfrak{N}_{i_0}^{n-1}$. Then the operator $P(x, D)$ (polynomial $P(x, \xi)$) has constant strength in Ω.*

PROOF. Let x and y be any two fixed points of Ω, and let $R(\xi)$ and $Q(\xi)$ respectively denote the polynomials $\widetilde{P}(x, \xi)^2$ and $\widetilde{P}(y, \xi)^2$ with constant coefficients. It suffices to prove that $R(\xi)$ and $Q(\xi)$ satisfy all of the conditions of Theorem 1.2′ for $\theta = 1$ and hence that inequality (0.2) is valid for them.

Condition 1) is verified in the same way as in the proof of Theorem 2.1.

Condition 2a) is obvious, since

$$R^{i_0, n-1}(\xi) = |P^{i_0, n-1}(x, \xi)|^2, \quad Q^{i_0, n-1}(\xi) = |P^{i_0, n-1}(y, \xi)|^2.$$

Condition 2b) follows from condition b).

Condition 2c′) follows from condition c).

---

(4) This condition means, in particular, that if $P^{i_0,n-1}(x^0, \xi^0) = 0$ at some point $(x^0, \xi^0) \in \Omega \times R_n^{(0)}$, then $P^{i_0,n-1}(x, \xi^0) = 0$ for all $x \in \Omega$. It is not difficult to see that the latter condition is necessary in order for the polynomial $P(x, \xi)$ to have constant strength.

(5) Clearly condition 2) is automatically satisfied if $P_{\lambda\lambda}^{i_0,n-1}(x, \xi) \equiv 0$, and this condition is satisfied in turn if, for example, $P(x, \xi) \equiv P^{i_0,n-1}(x, \xi)$.

Condition 2d′) follows from the conditions $\widetilde{\mathfrak{N}}(x) \equiv \widetilde{\mathfrak{N}}$ and $d_\lambda(x) \equiv d_\lambda$. Theorem 3.1 is proved.

§4. Differential operators of uniformly constant strength

DEFINITION 4.1. We will say that a differential operator $P(x, D)$ has *uniformly constant strength* in Ω if inequality (0.2) holds for it with a constant C not depending on the points $x, y \in \Omega$.

In the case when all of the completely proper faces of the c.p. $\mathfrak{N}(x) \equiv \mathfrak{N}$ of a polynomial $P(x, \xi)$ are P-regular, we will find conditions under which the operator $P(x, D)$ has uniformly constant strength in Ω. We note that in this case the polynomial $P(x, \xi)$ with c.p. $\mathfrak{N}(P)$ will be stronger than any polynomial $Q(\xi)$ with constant coefficients whose c.p. $\widetilde{\mathfrak{N}}(Q) \subseteq \widetilde{\mathfrak{N}}(P)$.

We cite an elementary lemma, a proof of which can be found, for example, in [2] or [3].

LEMMA 4.1. *Suppose given a collection of multi-indices $\{\alpha^k\}_1^N$ with a full c.p. $\mathfrak{N}$, and let e^k $(k = 1, \ldots, N_0)$ denote the vertices of $\mathfrak{N}$. Then the inequality*

$$|\xi^\nu| \leqslant C\left\{\sum_{k=1}^{N_0} |\xi^{e^k}| + 1\right\} \qquad \forall \xi \in R_n$$

is valid for any point $\nu \in \mathfrak{N}$, while the inequality

$$|\xi^\nu| \leqslant \delta \sum_{k=1}^{N_0} |\xi^{e^k}| + C(\delta) \qquad \forall \xi \in R_n,$$

in which $C(\delta) \to \infty$ as $\delta \to 0$, is valid for any nonprincipal point $\nu \in \mathfrak{N}$ and any $\delta > 0$.

Suppose a given differential operator $P(x, D) = \sum_\alpha \gamma_\alpha(x) D^\alpha$ with a c.p. $\mathfrak{N}(x)$ is such that

a) $\mathfrak{N}(x) \equiv \mathfrak{N}$; and

b) the coefficients $\gamma_\alpha(x)$ are defined in a bounded domain Ω and are bounded in $\overline{\Omega}$, with the completely proper coefficients, viz. the coefficients corresponding to the completely proper vertices of $\mathfrak{N}$, being continuous in $\overline{\Omega}$.

We note that the continuity in Ω of the coefficients of the polynomial $P(x, \xi)$ corresponding to the completely proper vertices of $\mathfrak{N}$, the condition $\mathfrak{N}(x) \equiv \mathfrak{N}$ and the compactness of $\overline{\Omega}$ imply the existence of positive numbers σ and M such that $\sigma \leqslant |\gamma_\alpha(x)| \leqslant M$ for all of the coefficients corresponding to the completely proper vertices of $\mathfrak{N}$ and for all points $x \in \overline{\Omega}$.

THEOREM 4.1. *Suppose a given differential operator $P(x, D)$ satisfies conditions* a) *and* b), *with all of the completely proper faces of the c.p. $\mathfrak{N}(x) \equiv \mathfrak{N}$ being P-regular. Then the operator $P(x, D)$ (polynomial $P(x, \xi)$) has uniformly constant strength in Ω.*

PROOF. It clearly suffices to prove the inequalities

$$c_1 \sum_{\mu \in \partial\widetilde{\mathfrak{M}}} |\xi^\mu| \leqslant \widetilde{P}(x, \xi)^2 \leqslant c_2 \sum_{\mu \in \partial\widetilde{\mathfrak{M}}} |\xi^\mu|, \tag{4.1}$$

where $\widetilde{\mathfrak{M}}$ is the c.p. for the polynomial $\widetilde{P}(x, \xi)^2$ and $\partial\widetilde{\mathfrak{M}}$ is the set of boundary points of $\widetilde{\mathfrak{M}}$.

The right inequality of (4.1) follows from conditions a) and b) and Lemma 4.1, so that it suffices to prove the left inequality.

Let x^0 be any point of Ω. According to Theorem 1.1 (when $\theta = 1$), for constant coefficients and for the polynomials

$$R(\xi) = \sum_{\mu \in \partial \widetilde{\mathfrak{M}}} |\xi^\mu|, \qquad Q(\xi) = \widetilde{P}(x^0, \xi)^2$$

we get

$$\sum_{\mu \in \partial \widetilde{\mathfrak{M}}} |\xi^\mu| \leqslant C(x^0) \widetilde{P}(x^0, \xi)^2 \qquad \forall\, \xi \in R_n. \tag{4.2}$$

We note that all of the principal coefficients $\widetilde{\gamma}_\alpha(x)$ of the polynomial $\widetilde{P}(x, \xi)^2$ are either squares of the completely proper coefficients of $P(x, \xi)$ or linear combinations of terms of the form $\gamma_{\alpha 1}(x)\gamma_{\alpha 2}(x)$, where $\gamma_{\alpha 1}(x)$ and $\gamma_{\alpha 2}(x)$ are completely proper coefficients. Therefore the principal coefficients $\widetilde{\gamma}_\alpha(x)$ of $\widetilde{P}(x, \xi)^2$ are continuous in $\overline{\Omega}$.

Let $\sigma > \theta$ be chosen so that $C(x^0)\sigma < 1$, and let $\rho = \rho(\sigma) > 0$ be chosen so that $|\widetilde{\gamma}_\alpha(x) - \widetilde{\gamma}_\alpha(x^0)| < \sigma/2$ for all of the principal coefficients $\widetilde{\gamma}_\alpha(x)$ and all points $x \in \overline{\Omega}$ for which $|x - x^0| \leqslant \rho$. Then

$$\begin{aligned} |\widetilde{P}(x, \xi)^2 - \widetilde{P}(x^0, \xi)^2| &\leqslant \sum_{\mu \in \partial \widetilde{\mathfrak{M}}} |\widetilde{\gamma}_\alpha(x) - \widetilde{\gamma}_\alpha(x^0)| \cdot |\xi^\mu| + \\ &+ C \sum_{\alpha \in \widetilde{\mathfrak{N}} \setminus \partial \widetilde{\mathfrak{M}}} |\xi^\alpha| \leqslant \frac{\sigma}{2} \cdot \sum_{\mu \in \partial \widetilde{\mathfrak{M}}} |\xi^\mu| + C \sum_{\alpha \in \widetilde{\mathfrak{N}} \setminus \partial \widetilde{\mathfrak{M}}} |\xi^\alpha|. \end{aligned} \tag{4.3}$$

On the other hand, Lemma 4.1 for $C\delta \leqslant \sigma/2$ implies

$$C \sum_{\alpha \in \widetilde{\mathfrak{M}} \setminus \partial \widetilde{\mathfrak{M}}} |\xi^\alpha| \leqslant \frac{\sigma}{2} \sum_{\mu \in \partial \widetilde{\mathfrak{M}}} |\xi^\mu| + C(\sigma). \tag{4.4}$$

It follows from (4.2)–(4.4) that

$$\begin{aligned} \sum_{\mu \in \partial \widetilde{\mathfrak{M}}} |\xi^\mu| &\leqslant C(x^0) |\widetilde{P}(x, \xi)^2 - \widetilde{P}(x^0, \xi)^2| + C(x^0) \widetilde{P}(x, \xi)^2 + C(\sigma) \\ &\leqslant C(x^0)\sigma \sum_{\mu \in \partial \widetilde{\mathfrak{M}}} |\xi^\mu| + C(x^0) \widetilde{P}(x, \xi)^2 + C'(\sigma) \end{aligned}$$

for $|x - x^0| \leqslant \rho$.

We transfer the term with σ to the left side and, making use of the choice of σ (viz. $C(x^0)\sigma < 1$), divide both sides of the inequality by $1 - C(x^0)\sigma$. This gives us the inequality

$$\sum_{\mu \in \partial \widetilde{\mathfrak{M}}} |\xi^\mu| \leqslant C_0 \widetilde{P}(x, \xi)^2 + C_0$$

for $|x - x^0| \leqslant \rho$.

Clearly, the set $\{U_\rho(x^0)\} \forall x^0 \in \Omega$ of domains

$$U_\rho(x^0) = \{x;\ x \in \Omega,\ |x - x^0| \leqslant \rho\} \tag{4.5}$$

covers Ω.

According to the Heine-Borel theorem, it is possible to select from this covering a finite subcovering $\{U^k\}_0^M$ of Ω. We take points $x^k \in U^k$ ($k = 0, 1, \ldots, M$) and write inequality (4.5) for each such x^k. Taking

$$C = \max_{0 \leqslant k \leqslant M} \{C_k(x^k),\ C_k\},$$

we get

$$\sum_{\mu \in \partial \widetilde{\mathfrak{M}}} |\xi^\mu| \leqslant C\{\widetilde{P}(x, \xi)^2 + 1\} \leqslant \widetilde{C}\widetilde{P}(x, \xi)^2 \quad \forall x \in \Omega, \quad \forall \xi \in R_n. \tag{4.6}$$

The latter inequality is obtained as a result of the fact that $\widetilde{P}(x, \xi)^2$ contains a nonzero constant term. It follows from (4.6) that the left inequality of (4.1) is valid for $c_1 = 1/\widetilde{C}$. Theorem 4.1 is proved.

I wish to express my gratitude to O. V. Besov for his unceasing interest in the present work.

BIBLIOGRAPHY

1. L. Hörmander, *Linear partial differential operators,* Die Grundlehren der math. Wissenschaften, Band **116**, Academic Press, New York; Springer-Verlag, Berlin, 1963. MR **28** #4221.

2. L. R. Volevič and S. G. Gindikin, *On a class of hypoelliptic polynomials,* Mat. Sb. **75 (117)** (1968), 400–416 = Math. USSR Sb. **4** (1968), 369–384. MR **37** #606a.

3. V. P. Mihaĭlov, *Behavior at infinity of a certain class of polynomials,* Trudy Mat. Inst. Steklov. **91** (1967), 59–80 = Proc. Steklov Inst. Math. **91** (1967), 61–82. MR **36** #3938.

4. G. G. Kazarjan, *Estimates of mixed derivatives in L_p by differential polynomials,* Trudy Mat. Inst. Steklov. **105** (1969), 66–76 = Proc. Steklov Inst. Math. **105** (1969), 79–91. MR **41** #4226.

5. ———, *Estimates of the L_p norms of derivatives by a nonregular set of differential operators,* Differencial'nye Uravnenija **5** (1969), 911–921 = Differential Equations **5** (1969), 662–670. MR **39** #5924.

6. G. C. Barozzi, *Sul prodotto di polinomi quasi-ellittici,* Boll. Un. Mat. Ital. (3) **20** (1965), 169–176. MR **34** #3099.

7. A. D. Aleksandrov, *Intrinsic geometry of convex surfaces,* OGIZ, Moscow, 1948; German transl., Akademie-Verlag, Berlin, 1955. MR **10**, 619; **17**, 74.

Translated by S. SMITH

Trudy Mat. Inst. Steklov.
131 (1974)

Proc. Steklov Inst. Math.
131 (1974)

THE METHOD OF FICTITIOUS DOMAINS FOR THE SECOND AND THIRD BOUNDARY VALUE PROBLEMS

UDC 517.51, 517.946

V. D. KOPČENOV

ABSTRACT. The second and third boundary value problems are investigated for a second-order elliptic differential operator in a bounded domain Ω_1 of n-dimensional space with a sufficiently smooth boundary S_1. The first boundary value problem is considered in a domain $\Omega \supset \bar{\Omega}_1$, in general, for a different elliptic operator; on the boundary S_1 of Ω_1 certain gluing conditions are imposed for the solution of this problem, which includes a small parameter $\epsilon > 0$. A theorem is proved to the effect that the two solutions are close to one another in Sobolev space, the estimate of the norm of their difference being dependent on ϵ. An example is given indicating that the estimate is sharp in the sense of the order with respect to ϵ.

Bibliography: 11 items.

1.1. We consider the boundary value problem

$$Lu = \sum_{i,k=1}^{n} \frac{\partial}{\partial x^i}\left(a_{ik}(x)\frac{\partial u}{\partial x_k}\right) - a(x)u = f(x) \quad (x \in \Omega_1), \tag{1.1}$$

$$\left(\sum_{i,k=1}^{n} a_{ik}\frac{\partial u}{\partial x_k}\cos(\widehat{n, x_i}) + \gamma(x)u\right)\Big|_{S_1} = 0 \tag{1.2}$$

in a bounded domain Ω_1 of n-dimensional space of points $x = (x_1, \ldots, x_n)$ with boundary $S_1 \in C^l$, where $a_{ik}(x) = a_{ki}(x)$, $a(x) \geqslant a_0 > 0$, $\gamma(x) \geqslant 0$ $(x \in S_1)$, while all the coefficients are continuous in $\bar{\Omega}_1$ together with their derivatives of order up to and including $l-1$, $\gamma \in C^{l-1}(S_1)$, and $f \in W_2^{l-2}(\Omega_1)$, $l \geqslant 2$. Also, let

$$\sum_{i,k=1}^{n} a_{ik}(x)\xi_i\xi_k \geqslant \beta\sum_{i=1}^{n}\xi_i^2, \tag{1.3}$$

where the constant $\beta > 0$ is independent of $x \in \Omega_1$ and $\xi = (\xi_1, \ldots, \xi_n)$; if $\gamma(x) \equiv 0$ on S_1, the problem (1.1)–(1.2) is the second boundary value problem; otherwise it is the third problem. Under these assumptions, the solution of the problem (1.1)–(1.2) belongs to the space $W_2^l(\Omega_1)$, while $\|u\|_{W_2^l(\Omega_1)} \leqslant c\|f\|_{W_2^{l-2}(\Omega_1)}$, where c is independent of u and f [1].

A device was used by V. K. Saul'ev in [2] and V. Ja. Rivkind in [3], whereby solution of the first boundary value problem for a second-order linear elliptic equation in a bounded domain with sufficiently smooth boundary is replaced by solution of the first boundary value problem in a domain encompassing the given domain, e.g. in a parallelepiped

---

AMS (MOS) subject classifications (1970). Primary 35J25, 35A35; Secondary 35A40, 35J20.

or sphere containing it. A similar device was employed by L. A. Ruhovec in [4] for the third boundary value problem.

This device for solving boundary value problems approximately is called the method of fictitious domains, and may be used for numerical solution of the above problems.

1.2. Let Ω be a sphere or parallelepiped with faces parallel to the coordinate planes, such that $\bar{\Omega}_1 \subset \Omega$. Denote the boundary of Ω by S_2, and the domain $\Omega \backslash \bar{\Omega}_1$ by Ω_2. We consider in Ω the boundary value problem

$$Lv_\varepsilon = f \qquad (x \in \Omega_1), \tag{1.4}$$

$$\Delta v_\varepsilon = 0 \qquad (x \in \Omega_2), \tag{1.5}$$

$$v_\varepsilon = 0 \qquad (x \in S_2), \tag{1.6}$$

$$v_{\varepsilon+} = v_{\varepsilon-} \qquad (x \in S_1), \tag{1.7}$$

$$\mu \left(\frac{\partial v_\varepsilon}{\partial \nu}\right)_+ + \gamma v_{\varepsilon+} = \varepsilon^2 \left(\frac{\partial v_\varepsilon}{\partial n}\right)_- \qquad (x \in S_1). \tag{1.8}$$

Here

$$\frac{\partial}{\partial \nu} = \frac{1}{\mu} \sum_{i,\,k=1}^{n} a_{ik} \cos(x_k, \widehat{\ } n) \frac{\partial}{\partial x_i}, \qquad \mu = \left[\sum_{i=1}^{n} \left(\sum_{k=1}^{n} a_{ik} \cos(x_k, \widehat{\ } n)\right)^2\right]^{1/2}.$$

(Henceforth, for brevity, the subscript ϵ will be omitted from the functions $v, \kappa, \ldots$.) We denote by v_+ and v_- the limits of v from the inside and from the outside of S_1; $\partial v/\partial \nu$ and $\partial v/\partial n$ are the derivatives with respect to the conormal and normal to S_1 respectively, and $\epsilon > 0$ is a small parameter. Under the same assumptions on the data of the problem (1.4)–(1.8) as in the case of the problem (1.1)–(1.2), $v \in W_2^{l-1}(\Omega_1 \cup \Omega_2)$ (see [5], [6]).

A problem closely similar to (1.4)–(1.8) (identical with it if $\gamma \equiv 0$) was discussed in [4]. It was shown that

$$\| u - v \|_{W_2^1(\Omega_1)} \leqslant \varepsilon c_1 \| f \|_{L_2(\Omega_1)}, \tag{1.9}$$

where the constant c_1 is independent of ϵ and f. We shall show here that a more exact estimate than (1.9) can be obtained for the problem (1.4)–(1.8); in fact:

THEOREM. *Let the boundary $S_1 \in C^l$, and let the coefficients a_{ik}, $a \in C^{l-1}(\Omega_1)$ and $\gamma \in C^{l-1}(S_1)$, while $f \in W_2^{l-2}(\Omega_1)$. Then*

$$\| u - v \|_{W_2^{l-1}(\Omega_1)} \leqslant \varepsilon^2 c_2 \| f \|_{W_2^{l-2}(\Omega_1)}, \tag{1.10}$$

where the constant c_2 depends on the data of the problem but not on ϵ and f. The estimate (1.10) *is exact in the sense of the order with respect to ϵ (see* §3).

It follows from the theorem that for $l = [n/2] + 4$

$$\| u - v \|_{C_2(\bar{\Omega}_1)} \leqslant \varepsilon^2 c_2 \| f \|_{C^{[n/2]+2}(\bar{\Omega}_1)}. \tag{1.11}$$

The estimate (1.11) follows easily from the Sobolev imbedding theorems [7].

1.3. Let $u \in W_2^l(\Omega_1)$ be the solution of the problem (1.1)–(1.2). Then $u \in W_2^{l-1/2}(S_1)$ [8], while

$$\| u \|_{W_2^{l-1/2}(S_1)} \leqslant c_3 \| u \|_{W_2^l(\Omega_1)} \leqslant c_4 \| f \|_{W_2^{l-2}(\Omega_1)}.$$

Let us further consider in Ω_2 the Dirichlet problem

$$\Delta\varkappa = 0 \qquad (x \in \Omega_2), \tag{1.12}$$

$$\varkappa = u \qquad (x \in S_1), \tag{1.13}$$

$$\varkappa = 0 \qquad (x \in S_2). \tag{1.14}$$

Note that $\kappa \in W_2^l(\Omega_2)$ [8], where

$$\|\varkappa\|_{W_2^l(\Omega_2)} \leqslant c_5 \|u\|_{W_2^{l-1/2}(S_1)} \leqslant c_6 \|f\|_{W_2^{l-2}(\Omega_1)},$$

and the constants c_5 and c_6 are independent of u and f.

1.4. We complete the definition of u by zero in Ω_2, and that of the function κ by zero in Ω_1, and then consider in Ω the function $w = u + \kappa - v$. The function w satisfies the differential problem

$$Lw = 0 \qquad (x \in \Omega_1), \tag{1.15}$$

$$\Delta w = 0 \qquad (x \in \Omega_2) \tag{1.16}$$

$$w_+ = w_- \qquad (x \in S_1), \tag{1.17}$$

$$\mu\left(\frac{\partial w}{\partial \nu}\right)_+ + \gamma w_+ = \varepsilon^2\left(\frac{\partial w}{\partial n}\right)_- + \varepsilon^2\varphi \qquad (x \in S_1), \tag{1.18}$$

$$w = 0 \qquad (x \in S_2), \tag{1.19}$$

where [8]

$$\varphi = -\frac{\partial\varkappa}{\partial \boldsymbol{n}}\bigg|_{S_1} \in W_2^{l-3/2}(S_1).$$

Let us write the variational equation for the problem (1.15)–(1.19). Let $\xi \in \overset{\circ}{W}{}_2^1(\Omega)$. Then the variational equations for the problems (1.1)–(1.2), (1.4)–(1.8), and (1.12)–(1.14) are

$$A(u, \xi) + (\gamma u, \xi)_{S_1} = (f, \xi)_{\Omega_1}, \tag{1.20}$$

$$A(v, \xi) + (\gamma v, \xi)_{S_1} + \varepsilon^2 \int_{\Omega_2} \sum_{i=1}^{n} \frac{\partial v}{\partial x_i}\frac{\partial \xi}{\partial x_i}\, dx = (f, \xi)_{\Omega_1}, \tag{1.21}$$

$$(\varphi, \xi)_{S_1} + \int_{\Omega_2} \sum_{i=1}^{n} \frac{\partial \varkappa}{\partial x_i}\frac{\partial \xi}{\partial x_i}\, dx = 0, \tag{1.22}$$

where

$$A(\omega, \xi) = \int_{\Omega_1} \left(\sum_{i,k=1}^{n} a_{ik} \frac{\partial \omega}{\partial x_i}\frac{\partial \xi}{\partial x_k} + a\omega\xi \right) dx,$$

and $(\ ,\ )_E$ is the scalar product in E.

The first term appears in (1.22) because ξ belongs to $\overset{\circ}{W}{}_2^1(\Omega)$ but not to $\overset{\circ}{W}{}_2^1(\Omega_2)$. On adding (1.20) and (1.22), multiplied by ϵ^2, then subtracting (1.21), we obtain the following variational equation for the function $w = u + \kappa - v$ in Ω:

$$\int_{\Omega_1} \sum_{i,k=1}^{n} a_{ik} \frac{\partial w}{\partial x_i}\frac{\partial \xi}{\partial x_k}\, dx + \int_{\Omega_1} aw\cdot\xi dx + \int_{S_1} \gamma w\cdot\xi dS + \varepsilon^2 \int_{S_1} \varphi\cdot\xi dS + \varepsilon^2 \int_{\Omega_2} \sum_{i=1}^{n} \frac{\partial w}{\partial x_i}\frac{\partial \xi}{\partial x_i}\, dx = 0. \tag{1.23}$$

Putting $\xi = w$ in (1.23), and using (1.17) and (1.19), we obtain the inequalities (see (1.2), [9]):

$$\beta \int_{\Omega_1} \left(\sum_{i=1}^{n} \left(\frac{\partial w}{\partial x_i} \right)^2 + a_0 w^2 \right) dx + \varepsilon^2 \int_{\Omega_2} \sum_{i=1}^{n} \left(\frac{\partial w}{\partial x_i} \right)^2 dx$$
$$\leqslant \varepsilon^2 \int_{S_1} |\varphi| \, |w| \, dS \leqslant c\varepsilon^2 \|\varphi\|_{L_2(S_1)} (\|w\|_{L_2(\Omega_1)} + \sqrt{D_{\Omega_1}(w)}). \tag{1.24}$$

The third term in (1.23) is dropped, since it is nonnegative with $\xi = w$, while $D_E(w)$ is the Dirichlet integral in E, and the constant c is independent of ϵ. Noting that

$$\|\varphi\|_{L_2(S_1)} = \left\| \frac{\partial \varkappa}{\partial n} \right\|_{L_2(S_1)} \leqslant c_1 \|\varkappa\|_{W_2^l(\Omega_2)} \leqslant c_2 \|f\|_{W_2^{l-2}(\Omega_1)},$$

we can strengthen (1.24):

$$\beta \int_{\Omega_1} \left(\sum_{i=1}^{n} \left(\frac{\partial w}{\partial x_i} \right)^2 + a_0 w^2 \right) dx + \varepsilon^2 \int_{\Omega_2} \sum_{i=1}^{n} \left(\frac{\partial w}{\partial x_i} \right)^2 dx$$
$$\leqslant c_3 \varepsilon^2 \|f\|_{W_2^{l-2}(\Omega_1)} \sqrt{\int_{\Omega_1} \left(\sum_{i=1}^{n} \left(\frac{\partial w}{\partial x_i} \right)^2 + w^2 \right) dx + \varepsilon^2 \int_{\Omega_2} \sum_{i=1}^{n} \left(\frac{\partial w}{\partial x_i} \right)^2 dx} \cdot \tag{1.25}$$

Let $\alpha = \min(\beta, a_0)$ and $c_4 = c_3/\alpha$; we then have from (1.25)

$$(D_{\Omega_1}(w))^{1/2} \leqslant c_4 \varepsilon^2 \|f\|_{W_2^{l-2}(\Omega_1)}, \tag{1.26}$$

$$\|w\|_{L_2(\Omega_1)} \leqslant c_4 \varepsilon^2 \|f\|_{W_2^{l-2}(\Omega_1)}, \tag{1.27}$$

$$(D_{\Omega_2}(\omega))^{1/2} \leqslant c_4 \varepsilon \|f\|_{W_2^{l-2}(\Omega_1)}. \tag{1.28}$$

In addition, since $\Delta w = 0$ in Ω_2 and $w = 0$ on S_2, we have [9]

$$\|w\|_{L_2(\Omega_2)} \leqslant c_5 \sqrt{D_{\Omega_2}(\omega)} \leqslant c_6 \varepsilon \|f\|_{W_2^{l-2}(\Omega_1)}, \tag{1.29}$$

where the constant c_6 is independent of ϵ and f.(1)

2.1. For what follows, we need to show that the highest generalized derivatives of w are square-summable. We shall first perform these estimates close to the surface S_1. For this, we specify a point $Q \in S_1$. Taking Q as center, we draw a sphere of sufficiently small radius r, such that the piece (open set) σ cut by it from S_1 projects one-to-one onto one of the $(n-1)$-dimensional coordinate planes. We shall assume that this coordinate plane is the $(x_1, \ldots, x_{n-1})$ plane. Then σ is given by the equation

$$x_n = \chi(x_1, \ldots, x_{n-1}) \qquad (x_1, \ldots, x_{n-1}) \in \sigma',$$

where σ' is an $(n-1)$-dimensional (open) domain (the projection of σ onto the $x_0 = 0$ plane), and the function χ belongs to the class $C^l(\overline{\sigma'})$.

Inside the sphere we introduce local coordinates $(y_1, \ldots, y_n)$, connected with the old coordinates by the equations

$$y_1 = x_1, \ldots, \quad y_{n-1} = x_{n-1}, \quad y_n = x_n - \chi(x_1, \ldots, x_{n-1}). \tag{2.1}$$

Let H_ρ be the circular cylinder of radius $\rho = r/2$ such that the $y_n = 0$ plane is perpendicular to its axis, and its axis passes through the point Q. Denote by H_{ρ,a_1,a_2} $(a_1 < a_2)$ the part of H_ρ lying between the $y_n = a_1$ and $y_n = a_2$ planes. Since the surface S_1 is smooth, there

---

(1) Note that the theorem, with $l = 2$, follows from (1.26) and (1.27).

exists a sufficiently small positive number a such that one of the finite cylinders $H_{\rho,0,a}$ or $H_{\rho,-a,0}$ belongs entirely to the domain Ω_1, while the other belongs to Ω_2. For clarity, we shall assume that the first cylinder belongs to Ω_1. Its points have coordinate $y_n > 0$. Note that the transformation (2.1) has a Jacobian equal to unity.

2.2. Let us write the integral identity (1.23) in the cylinder $H_{\rho-\delta,-a+\delta,a-\delta}$ $(0<\delta<a,\ \rho)$ in $(y_1, \ldots, y_n)$ coordinates, assuming that ξ vanishes outside the cylinder $H_{\rho-\delta,-a+\delta,a-\delta}$:

$$\int\limits_{H_{\rho-\delta,\,0,\,a-\delta}} \sum_{i,\,k=1}^{n} b_{ik} \frac{\partial w}{\partial y_i} \frac{\partial \xi}{\partial y_k} dy + \int\limits_{H_{\rho-\delta,\,0,\,a-\delta}} aw\xi dy + \int\limits_{\sigma_{\rho-\delta}} \gamma w\xi ds$$
$$+ \varepsilon^2 \int\limits_{\sigma_{\rho-\delta}} \varphi\xi ds + \varepsilon^2 \int\limits_{H_{\rho-\delta,\,-a+\delta,\,0}} \sum_{i,\,k=1}^{n} c_{ik} \frac{\partial w}{\partial y_i} \frac{\partial \xi}{\partial y_k} dy = 0, \tag{2.2}$$

where $\sigma_{\rho-\delta}$ is the part of the surface $y_n = 0$ belonging to $H_{\rho-\delta,-a+\delta,a-\delta}$, and

$$b_{ik} = \sum_{j,\,l=1}^{n} a_{ik} \frac{\partial y_i}{\partial x_j} \frac{\partial y_k}{\partial x_l}, \qquad \sum_{i,\,k=1}^{n} b_{ik}\xi_i\xi_k > \beta_1 \sum_{i=1}^{n} \xi_i^2,$$
$$c_{ik} = \sum_{j,\,l=1}^{n} \frac{\partial y_i}{\partial x_j} \frac{\partial y_k}{\partial x_l}, \qquad \sum_{i,\,k=1}^{n} c_{ik}\xi_i\xi_k > \beta_2 \sum_{i=1}^{n} \xi_i^2 \tag{2.3}$$

($\beta_1, \beta_2 > 0$ are constants).

Let h_j be the vector of length $h_j < \delta$ directed along the y_j axis ($j = 1, \ldots, n-1$). Let us agree to denote by ψ_{h_j} the separated first difference of ψ with respect to the variable y_j with the step h_j. Then, from (2.2),

$$\int\limits_{H_{\rho-\delta,\,0,\,a-\delta}} \sum_{k=1}^{n} \frac{\partial \xi}{\partial y_k} \left(\sum_{i=1}^{n} b_{ik} \frac{\partial w}{\partial y_i} + \frac{\partial w(y+\mathbf{h}_j)}{\partial y_i} b_{ikh_j} \right) dy$$
$$+ \int\limits_{H_{\rho-\delta,\,0,\,a-\delta}} \xi\,(aw_{h_j} + w(y+\mathbf{h}_j)\, a_{h_j})\, dy + \int\limits_{\sigma_{\rho-\delta}} \xi\,(\gamma w_{h_j} + w(y+\mathbf{h}_j)\,\gamma_{h_j})\, ds \tag{2.4}$$
$$+ \varepsilon^2 \int\limits_{\sigma_{\rho-\delta}} \varphi_{h_j}\xi ds + \varepsilon^2 \int\limits_{H_{\rho-\delta,\,-a+\delta,\,0}} \sum_{k=1}^{n} \frac{\partial \xi}{\partial y_k} \left(c_{ik} \frac{\partial w_{h_j}}{\partial y_i} + \frac{\partial w(y+\mathbf{h}_j)}{\partial y_i} c_{ikh_j} \right) dy = 0.$$

Now let

$$\xi = \eta w_{h_j}, \tag{2.5}$$

where η is a function, continuous and nonnegative in $H_{\rho-\delta,-a+\delta,a-\delta}$, such that

$$\eta = \begin{cases} 1, & \text{if } \ y \in H_{\rho-2\delta,\,-a+2\delta,\,a-2\delta} \\ 0, & \text{if } \ y \notin H_{\rho-\delta,\,-a+\delta,\,a-\delta} \end{cases} \tag{2.6}$$

and we have

$$\left(\frac{\partial \eta}{\partial y_i}\right)^2 \leqslant c(\delta)\,\eta \qquad (i = 1, \ldots, n). \tag{2.7}$$

The equation (2.4) then becomes

$$\int\limits_{H_{\rho-\delta,\,0,\,a-\delta}} \eta \sum_{i,k=1}^{n} b_{ik} \frac{\partial w_{h_j}}{\partial y_i} \frac{\partial w_{h_j}}{\partial y_k} dy + \int\limits_{H_{\rho-\delta,\,0,\,a-\delta}} \eta \sum_{i,k=1}^{n} \frac{\partial w_{h_j}}{\partial y_k} \frac{\partial w(y+\mathbf{h}_j)}{\partial y_i} b_{ikh_j} dy$$
$$+ \int\limits_{H_{\rho-\delta,\,0,\,a-\delta}} w_{h_j} \sum_{i,k=1}^{n} b_{ik} \frac{\partial w_{h_j}}{\partial y_i} \frac{\partial \eta}{\partial y_i} dy$$
$$+ \int\limits_{H_{\rho-\delta,\,0,\,a-\delta}} w_{h_j} \sum_{i,k=1}^{n} b_{ikh_j} \frac{\partial w(y+\mathbf{h}_j)}{\partial y_i} \frac{\partial \eta}{\partial y_k} dy + \int\limits_{H_{\rho-\delta,\,0,\,a-\delta}} a\eta w_{h_j}^2 dy$$
$$+ \int\limits_{H_{\rho-\delta,\,0,\,a-\delta}} \eta\omega_{h_j} w(y+\mathbf{h}_j)\, a_{h_j} dy + \int\limits_{\sigma_{\rho-\delta}} \gamma\eta w_{h_j}^2 d\sigma \tag{2.8}$$
$$+ \int\limits_{\sigma_{\rho-\delta}} \eta w_{h_j} w(y+\mathbf{h}_j)\,\gamma_{h_j} d\sigma + \varepsilon^2 \int\limits_{\sigma_{\rho-\delta}} \eta\varphi_{h_j} w_{h_j} d\sigma + \varepsilon^2 \int\limits_{H_{\rho-\delta,\,-a+\delta,\,0}} \eta \sum_{i,k=1}^{n} c_{ik} \frac{\partial w_{h_j}}{\partial y_i}$$
$$\times \frac{\partial w_{h_j}}{\partial y_k} dy + \varepsilon^2 \int\limits_{H_{\rho-\delta,\,-a+\delta,\,0}} \eta \sum_{i,k=1}^{n} c_{ikh_j} \frac{\partial w_{h_j}}{\partial y_k} \frac{\partial w(y+\mathbf{h}_j)}{\partial y_i} dy$$
$$+ \varepsilon^2 \int\limits_{H_{\rho-\delta,\,-a+\delta,\,0}} w_{h_j} \sum_{i,k=1}^{n} c_{ik} \frac{\partial w_h}{\partial y_i} \frac{\partial \eta}{\partial y_k} dy + \varepsilon^2 \int\limits_{H_{\rho-\delta,\,-a+\delta,\,0}} w_{h_j} \sum_{i,k=1}^{n} c_{ikh_j}$$
$$\times \frac{\partial w(y+\mathbf{h}_j)}{\partial y_i} \frac{\partial \eta}{\partial y_k} dy = J_1 + J_2 + \ldots + J_{13} = 0.$$

2.3. We shall find lower bounds for the integrals J_1, J_5, J_7, J_{10}, and upper bounds for $J_2, J_3, J_4, J_6, J_8, J_9, J_{11}, J_{12}, J_{13}$:

$$J_1 \geqslant \beta_1 \int\limits_{H_{\rho-\delta,\,0,\,a-\delta}} \eta \sum_{i=1}^{n} \left(\frac{\partial w_{h_j}}{\partial y_i}\right)^2 dy = \beta_1 \Lambda_{1h_j}^2.$$

We introduce the notation

$$M(\psi) = \|\psi\|_{C^{l-2}(\overline{\Omega}_1)},$$

if ψ depends on i and k, the max is taken with respect to all i and k. Further, from (1.26)

$$|J_2| \leqslant n^{3/2} M(b_{ik}) \sqrt{\int\limits_{H_{\rho-\delta,\,0,\,a-\delta}} \eta \sum_{i=1}^{n} \left(\frac{\partial w(y+\mathbf{h}_j)}{\partial y_i}\right)^2 dy}\ \Lambda_{1h_j} \leqslant A_1 \varepsilon^2 \Lambda_{1h_j},$$

where A_1, and later $A_2, A_3, \ldots$, are constants independent of h_j and ϵ,

$$|J_3| \leqslant c_1(\delta)\, n^{3/2} M(b_{ik}) \sqrt{\int\limits_{H_{\rho-\delta,\,0,\,a-\delta}} w_{h_j} dy}\ \Lambda_{1h_j} \leqslant c_1(\delta)\, n^{3/2} M(b_{ik})$$
$$\times \sqrt{D_{H_{\rho,\,0,\,a}}(w)}\ \Lambda_{1h_j} \leqslant A_2 \varepsilon^2 \Lambda_{1h_j},$$

where $c_1(\delta) = \sqrt{c(\delta)}$ (see (2.7)); and

$$|J_4| \leqslant c_1(\delta)\, n^{3/2} M(b_{ik}) \sqrt{\int\limits_{H_{\rho,\,0,\,a}} \sum_{i=1} \left(\frac{\partial w}{\partial y_i}\right)^2 dy} \sqrt{\int\limits_{H_{\rho-\delta,\,0,\,a-\delta}} \eta w_{h_j}^2 dy} \leqslant A_3 \varepsilon^4,$$

$$J_5 \geqslant 0,$$

$$|J_6| \leqslant n^{3/2} M(a) \sqrt{\int\limits_{H_{\rho-\delta,\,0,\,a}} \eta \sum_{i=1} \left(\frac{\partial w}{\partial y_i}\right)^2 dy} \sqrt{\int\limits_{H_{\rho,\,0,\,a}} w^2 dy} \leqslant A_4 \varepsilon^4$$

(see (1.26) and (1.27)), while

$$J_7 \geqslant 0.$$

We have (see Lemma 3 of [10])

$$\| w_{h_j} \|_{L_2(\sigma_{\rho-\delta})} \leqslant \left\| \frac{\partial w}{\partial y_i} \right\|_{L_2(\sigma_\rho)}. \tag{2.9}$$

We introduce the notation

$$N = \| \varphi \|_{W^{l-2}(\sigma_\rho)}, \quad L = \| \gamma \|_{C^{l-2}(S_1)}$$

and find bounds for J_8 and J_9:

$$|J_8| \leqslant L \| w \|_{L_2(\sigma_\rho)} \left(\sqrt{\int_{H_{\rho-\delta,\,0,\,a-\delta}} \eta w_{h_j}^2 dy} + \sqrt{\int_{H_{\rho-\delta,\,0,\,a-\delta}} \sum_{i=1}^n \left(\frac{\partial(\eta w_{h_j})}{\partial y_i} \right)^2 dy} \right)$$

$$\leqslant A_5 \varepsilon^2 \left(\sqrt{\int_{H_{\rho-\delta,\,0,\,a-\delta}} \eta w_{h_j}^2 dy} + \sqrt{\int_{H_{\rho-\delta,\,0,\,a-\delta}} \eta \sum_{i=1}^n \left(\frac{\partial w_{h_j}}{\partial y_i} \right)^2 dy} \right)$$

$$\leqslant A_6 (\varepsilon^4 + \varepsilon^2 \Lambda_{1h_j}),$$

$$|J_9| \leqslant N \varepsilon^2 \left(\sqrt{\int_{H_{\rho-\delta,\,0\,a-\delta}} \eta w_{h_j}^2 dy} + \sqrt{\int_{H_{\rho-\delta,\,0,\,a-\delta}} \sum_{i=1}^n \left(\frac{\partial(\eta w_{h_j})}{\partial y_i} \right)^2 dy} \right)$$

$$\leqslant A_7 (\varepsilon^4 + \varepsilon^2 \Lambda_{1h_j}),$$

$$J_{10} \geqslant \varepsilon^2 \beta_2 \int_{H_{\rho-\delta,\,-a+\delta,\,0}} \eta \sum_{i=1}^n \left(\frac{\partial w_{h_j}}{\partial y_i} \right)^2 dy = \beta_2 \varepsilon^2 \Lambda_{2h_j}^2,$$

$$|J_{11}| \leqslant \varepsilon^2 n^{3/2} M(c_{ik}) \sqrt{\int_{H_{\rho-\delta,\,-a+\delta,\,0}} \eta \sum_{i=1}^n \left(\frac{\partial w(y + \mathbf{h}_j)}{\partial y_i} \right)^2 dy}\; \Lambda_{2h_j} \leqslant A_8 \varepsilon^3 \Lambda_{2h_j},$$

(see (1.28))

$$|J_{12}| \leqslant \varepsilon^2 c_1(\delta) n^{3/2} M(c_{ik}) \sqrt{\int_{H_{\rho-\delta,\,-a+\delta,\,0}} w_{h_j}^2 dy}\; \Lambda_{2h_j} \leqslant A_9 \varepsilon^3 \Lambda_{2h_j},$$

$$|J_{13}| \leqslant \varepsilon^2 c_1(\delta) n^{3/2} M(c_{ik}) \sqrt{\int_{H_{\rho,\,-a,\,0}} \sum_{i=1}^n \left(\frac{\partial w}{\partial y_i} \right)^2 dy} \sqrt{\int_{H_{\rho-\delta,\,-a+\delta,\,0}} \eta w_{h_j} dy}$$

$$\leqslant A_{10} \varepsilon^4.$$

It then follows from (2.8) that

$$\Lambda_{1h_j}^2 + \varepsilon^2 \Lambda_{2h_j}^2 \leqslant \frac{1}{\beta_3} (A_{11} \varepsilon^2 \Lambda_{1h_j} + A_{12} \varepsilon^3 \Lambda_{2h_j} + A_{13} \varepsilon^4)$$

$$\leqslant A_{14} \varepsilon^2 (\Lambda_{1h_j} + \varepsilon \Lambda_{2h_j}) + A_{15} \varepsilon^4 \leqslant A_{16} \varepsilon^2 \sqrt{\Lambda_{1h_j}^2 + \varepsilon^2 \Lambda_2^2 h_j} + A_{15} \varepsilon^4, \tag{2.10}$$

where $\beta_3 = \min(\beta_1, \beta_2)$.

Let

$$X = \frac{1}{\varepsilon^2} \sqrt{\Lambda_{1h_j}^2 + \varepsilon^2 \Lambda_{2h_j}^2};$$

we then obtain from (2.10)

$$X^2 - A_{16} X - A_{15} \leqslant 0.$$

Hence we always have $X \leqslant A_{17}$, where A_{17} is the positive root of the equation $X^2 - A_{16}X - A_{15} = 0$. It immediately follows from this that

$$\sqrt{\int\limits_{H_{\rho-\delta,\,0,\,a-\delta}} \eta \sum_{i=1}^{n} \left(\frac{\partial w_{h_j}}{\partial y_i}\right)^2 dy + \varepsilon^2 \int\limits_{H_{\rho-\delta,\,-a+\delta,\,0}} \eta \sum_{i=1}^{n} \left(\frac{\partial w_{h_j}}{\partial y_i}\right)^2 dy} \leqslant A_{17}\varepsilon^2, \tag{2.11}$$

where A_{17} is a constant independent of w, ϵ, and h_j.

It can easily be seen that the following bound holds for the constant A_{17} (see (1.26)–(1.29)):

$$A_{17} \leqslant B \| f \|_{W_1^{l-2}(\Omega_1)}, \tag{2.12}$$

where B is a constant independent of f.

2.4. On taking the integrals on the left side of (2.11) with respect to the lesser of the cylinders $H_{\rho-2\delta,0,a-2\delta}$ and $H_{\rho-2\delta,-a+2\delta,0}$, and recalling Lemma 1 of [10], we get

$$\left\| \frac{\partial^2 w}{\partial y_i \partial y_k} \right\|_{L_2(H_{\rho-2\delta,\,0,\,a-2\delta})} \leqslant A_{17}\varepsilon^2 \qquad (i = 1, \ldots, n;\; k = 1, \ldots, n-1),$$

$$\left\| \frac{\partial^2 w}{\partial y_i \partial y_k} \right\|_{L_2(H_{\rho-2\delta,\,-a+2\delta,\,0})} \leqslant A_{17}\varepsilon \qquad (i = 1, \ldots, n;\; k = 1, \ldots, n-1).$$

Let λ satisfy the inequalities $0 < \lambda < a, \rho$. Using induction, we have

$$\| D^m w \|_{L_2(H_{\rho-\lambda,\,0,\,a-\lambda})} \leqslant A\varepsilon^2 \quad (m = (m_1, \ldots, m_n);\; m_n = 0, 1;\; 2 \leqslant |m| \leqslant l-1), \tag{2.13}$$

$$\| D^m w \|_{L_2(H_{\rho-\lambda,\,-a+\lambda,\,0})} \leqslant A\varepsilon \quad (m = (m_1, \ldots, m_n);\; m_n = 0, 1;\; 2 \leqslant |m| \leqslant l-1), \tag{2.14}$$

where the constant A is independent of ϵ.

By means of the differential problem, the derivatives $D^m w$ with $m_n = k \geqslant 2$ can be expressed and estimated in terms of $D^m w$ with $m_n = k-1, k-2, \ldots$ (see [11]). In this way the estimates (2.13) and (2.14) can be extended to any derivatives of w of the form $D^m w$ $(2 \leqslant |m| \leqslant l-1, y \in H_{\rho-\lambda,-a+\lambda,a-\lambda})$.

We return to the old coordinates and note that, by the Borel lemma, the surface S_1 can be covered by a finite number of domains of the type $H_{\rho-\lambda,-a+\lambda,a-\lambda}$. All in all, we find that there exist a bounded domain O containing the surface S_1 (an n-dimensional neighborhood of S_1), and a constant A such that, for any partial derivative of order m $(0 \leqslant |m| \leqslant l-1)$ of w, we have the bounds

$$\| D^m w \|_{L_2(\Omega_1 \cap O)} \leqslant A\varepsilon^2, \tag{2.15}$$

$$\| D^m w \|_{L_2(\Omega_2 \cap O)} \leqslant A\varepsilon, \tag{2.16}$$

where the constant A is independent of w and ϵ.

2.5. Bounds of the type (2.15) may be similarly obtained, for the same m, strictly inside the domain Ω_1; here it is unnecessary to transform to local coordinates. If it is further noted that (1.26)–(1.29) and (2.12) are used here as initial estimates, we finally obtain

$$\| D^m w \|_{L_2(\Omega_1)} \leqslant \varepsilon^2 c \| f \|_{W_2^{l-2}(\Omega_1)} \qquad (0 \leqslant |m| \leqslant l-1), \tag{2.17}$$

where the constant c is independent of f and ϵ. The theorem of §1.2 follows readily from (2.17).

3. If Ω is a sphere, it can easily be shown that the estimate (1.10) is sharp in the sense of the order with respect to ϵ. Let Ω_1 be the open sphere of radius $r = 1$ and center at the origin, and Ω the sphere of radius $r = 2$ and center at the origin. We consider the following problem of the type (1.1)–(1.2) ($n = 3$):

$$\Delta u - u = 1 \qquad (x \in \Delta_1), \tag{3.1}$$

$$\left(\sum_{i=1}^{n} \frac{\partial u}{\partial x_i} \cos(\widehat{n, x_i}) + u\right)\Big|_{S_1} = 0. \tag{3.2}$$

In the light of the symmetry, we can conveniently write the problem (3.1)–(3.2) in a spherical coordinate system:

$$\frac{1}{r^2}\frac{\partial}{\partial r}\left(r^2 \frac{\partial u}{\partial r}\right) - u = 1 \qquad (r < 1),$$

$$\frac{\partial u}{\partial r} + u = 0 \qquad (r = 1).$$

The solution u can be written explicitly:

$$u = \frac{1}{e}\left(\frac{e^r - e^{-r}}{r}\right) - 1.$$

Let us write a problem of the type (1.4)–(1.8) in spherical coordinates:

$$\frac{1}{r^2}\frac{\partial}{\partial r}\left(r^2 \frac{\partial v}{\partial r}\right) - v = 1 \qquad (r < 1),$$

$$\frac{\partial}{\partial r}\left(r^2 \frac{\partial v}{\partial r}\right) = 0 \qquad (1 < r < 2),$$

$$v_+ = v_- \qquad (r = 1),$$

$$\left(\frac{\partial v}{\partial r}\right)_+ + v_+ = \varepsilon^2 \left(\frac{\partial v}{\partial r}\right)_- \qquad (r = 1),$$

$$v = 0 \qquad (r = 2).$$

The solution v can again be written explicitly:

$$v = \frac{1 + 2\varepsilon^2}{2\varepsilon^2 (e^{-1} - e) - e}\left(\frac{e^{-r} - e^r}{r}\right) - 1 \qquad (r < 1),$$

$$v = \frac{4e^2 - 2}{e^2 - 2\varepsilon^2 (1 - e^2)}\left(\frac{1}{r} - \frac{1}{2}\right) \qquad (1 < r < 2).$$

We then obtain, for the difference $u - v$,

$$u - v = \frac{2\varepsilon^2}{2\varepsilon^2 (e - e^3) - e^3}\left(\frac{e^{-r} - e^r}{r}\right) \qquad (r < 1),$$

which in fact proves our assertion.

In conclusion, the author thanks V. I. Burenkov for advice and assistance in writing the article, and S. M. Nikol'skiĭ and E. A. Volkov for discussion of the results.

BIBLIOGRAPHY

1. O. A. Ladyženskaja, *A simple proof of the solvability of the fundamental boundary problems and of a problem on eigenvalues for linear elliptic equations,* Vestnik Leningrad. Univ. **10** (1955), no. 11, 23–29. (Russian) MR **17**, 855.

2. V. K. Saul'ev, *Solution of certain boundary-value problems on high-speed computers by the fictitious-domain method,* Sibirsk. Mat. Ž. **4** (1963), 912–925. (Russian) MR **29** #1748.

3. V. Ja. Rivkind, *An approximate method of solving the Dirichlet problem and estimates of the speed of convergence of solutions of the difference equations to solutions of elliptic equations with discontinuous coefficients,* Vestnik Leningrad. Univ. Ser. Mat. Meh. Astronom. **19** (1964), no. 3, 37–52. (Russian) MR **30** #5500.

4. L. A. Ruhovec, *A remark on the method of fictitious domains,* Differencial'nye Uravnenija **3** (1967), 698–701 = Differential Equations **3** (1967), 357–359. MR **35** #3253.

5. E. A. Volkov, *Differentiability properties of solutions of boundary value problems for the Laplace equation on a polygon,* Trudy Mat. Inst. Steklov. **77** (1965), 113–142 = Proc. Steklov Inst. Math. **77** (1965), 127–159. MR **33** #305.

6. V. D. Kopčenov, *The approximation of the solution of the Dirichlet problem by the method of fictitious domains,* Differencial'nye Uravnenija **4** (1968), 151–164 = Differential Equations **4** (1968), 79–85. MR **36** #6792.

7. S. L. Sobolev, *Applications of functional analysis in mathematical physics,* Izdat. Leningrad. Gos. Univ., Leningrad, 1950; English transl., Transl. Math. Monographs, vol. 7, Amer. Math. Soc., Providence, R. I., 1963. MR **14**, 565; **29** #2624.

8. L. N. Slobodeckiĭ, *Generalized Sobolev spaces and their application to boundary problems for partial differential equations,* Leningrad. Gos. Ped. Inst. Učen. Zap. **197** (1958), 54–112; English transl., Amer. Math. Soc. Transl. (2) **57** (1966), 207–276. MR **34** #3075.

9. S. M. Nikol'skiĭ, *A variational problem of Hilbert,* Izv. Akad. Nauk SSSR Ser. Mat. **22** (1958), 599–630. (Russian) MR **20** #7235.

10. S. V. Uspenskiĭ, *A variational problem of Hilbert,* Dokl. Akad. Nauk SSSR **127** (1959), 526–528. (Russian) MR **21** #6554.

11. V. D. Kopčenov, *A variational problem with a small parameter.* II, Differencial'nye Uravnenija **2** (1966), 967–987 = Differential Equations 2 (1966), 504–514. MR **34** #1668.

Translated by D. E. BROWN

Trudy Mat. Inst. Steklov.
131 (1974)

Proc. Steklov Inst. Math.
131 (1974)

SOME WEIGHTED IMBEDDING THEOREMS IN A REGION WITH NONSMOOTH BOUNDARY

UDC 517.518.22

A. F. KOČARLI

ABSTRACT. In this paper, weighted imbedding theorems are obtained for the Sobolev classes $W^1_{p,\alpha}(\Omega)$ defined on a region Ω whose boundary is not smooth. All considerations in the paper are carried out for unbounded regions (of a well-defined class), but the results obtained can be used as a basis to establish appropriate estimates for more general regions, in particular for bounded regions. The theorems include the case of so-called imbedding theorems with a limiting exponent.

Bibliography: 11 items.

The aim of this paper is to obtain weighted imbedding theorems for the Sobolev class $W^1_{p,\alpha}(\Omega)$ [1], defined on a region Ω with a nonsmooth boundary. All considerations are carried out for unbounded regions (of well-defined class), but the results obtained can be used as a basis to establish appropriate estimates for more general regions, in particular for bounded regions (of a specific class).

Note also that the theorems given below include the case of so-called imbedding theorems with a limiting exponent. For the proofs of the theorems, we use the integral representation of a function in terms of its derivatives obtained by V. P. Il'in [2](1) (see [3] also), results of Hardy and Littlewood concerning the boundedness of an integral operator of potential type from L_p into L_q, results of Zygmund and Calderón concerning the boundedness of singular operators from L_p into L_q and generalizations of these results, and, in addition, the inequalities of Minkowski, Hölder, Hardy, and others.

Basic definitions

Let Ω denote a region in n-dimensional euclidean space E^n of the form

$$\Omega = \{x : x' = (x_1, \ldots, x_{n-1}) \in E^{n-1}, \ \varphi(x') < x_n < \infty\}, \tag{1}$$

where the function $\varphi(x')$ describing the equation of the boundary satisfies a Hölder condition

$$|\varphi(x') - \varphi(y')|^{l_n} \leqslant M\left(\sum_{i=1}^{n-1} |x_i - y_i|^{l_i}\right) \tag{2}$$

(the l_i are nonnegative integers for $i = 1, \ldots, n$).

Note that if certain $l_j > l_n$, then it is easy to see [4] that $\varphi(x')$ does not depend on the x_j. In the case where the l_j are equal, the function $\varphi(x')$ describing the equation of the boundary $\partial\Omega$ of the region Ω will satisfy a Lipschitz condition of the first order, i.e.

AMS (MOS) subject classifications (1970). Primary 26A54, 26A84, 46E35.

(1) The method of integral representations of functions had its beginning in the work of Sobolev [1].

$$|\varphi(x') - \varphi(y')| \leqslant M\left(\sum_{i=1}^{n-1} |x_i - y_i|\right).$$

For example, in the case of unequal l_i ($l_i \leqslant l_n$ for $i = 1, \ldots, n-1$), regions with degenerate corners are permitted.

Let $L_{p,\alpha}(\Omega)$ denote the set of all measurable functions $f(x)$ defined on the region Ω with finite norm

$$\|f\|_{L_{p,\alpha}(\Omega)} = \left\{\int_\Omega |\rho^\alpha(x, \partial\Omega) f(x)|^p dx\right\}^{1/p}, \quad 1 \leqslant p < \infty, \tag{3}$$

$$\|f\|_{L_{\infty,\alpha}(\Omega)} = \operatorname*{ess\,sup}_{x\in\Omega} |\rho^\alpha(x, \partial\Omega) f(x)|, \tag{4}$$

where

$$\rho(x, \partial\Omega) = \inf_{y\in\partial\Omega}\left(\sum_{i=1}^{n} |x_i - y_i|^{l_i/l_n}\right).$$

We shall refer to $\rho(x, \partial\Omega)$ as a generalized distance. Note that when the l_i ($i = 1, \ldots, n$) are equal, the generalized distance is equivalent to the ordinary distance from the boundary of the region.

We shall say that a function $f(x)$ defined on the region Ω belongs to the weighted spaces $L^{\mathbf{l}}_{p,\alpha}(\Omega)$ or $W^{\mathbf{l}}_{p,\alpha}(\Omega)$ (for the vector $\mathbf{l} = (l_1, \ldots, l_n)$ with positive integral coordinates and $1 \leqslant p \leqslant \infty$) if $f(x)$ has generalized (Sobolev) derivatives

$$D_i^{l_i} f(x) = \frac{\partial^{l_i}}{\partial x_i^{l_i}} f(x) \qquad (i = 1, \ldots, n)$$

on Ω and if the corresponding seminorm

$$\|f\|_{L^{\mathbf{l}}_{p,\alpha}(\Omega)} = \sum_{i=1}^{n} \|D_i^{l_i} f\|_{L_{p,\alpha}(\Omega)} \tag{5}$$

or the norm

$$\|f\|_{W^{\mathbf{l}}_{p,\alpha}(\Omega)} = \|f\|_{L_{p,\alpha}(\Omega)} + \|f\|_{L^{\mathbf{l}}_{p,\alpha}(\Omega)} \tag{6}$$

is finite.

Let us consider certain geometric properties of a region.

As is well known, a region with a boundary satisfying a Hölder condition also satisfies a so-called "horn condition"; i.e., if x is an arbitrary point of the region Ω and y is an arbitrary point of the horn $R(l)$, where $R(l) = \{y\colon y_i > 0, 0 < a_i h < y_i^{l_i} < (a_i + \epsilon)h$ $(i = 1, \ldots, n), 0 < h < r < \infty\}$, then $x + y \in \Omega$. This property of a region is established in the following lemma.

LEMMA 1. *Let $R(l) = \{y\colon y_i > 0, 0 < a_i h < y_i^{l_i} < (a_i + \epsilon)h$ $(i = 1, \ldots, n), 0 < h < r < \infty, a_n > M(\Sigma_{i=1}^{n-1}(a_i + \epsilon))\}$. Then, for a region Ω satisfying* (1) *and* (2) *and for an arbitrary point $x \in \Omega$, $x + R(l) \subset \Omega$ (here the $+$ sign denotes the vector sum of the two sets).*

PROOF. Let $x \in \Omega$ and $y \in R(l)$. Then $x + y$ is an arbitrary point of the horn $R(l)$ with vertex at the point x, and, by the definition of the region Ω, $x_n - \varphi(x') > 0$. By (2) and the properties of a horn we obtain, by carrying out elementary transformations, the following chain of inequalities:

$$x_n + y_n > x_n + a_n^{\sigma_n} h^{\sigma_n} \geqslant x_n + M^{\sigma_n} \left(\sum_{i=1}^{n-1} |a_i + \varepsilon| \right)^{\sigma_n} h^{\sigma_n}$$
$$> x_n + M^{\sigma_n} \left(\sum_{i=1}^{n-1} |y_i|^{l_i} \right)^{\sigma_n} = x_n + M^{\sigma_n} \left(\sum_{i=1}^{n-1} |(x_i + y_i) - x_i|^{l_i} \right)^{\sigma_n}$$
$$> x_n + |\varphi(x' + y') - \varphi(x')| \geqslant x_n + \varphi(x' + y') - \varphi(x')$$
$$> \varphi(x') + \varphi(x' + y') - \varphi(x') = \varphi(x' + y')$$
$$\left(\sigma_i = \frac{1}{l_i}, \quad i = 1, \ldots, n \right).$$

Thus $x_n + y_n > \varphi(x' + y')$, which was to be shown.

Let $\pi(x) = x_n - \varphi(x')$. A region Ω satisfying (1) and (2) is characterized in this case by the condition $\pi(x) > 0$; moreover, for $x \in \Omega$, $\pi(x)$ is the distance along the vertical of the point $x = (x', x_n)$ from the boundary of the region.

LEMMA 2. *Let Ω satisfy* (1) *and* (2). *Then, for each point $x \in \Omega$, $\rho(x, \partial\Omega)$ is equivalent to $\pi(x)$.*

PROOF. Let x be an arbitrary point of Ω and let y be an arbitrary point of the boundary. Without loss of generality, we can assume that the origin is taken at the point $\varphi(x') = 0$ so that $\pi(x) = x_n$. Since

$$\rho(x, \partial\Omega) = \inf_{y \in \partial\Omega} \left(\sum_{i=1}^{n} |x_i - y_i|^{l_i/l_n} \right)$$

and since the point y, in particular, can assume the value $y = (0, \ldots, 0)$, it is clear that $\pi(x) \geqslant \rho(x, \partial\Omega)$.

Now let us establish the reverse inequality. It is clear that the worst case will occur when $\rho(x, \partial\Omega)$ takes its smallest value, i.e. when the boundary satisfies the equation $x_n = \varphi(x')$ in a neighborhood of the origin. Moreover,

$$|\varphi(x') - \varphi(y')|^{l_n} \leqslant M \left(\sum_{i=1}^{n-1} |x_i - y_i|^{l_i} \right),$$

where the y_i are positive. In this case we have

$$\rho(x, \partial\Omega) = \left(\sum_{i=1}^{n} |x_i - y_i|^{l_i/l_n} \right) = \left(\sum_{i=1}^{n-1} |-y_i|^{l_i/l_n} + |x_n - y_n| \right)$$
$$\geqslant C_1 \left(\sum_{i=1}^{n-1} |-y_i|^{l_i} \right)^{1/l_n} + |x_n - y_n| \geqslant C_2 |-\varphi(y')|$$
$$+ |x_n - \varphi(y')| = C_2 \varphi(y') + |x_n - \varphi(y')| \geqslant C_3 \pi(x),$$

if $x_n = \pi(x) \geqslant \varphi(y')$ and $C_3 = \min\{C_2, 1\}$.

Now let $x_n < \varphi(y')$. Then it follows that

$$\rho(x, \partial\Omega) \geqslant C_2 \varphi(y') + \varphi(y') - x_n \geqslant C_2 \pi(x).$$

Thus $\rho(x, \partial\Omega) \geqslant C\pi(x)$; hence, since $y \in \partial\Omega$ was arbitrary, $\rho(x, \partial\Omega) \geqslant C\pi(x)$, and the equivalence is established.

REMARK 1. It is clear from (3)–(6) that a power of the generalized distance $\rho(x, \partial\Omega)$

can be used as a weight. On the basis of Lemma 2, the norms can be replaced by an equivalent one if $\rho(x, \partial\Omega)$ is replaced in them by $\pi(x)$, the distance along the vertical to the boundary. To simplify calculations in what follows, an equivalent norm of the necessary weighted spaces will be used when $\pi(x)$ is taken instead of $\rho(x, \partial\Omega)$.

Let us recall the integral representation of a function in terms of its derivatives from the paper [3]:

$$f(x) = f_{r\sigma}(x) + \int_0^r \sum_{i=1}^n \int_{E^n} h^{-|\sigma|}\Phi_i\left(\frac{y}{h^\sigma}\right) D^{l_i} f(x+y)\,dy\,dh. \tag{7}$$

Here $\sigma = (\sigma_1, \ldots, \sigma_n)$, $\sigma_i = 1/l_i$ $(i = 1, \ldots, n)$, $|\sigma| = \Sigma_{i=1}^n \sigma_i$, $h > 0$, $h^\sigma = (h^{\sigma_1}, \ldots, h^{\sigma_n})$, $y/h^\sigma = (y_1/h^{\sigma_1}, \ldots, y_n/h^{\sigma_n})$, $f_{r\sigma}(x) = r^{-|\sigma|} \int_{E^n} \Phi_0(y/r^\sigma) f(x+y)\,dy$, where the last quantity is the average of the function $f(x)$ with respect to the horn $R(l)$ with a smooth kernel having compact support, the $\Phi_i(x) \in C_0^\infty(E^n)$ are concentrated in an arbitrary preassigned cube from the first orthant, and for any α

$$\int D^\alpha \Phi_i(x)\,dx = 0 \quad (i = 1, \ldots, n).$$

In view of the arbitrariness of the supports of the Φ_i, the actual integration in the right side of (7) is carried out only over the points $y \in R(l)$: $R(l) = \{y\colon y_i > 0,\ 0 < a_i h < y_i^{l_i} < (a_i + \epsilon)h\ \ (i = 1, \ldots, n),\ 0 < h < r < \infty\}$.

Note that for equal values of l_i $(l_i = l)$, the horn $R(l)$ becomes the cone $V(r) = \{y\colon y_i > 0,\ 0 < a_i h < y_i < (a_i + \epsilon)h\ (i = 1, \ldots, n),\ 0 < h < r < \infty\}$.

In this case the following representation will be used in place of (7):

$$f(x) = f_r(x) + \int_0^r \sum_{i=1}^n \int_{E^n} h^{-1-n+l}\,\Phi_i\left(\frac{y}{h}\right) D_i^l f(x+y)\,dy\,dh. \tag{7$'$}$$

Basic results

Everywhere in what follows, we shall assume that the region Ω satisfies (1) and (2) and that the parameters of the horn $R(l)$ are chosen so that for an arbitrary point $x \in \partial\Omega$ we have $\partial\Omega + R(l) \subset \Omega$, where the symbol $+$ denotes the arithmetic sum of the two sets, and moreover the horn $R(l)$ has the form

$$R(l) = \Big\{y : y_i > 0,\ 0 < a_i h < y_i^{l_i} < (a_i + \varepsilon)h\ \ (i = 1, \ldots, n),$$
$$0 < h < r < \infty,\ a_n > M\Big(\sum_{i=1}^{n-1}(a_i + \varepsilon)\Big)\Big\}.$$

In addition, by Lemma 2 and Remark 1, when $\pi(x)$ is used in place of $\rho(x, \partial\Omega)$, we shall consider the norms of the weighted spaces to be equivalent.

THEOREM 1. *Let $f(x) \in W_{p,\alpha}^1(\Omega)$, $1 < p < \infty$, $\kappa = 1 - \Sigma_1^n \nu_i/l_i - (\alpha - \beta)/l_n \geqslant 0$, where $\alpha - \beta \geqslant 0$, $\alpha \geqslant \beta - 1/p$, and $\nu = (\nu_1, \ldots, \nu_n)$ where the ν_i $(i = 1, \ldots, n)$ are integers. Then $D^\nu f(x) \in L_{p,\beta}(\Omega)$ and*

$$\| D^\nu f \|_{L_{p,\beta}(\Omega)} \leqslant C \| f \|_{W_{p,\alpha}^1(\Omega)},$$

where C is a constant not depending on $f(x)$.

PROOF. First consider the limiting case when κ in the hypothesis of the theorem is equal to zero. In the author's view, this case is the most interesting and the most difficult to prove.

Since $\alpha - \beta \geqslant 0$ by hypothesis, it is clear that $\alpha - \beta$ can be represented in the form $\alpha - \beta = k + \gamma$, where k is a nonnegative integer and $0 \leqslant \gamma < 1$. Let $\alpha - \beta = k + \gamma$, and fix this k.

From (7), integrating k times with respect to y_n, we obtain by the finiteness of the kernel Φ_i that[2]

$$D^{\nu} f_{\varepsilon}(x) = C \int_{E^n} \Phi_0^{(\bar{\nu})}\left(\frac{y}{h^{\sigma}}\right) \int_{y_n}^{\infty} \int_{t_{k-1}}^{\infty} \cdots \int_{t_1}^{\infty} f(x' + y', x_n + t_0)\, dt_0 \ldots dt_{k-1} dy$$
$$+ \sum_{i=1}^{n} \int_{\varepsilon}^{r} \int_{E^n} h^{-|\sigma|-(\sigma,\,\nu)-k\sigma_n} \Phi_i^{(\bar{\nu})}\left(\frac{y}{h^{\sigma}}\right) \int_{y_n}^{\infty} \int_{t_{k-1}}^{\infty} \int_{t_1}^{\infty} D_i^{l_i} f(x' + y', x_n + t_0)\, dt_0 \ldots dt_{k-1} dy dh, \tag{8}$$

where $\bar{\nu} = (\nu_1, \ldots, \nu_{n-1}, \nu_n + k)$.

It is clear that for $k = 0$, representation (8) reduces to (7).

Because of the assumption that κ is equal to zero, with a simple calculation of the powers of h in (8) we obtain

$$D^{\nu} f_{\varepsilon}(x) = C \int_{E^n} \Phi_0^{(\bar{\nu})}\left(\frac{y}{h^{\sigma}}\right) F_0(x + y)\, dy$$
$$+ \sum_{i=1}^{n} \int_{\varepsilon}^{r} \int_{E^n} h^{-1-|\sigma|-k\sigma_n + \frac{1}{l_n}(\alpha-\beta)} \Phi_i^{(\bar{\nu})}\left(\frac{y}{h^{\sigma}}\right) F_i(x + y)\, dy dh, \tag{9}$$

where

$$F_0(x) = \int_{x_n}^{\infty} \int_{t_{k-1}}^{\infty} \cdots \int_{t_1}^{\infty} f(x', t_0)\, dt_0 \ldots dt_{k-1},$$

$$F_i(x) = \int_{x_n}^{\infty} \int_{t_{k-1}}^{\infty} \cdots \int_{t_1}^{\infty} D_i^{l_i} f(x', t_0)\, dt_0 \ldots dt_{k-1} \qquad (i = 1, \ldots, n),$$

and for $k = 0$

$$F_0(x) = f(x), \quad F_i(x) = D_i^{l_i} f(x) \qquad (i = 1, \ldots, n).$$

We shall prove the theorem initially for $\gamma = 0$, i.e. for $\alpha - \beta = k$, where k is a nonnegative integer. In this case, from (9) we will have

$$\| D^{\nu} f_{\varepsilon} \|_{L_{p,\,\beta}(\Omega)} \leqslant C \left\| \int_{E^n} \Phi_0^{(\bar{\nu})}\left(\frac{y}{h^{\sigma}}\right) F_0(x + y)\, dy \right\|_{L_{p,\,\beta}(\Omega)}$$
$$+ \sum_{i=1}^{n} \left\| \int_{\varepsilon}^{r} \int_{E^n} h^{-1-|\sigma|} \Phi_i^{(\bar{\nu})}\left(\frac{y}{h^{\sigma}}\right) F_i(x + y)\, dy dh \right\|_{L_{p,\,\beta}(\Omega)}. \tag{10}$$

(2) As we shall see later on, an application of the Hardy inequality shows that all of these improper integrals have meaning.

In what follows, all constants will be denoted by C without any subscripts.

The following bounds are obtained from bounds of the integral operators in the right side of (10):

$$M_0 = \left\| \int_{E^n} \Phi_0^{(\bar{\nu})}\left(\frac{y}{h^\sigma}\right) F_0(x+y)\,dy \right\|_{L_{p,\beta}(\Omega)} \leqslant C \|f\|_{L_{p,\alpha}(\Omega)}, \tag{11}$$

$$M = \sum_{i=1}^{n} \left\| \int_{\varepsilon}^{r} \int_{E^n} h^{-1-|\sigma|} \Phi_i^{(\bar{\nu})}\left(\frac{y}{h^\sigma}\right) F_i(x+y)\,dy\,dh \right\|_{L_{p,\beta}(\Omega)} \leqslant C \|f\|_{L^1_{p,\alpha}(\Omega)}, \tag{12}$$

moreover, C does not depend on $f(x)$ or ϵ.

Let us establish (12). Performing elementary transformations, we obtain

$$M \leqslant \sum_{i=1}^{n} \left\| \int_{\varepsilon}^{r} \int_{E^n} h^{-1-|\sigma|} \Phi_i^{(\bar{\nu})}\left(\frac{y}{h^\sigma}\right) \pi^{\alpha-k}(x+y) F_i(x+y)\,dy\,dh \right\|_{L_p(\Omega)}$$

$$+ \sum_{i=1}^{n} \left\| \int_{\varepsilon}^{r} \int_{E^n} h^{-1-|\sigma|} \Phi_i^{(\bar{\nu})}\left(\frac{y}{h^\sigma}\right) \left[\frac{\pi^\beta(x) - \pi^{\alpha-k}(x+y)}{\pi^{\alpha-k}(x+y)}\right]\right.$$

$$\left. \times\, \pi^{\alpha-k}(x+y) F_i(x+y)\,dy\,dh \right\|_{L_p(\Omega)} = \sum_{i=1}^{n} (M_i^{(1)} + M_i^{(2)}).$$

Define the function $F_i'(x) = \pi^{\alpha-k}(x) F_i(x)$ for $x \in \Omega$, $F_i'(x) = 0$ for $x \in E^n \backslash \Omega$. Then

$$M_i^{(1)} \leqslant \left\| \int_{\varepsilon}^{r} \int_{E^n} h^{-1-|\sigma|} \Phi_i^{(\bar{\nu})}\left(\frac{y}{h^\sigma}\right) F_i'(x+y)\,dy\,dh \right\|_{L_p(E^n)} \leqslant C \| F_i' \|_{L_p(E^n)}, \tag{13}$$

where C does not depend on F_i', ϵ, or r $(0 < \epsilon < r < \infty)$. This bound is based on the following theorem, generalizing the well-known Calderón-Zygmund bound for singular integral operators.

THEOREM. *Let $\Phi(x) \in C_0^\infty(E^n)$ be a function concentrated on the interior of the first orthant and let $\int_{E^n} \Phi(x)\,dx = 0$. For $1 < p < \infty$, and for $u(x) \in L_p(E^n)$, set*

$$v_{\varepsilon r} = \int_{\varepsilon}^{r} \int_{E^n} h^{-1-|\sigma|} \Phi\left(\frac{y}{h^\sigma}\right) u(x+y)\,dy\,dh.$$

Then

1) $\| v_{\varepsilon r} \|_{L_p(E^n)} \leqslant C_p \| u \|_{L_p(E^n)}$;

2) $v_{\varepsilon r}(x) \to v_{0r}(x)$ *in* $L_p(E^n)$ *as* $\varepsilon \to 0$.

The proof of this theorem can be found in [5] and [6].

Let us establish a bound for $M_i^{(2)}$. Note that the following bound holds for the kernel:

$$\int_{\varepsilon}^{r} h^\tau \left| \Phi_i^{(\bar{\nu})}\left(\frac{y}{h^\sigma}\right) \right| dh \leqslant C\chi(y; R) \left(\sum_{i=1}^{n} |y_i|^{l_i} \right)^{\tau+1}, \tag{14}$$

where $\chi(y; R)$ is the characteristic function of the horn $R(l)$ and C does not depend on ϵ.

Bound (14) comes from the inequality

$$\int_{\varepsilon}^{r} h^{\tau}\left|\Phi_i^{(\bar{\nu})}\left(\frac{y}{h^{\sigma}}\right)\right| dh \leqslant C\chi(y;R)\int_{A}^{B} h^{\tau}\, dh = C\chi(y;R)\frac{1}{\tau+1} h^{\tau+1}\Big|_A^B$$

$$= C\chi(y;R)\frac{1}{\tau+1}\{(\min_i y_i^{l_i})^{\tau+1} - (\max_i y_i^{l_i})^{\tau+1}\}$$

$$\left(\text{here } A = \max_i \frac{y_i^{l_i}}{a_i+\varepsilon} \text{ and } B = \min_i \frac{y_i^{l_i}}{a_i}\right)$$

since, in view of the choice of the support of the kernel, the actual integration with respect to h is carried out for

$$\frac{y_i^{l_i}}{a_i+\varepsilon} \leqslant h \leqslant \frac{y_i^{l_i}}{a_i} \qquad (i = 1, \dots, n).$$

Hence

$$\int_{\varepsilon}^{r} h^{\tau}\left|\Phi_i^{(\bar{\nu})}\left(\frac{y}{h^{\sigma}}\right)\right| dh \leqslant C\chi(y;R)\frac{1}{\tau+1}\begin{cases}(\min_i y_i^{l_i})^{\tau+1}, & \text{if } \tau+1>0;\\ (\max_i y_i^{l_i})^{\tau+1}, & \text{if } \tau+1<0.\end{cases}$$

But

$$(\min_i y_i^{l_i})^{\tau+1} \leqslant C\left(\sum_{i=1}^{n} |y_i|^{l_i}\right)^{\tau+1},$$

$$(\max_i y_i^{l_i})^{\tau+1} \leqslant C\left(\sum_{i=1}^{n} |y_i|^{l_i}\right)^{\tau+1}.$$

Bound (14) is established.

Note that in the isotropic case ($l_1 = l$, $i = 1, \dots, n$) the bound (14) can be rewritten in the following way:

$$\int_{\varepsilon}^{r} h^{\tau}\left|\Phi_i^{(\bar{\nu})}\left(\frac{y}{h}\right)\right| dh \leqslant C\chi(y;V)\left(\sum_{i=1}^{n} |y_i|\right)^{\tau+1},$$

where $\chi(y;\ V)$ is the characteristic function of the cone $V(r)$.

Taking (14) into account, we find that

$$M_i^{(2)} \leqslant \left\| \int_{E^n} \chi(y;R)\left(\sum_{i=1}^{n} |y_i|^{l_i}\right)^{-|\sigma|}\left|1-\frac{\pi^{\beta}(x)}{\pi^{\alpha-k}(x+y)}\right| |F_i'(x+y)|\, dy\right\|_{L_p(E^n)}. \tag{15}$$

Since the integration with respect to y in the last expression is actually carried out over the horn $R(l)$, on replacing $x + y$ by y and extending the region of integration, we obtain

$$M_i^{(2)} \leqslant \left\| \int_{E^n} \chi(y-x;R)\left(\sum_{i=1}^{n} |y_i-x_i|^{l_i}\right)^{-|\sigma|}\left|1-\frac{\pi^{\beta}(x)}{\pi^{\alpha-k}(y)}\right| \cdot |F_i'(y)|\, dy\right\|_{L_p(E^n)}.$$

Let

$$I(x) = \int_{E^n} \chi(y - x; R) \left(\sum_{i=1}^{n} | y_i - x_i |^{l_i} \right)^{-|\sigma|} \left| 1 - \frac{\pi^\beta(x)}{\pi^{\alpha-k}(y)} \right| \cdot | F_i'(y) | \, dy.$$

By the Fubini theorem,

$$I(x) = \int_{E^{n-1}} dy' \int_{-\infty}^{\infty} \chi(y' - x', y_n - x_n; R) \left(\sum_{i=1}^{n-1} | y_i - x_i |^{l_i} + | y_n - x_n |^{l_n} \right)^{-|\sigma|}$$
$$\times \left| 1 - \frac{(x_n - \varphi(x'))^\beta}{(y_n - \varphi(y'))^{\alpha-k}} \right| \, | F_i'(y', y_n) | \, dy_n.$$

Performing the substitution $\eta_n = y_n - \varphi(y')$ in the inside integral, we obtain

$$I(x) = \int_{E^{n-1}} dy' \int_{-\infty}^{\infty} \chi(y' - x', \eta_n + \varphi(y') - x_n; R)$$
$$\times \left(\sum_{i=1}^{n-1} | y_i - x_i |^{l_i} + | \eta_n + \varphi(y') - x_n |^{l_n} \right)^{-|\sigma|} \left| 1 - \frac{(x_n - \varphi(x'))^\beta}{\eta_n^{\alpha-k}} \right|$$
$$\times | F'(y', \eta_n + \varphi(y')) | \, d\eta_n.$$

Redesignating y_i by η_i $(i = 1, \ldots, n-1)$, we get

$$I(x) = \int_{E^{n-1}} d\eta' \int_{-\infty}^{\infty} \chi(\eta' - x', \eta_n + \varphi(\eta') - x_n; R)$$
$$\times \left(\sum_{i=1}^{n-1} | \eta_i - x_i |^{l_i} + | \eta_n + \varphi(\eta') - x_n |^{l_n} \right)^{-|\sigma|}$$
$$\times \left| 1 - \frac{(x_n - \varphi(x'))^\beta}{\eta_n^{\alpha-k}} \right| \cdot | F_i'(\eta', \eta_n + \varphi(\eta')) | \, d\eta_n.$$

In like manner, we substitute $\xi_n = x_n - \varphi(x')$, then redesignate $\xi_i = x_i$ $(i = 1, \ldots, n-1)$ in order to calculate the L_p norm of I, and we find that

$$M_i^{(2)} \leqslant \Big\{ \int_{E^{n-1}} d\xi' \int_{-\infty}^{\infty} \Big| \int_{E^{n-1}} d\eta' \int_{-\infty}^{\infty} \chi(\eta' - \xi', \eta_n + \varphi(\eta') - \xi_n - \varphi(\xi'); R)$$
$$\times \left(\sum_{i=1}^{n-1} | \eta_i - \xi_i |^{l_i} + | \eta_n + \varphi(\eta') - \xi_n - \varphi(\xi') |^{l_n} \right)^{-|\sigma|}$$
$$\times \left| 1 - \frac{\xi_n^\beta}{\eta_n^{\alpha-k}} \right| \cdot | \bar{F}_i'(\eta', \eta_n) | \, d\eta_n \Big|^p d\xi_n \Big\}^{1/p},$$

where $\bar{F}_i'(\eta', \eta_n) = F_i'(\eta', \eta_n + \varphi(\eta'))$.

We shall prove several assertions.

By the Fubini theorem,

$$\Big\{ \int_{E^n} | f(\eta', \eta_n + \varphi(\eta')) |^p \, d\eta \Big\}^{1/p} = \Big\{ \int_{E^{n-1}} d\eta' \int_{-\infty}^{\infty} | f(\eta', \eta_n + \varphi(\eta')) |^p \, d\eta_n \Big\}^{1/p}$$
$$= \Big\{ \int_{E^{n-1}} d\eta' \int_{-\infty}^{\infty} | f(\eta', \eta_n) |^p \, d\eta_n \Big\}^{1/p} = \Big\{ \int_{E^n} | f(\eta) |^p \, d\eta \Big\}^{1/p}.$$

Later on, we shall show that

$$C_1\Big(\sum_{i=1}^{n}|\eta_i-\xi_i|^{l_i}\Big)\leqslant\Big(\sum_{i=1}^{n-1}|\eta_i-\xi_i|^{l_i}+|\eta_n+\varphi(\eta')-\xi_n-\varphi(\xi')|^{l_n}\Big) \tag{16}$$
$$\leqslant C_2\Big(\sum_{i=1}^{n}|\eta_i-\xi_i|^{l_i}\Big).$$

We have that

$$\sum_{i=1}^{n-1}|\eta_i-\xi_i|^{l_i}+|\eta_n+\varphi(\eta')-\xi_n-\varphi(\xi')|^{l_n}$$
$$\leqslant\sum_{i=1}^{n-1}|\eta_i-\xi_i|^{l_i}+2^{l_n}|\eta_n-\xi_n|^{l_n}+2^{l_n}|\varphi(\eta')-\varphi(\xi')|^{l_n}$$
$$\leqslant\sum_{i=1}^{n-1}|\eta_i-\xi_i|^{l_i}+2^{l_n}|\eta_n-\xi_n|^{l_n}+2^{l_n}M\Big(\sum_{i=1}^{n-1}|\eta_i-\xi_i|^{l_i}\Big)$$
$$\leqslant(2^{l_n}M+1)\sum_{i=1}^{n}|\eta_i-\xi_i|^{l_i}=C_2\Big(\sum_{i=1}^{n}|\eta_i-\xi_i|^{l_i}\Big).$$

We also obtain the opposite inequality:

$$\sum_{i=1}^{n}|\eta_i-\xi_i|^{l_i}=\sum_{i=1}^{n-1}|\eta_i-\xi_i|^{l_i}+|\eta_n-\xi_n+\varphi(\eta')-\varphi(\xi')-\varphi(\eta')+\varphi(\xi')|^{l_n}$$
$$\leqslant\sum_{i=1}^{n-1}|\eta_i-\xi_i|^{l_i}+2^{l_n}|\eta_n-\xi_n+\varphi(\eta')-\varphi(\xi')|^{l_n}+2^{l_n}M\Big(\sum_{i=1}^{n=1}|\eta_i-\xi_i|^{l_i}\Big)$$
$$\leqslant(2^{l_n}M+1)\Big\{\sum_{i=1}^{n-1}|\eta_i-\xi_i|^{l_i}+|\eta_n+\varphi(\eta')-\xi_n-\varphi(\xi')|^{l_n}\Big\}.$$

Thus the equivalence is established.

We shall prove that

$$\chi(\eta'-\xi',\ \eta_n+\varphi(\eta')-\xi_n-\varphi(\xi');\ R)\leqslant\chi(\eta'-\xi',\ \eta_n-\xi_n;\ E^n_+), \tag{17}$$

where $E^n_+=\{x=(x',x_n)\colon x'\in E^{n-1},\,x_n>0\}$, i.e. we shall prove that $\eta_n-\xi_n>0$ follows from the left side of the last inequality.

From the definition of $R(l)$, we have

$$0<a_i^{1/l_i}h^{1/l_i}<\eta_i-\xi_i<(a_i+\varepsilon)^{1/l_i}h^{1/l_i}\qquad(i=1,\dots,n-1), \tag{*}$$
$$0<a_n^{1/l_n}h^{1/l_n}<\eta_n+\varphi(\eta')-\xi_n-\varphi(\xi')<(a_n+\varepsilon)^{1/l_n}h^{1/l_n}. \tag{**}$$

From the left side of (**) we obtain

$$a_n^{1/l_n}h^{1/l_n}-\varphi(\eta')+\varphi(\xi')<\eta_n-\xi_n.$$

Consequently it is necessary to show that

$$a_n^{1/l_n}h^{1/l_n}-\varphi(\eta')+\varphi(\xi')>0.$$

By (2), (*), and the condition $a_n\geqslant(\Sigma_{i=1}^{n-1}|a_i+\epsilon|)$ we have

$$a_n^{1/l_n} h_n^{1/l_n} - \varphi(\eta') + \varphi(\xi') > a^{1/l_n} h^{1/l_n} - |\varphi(\eta') - \varphi(\xi')|$$

$$> a_n^{1/l_n} h^{1/l_n} - M^{1/l_n} h^{1/l_n} \Big(\sum_{i=1}^{n-1} |a_i + \varepsilon| \Big)^{1/l_n}$$

$$= h^{1/l_n} \Big\{ a_n^{1/l_n} - M^{1/l_n} \Big(\sum_{i=1}^{n-1} |a_i + \varepsilon| \Big)^{1/l_n} \Big\} > 0.$$

Note that in the isotropic case ($l_i = l$), the last calculation and argument can be simplified somewhat. For this we substitute $\eta_i = y_i$ $(i = 1, \dots, n-1)$ and $\eta_n = y_n - \varphi(y')$ directly. Since the boundary function $\varphi(x')$ is differentiable almost everywhere, the Jacobian of this mapping is equal to one almost everywhere.

On the basis of what has been said,

$$M_i^{(2)} \leqslant \Big\{ \int_{E^{n-1}} d\xi' \int_{-\infty}^{\infty} \Big| \int_{E^{n-1}} d\eta' \int_{-\infty}^{\infty} \chi(\eta - \xi; E_+^n)$$

$$\times \Big(\sum_{i=1}^{n} |\eta_i - \xi_i|^{l_i} \Big)^{-|\sigma|} \Big| 1 - \frac{\xi_n^\beta}{\eta_n^{\alpha - k}} \Big| \cdot |\bar{F}_i'(\eta', \eta_n)| \, d\eta_n \Big|^p d\xi_n \Big\}^{1/p}.$$

Changing the order of integration, we obtain

$$M_i^{(2)} \leqslant \Big\| \int_{E^n} \chi(\eta - \xi; E_+^n) \Big(\sum_{i=1}^{n} |\eta_i - \xi_i|^{l_i} \Big)^{-|\sigma|} \Big| 1 - \frac{\xi_n^\beta}{\eta_n^{\alpha-k}} \Big| \cdot |\bar{F}_i'(\eta', \eta_n)| \, d\eta \Big\|_{L_p(E^n)}.$$

In the right side of the last expression, we apply the one-dimensional bound of the convolution

$$\|f * g\|_p \leqslant \|g\|_{L_1} \|f\|_p, \qquad 1 \leqslant p \leqslant \infty,$$

first with respect to the variable ξ_1, then with respect to ξ_2, and so on up to ξ_{n-1}. We obtain

$$M_i^{(2)} \leqslant \Big\| \int_{\xi_n}^{\infty} |\eta_n - \xi_n|^{-1} \Big| 1 - \frac{\xi_n^\beta}{\eta_n^{\alpha-k}} \Big| \, \|\bar{F}_i'(\cdot, \eta_n)\|_{L_p(E^{n-1})} \, d\eta_n \Big\|_{L_p(E')}.$$

If we substitute $\eta_n = \lambda \xi_n$ and apply the generalized Minkowski inequality, we find that

$$M_i^{(2)} \leqslant C \|\bar{F}_i'\|_{L_p(E^n)} \int_1^{\infty} \frac{|1 - \lambda^{-(\alpha-k)}|}{|\lambda - 1| \lambda^{1/p}} \, d\lambda. \tag{18}$$

By the conditions of the theorem, the integral with respect to λ in (18) converges, and thus

$$M_i^{(2)} \leqslant C \|\bar{F}_i'\|_{L_p(E^n)} = C \|F_i'\|_{L_p(E^n)}.$$

Now consider

$$\|F_i'\|_{L_p(E^n)} = \Big\| \pi^{\alpha-k}(x) \int_{x_n}^{\infty} \int_{t_{k-1}}^{\infty} \dots \int_{t_1}^{\infty} D_i^{l_i} f(x', t_0) \, dt_0 \dots dt_{k-1} \Big\|_{L_p(E^n)}.$$

By the Fubini theorem,

$$\|F_i'\|_{L_p(E^n)} = \Big\{ \int_{E^{n-1}} dx' \int_{-\infty}^{\infty} \Big| (x_n - \varphi(x'))^{\alpha-k} \int_{x_n}^{\infty} \int_{t_{k-1}}^{\infty} \dots \int_{t_1}^{\infty} D_i^{l_i} f(x', t_0) \, dt_0 \dots dt_{k-1} \Big|^p dx_n \Big\}^{1/p}.$$

If we substitute $\xi_n = x_n - \varphi(x')$ in the last equation, redesignate x_i by ξ_i $(i = 1, \ldots, n-1)$, interchange the order of integration, and finally substitute $t_{k-j} = u_{k-j} + \varphi(\xi')$ $(j = 1, \ldots, k)$, we obtain

$$\| F_i' \|_{L_p(E^n)} = \left\| \xi_n^{\alpha-k} \int_{\xi_n}^{\infty} \int_{u_{k-1}}^{\infty} \ldots \int_{u_1}^{\infty} D_i^{l_i} f(\xi', u_0 + \varphi(\xi'))\, du_0\, du_1 \ldots du_{k-1} \right\|_{L_p(E^n)}.$$

From the conditions of the theorem we can apply the Hardy inequality

$$\| x^{-\gamma} F(x) \|_p \leqslant C \| x^{-\gamma+1} f(x) \|_p, \qquad 1 \leqslant p \leqslant \infty;$$

$$F(x) = \int_x^{\infty} f(\xi)\, d\xi, \qquad \gamma < 1/p; \qquad F(x) = \int_0^x f(\xi)\, d\xi, \qquad \gamma > 1/p,$$

k times to find that

$$\| F_i' \|_{L_p(E^n)} \leqslant C \, \| \xi_n^{\alpha} D_i^{l_i} f(\xi', \xi_n + \varphi(\xi')) \|_{L_p(E^n)} = \| \pi^{\alpha}(x) D_i^{l_i} f(x) \|_{L_p(\Omega)}. \tag{19}$$

Thus, for $\alpha - \beta = k$ $(k = 0, 1, 2, \ldots)$, on collecting bounds we obtain

$$M \leqslant C \sum_{i=1}^{n} \| \pi^{\alpha} D_i^{l_i} f \|_{L_p(\Omega)} = C \| f \|_{L^1_{p,\alpha}(\Omega)}.$$

Now let $\alpha - \beta = k + \gamma$ $(k = 0, 1, 2, \ldots; 0 < \gamma < 1)$. In this case we have from (8) that

$$\| D^{\nu} f_{\varepsilon} \|_{L_{p,\beta}(\Omega)} \leqslant \left\| \int_{E^n} \Phi_0^{(\bar{\nu})} \left(\frac{y}{h^{\sigma}} \right) F_0(x+y)\, dy \right\|_{L_{p,\beta}(\Omega)}$$
$$+ \sum_{i=1}^{n} \left\| \int_{\varepsilon}^{r} \int_{E^n} h^{-1-|\sigma|+\gamma/l_n} \Phi_i^{(\bar{\nu})} \left(\frac{y}{h^{\sigma}} \right) F_i(x+y)\, dy\, dh \right\|_{L_{p,\beta}(\Omega)}.$$

Let

$$N = \sum_{i=1}^{n} \left\| \int_{\varepsilon}^{r} \int_{E^n} h^{-1-|\sigma|-\gamma/n} \Phi_i^{(\bar{\nu})} \left(\frac{y}{h^{\sigma}} \right) F_i(x+y)\, dy\, dh \right\|_{L_{p,\beta}(\Omega)}.$$

Then

$$N \leqslant \sum_{i=1}^{n} \left\| \int_{\varepsilon}^{r} \int_{E^n} h^{-1-|\sigma|-\gamma/l_n} \Phi_i^{(\bar{\nu})} \left(\frac{y}{h^{\sigma}} \right) \frac{\pi^{\beta}(x)}{\pi^{\alpha-k}(x+y)} F_i'(x+y)\, dy\, dh \right\|_{L_p(E^n)}.$$

A literal repetition of the proof of the bound for $M_i^{(2)}$ leads us to the following analog of inequality (18):

$$N \leqslant C \sum_{i=1}^{n} \| F_i' \|_{L_p(E^n)} \int_1^{\infty} \frac{|\lambda - 1|^{-1+\alpha-\beta-k}}{\lambda^{\alpha-k} \lambda^{1/p}}\, d\lambda.$$

From this, (19), and the conditions of the theorem we obtain

$$N \leqslant C \sum_{i=1}^{n} \| F_i' \|_{L_p(E^n)} \leqslant C \| f \|_{L^1_{p,\alpha}(\Omega)}. \tag{20}$$

Similar calculations and arguments lead to the bound

$$\left\| \int_{E^n} \Phi_0^{(\bar{\nu})} \left(\frac{y}{h^{\sigma}} \right) F_0(x+y)\, dy \right\|_{L_{p,\beta}(\Omega)} \leqslant C \| f \|_{L_{p,\alpha}(\Omega)}, \tag{21}$$

if we consider that

$$|\Phi_0(y)| \leqslant C\chi(y;\ R)\Big(\sum_{i=1}^{n} |y_i|^{l_i}\Big)^{-|\sigma|},$$

where $\chi(y; R)$ is the characteristic function of the horn $R(l)$ and the last bound follows from the form of the kernel in the average.

Thus, collecting bounds, we obtain

$$\| D^{\nu} f_{\varepsilon} \|_{L_{p,\beta}(\Omega)} \leqslant C\{ \| f \|_{L_{p,\alpha}(\Omega)} + \| f \|_{L^1_{p,\alpha}(\Omega)} \}.$$

To complete the proof of the theorem, it is necessary to establish two facts: first, we must show that $D^{\nu} f_{\epsilon}$ converges to some element of $L_{p,\beta}(\Omega)$ as $\epsilon \longrightarrow 0$, and second, we must prove that this limiting element is the generalized derivative $D^{\nu} f$ of the function $f(x)$ to which $f_{\epsilon}(x)$ converges as $\epsilon \longrightarrow 0$.

To establish the convergence of $D^{\nu} f_{\epsilon}$ to some element of $L_{p,\beta}(\Omega)$ as $\epsilon \longrightarrow 0$, we shall show that the sequence $\{D^{\nu} f_{\epsilon}\}$ is fundamental in the $L_{p,\beta}(\Omega)$ norm. For $0 < \epsilon < \eta$, $k = 0$, $1, 2, \ldots$, and $\gamma = 0$, we have that

$$\| D^{\nu} f_{\varepsilon} - D^{\nu} f_{\eta} \|_{L_{p,\beta}(\Omega)} \leqslant \sum_{i=1}^{n} (M_i^{(1)} + M_i^{(2)}).$$

Each term of the first sum on the right is bounded by $C \| \pi^{\alpha} D_i^{l_i} f \|_{L_p(\Omega)} = C'$ and tends to zero as $\epsilon \longrightarrow 0$ and $\eta \longrightarrow 0$ on the basis of an appropriate theorem about singular integrals in L_p. Each term of the second sum on the right is also bounded by C'. By Lebesgue's dominated convergence theorem (as $\eta \longrightarrow 0$), we conclude that each term of this second sum tends to zero. Consequently, in the case $\gamma = 0$, the sequence $\{D^{\nu} f_{\epsilon}\}$ is fundamental as $\epsilon \longrightarrow 0$.

For $\alpha - \beta = k + \gamma$ ($k = 0, 1, 2, \ldots$; $0 < \gamma < 1$), we have for $0 < \epsilon < \eta$

$$\| D^{\nu} f_{\varepsilon} - D^{\nu} f_{\eta} \|_{L_{p,\beta}(\Omega)} \leqslant \sum_{i=1}^{n} \Big\| \int_{\varepsilon}^{r} \int_{E^n} h^{-|\sigma|-1+\gamma/l_n} \Phi_i^{(\bar{\nu})}\Big(\frac{y}{h^{\sigma}}\Big) \frac{\pi^{\beta}(x)}{\pi^{\alpha-k}(x+y)} F_i'(x+y)\, dy\, dh \Big\|_{L_p(E^n)}.$$

As was shown above, each term on the right side is bounded by C'. Thus, using Lebesgue's dominated convergence theorem again, we conclude that

$$\| D^{\nu} f_{\varepsilon} - D^{\nu} f_{\eta} \|_{L_{p,\beta}(\Omega)} \to 0 \quad \text{as} \quad \eta \to 0$$

For the case $\alpha - \beta = k + \gamma$, $k = 0, 1, \ldots$, $0 < \gamma < 1$, the sequence $\{D^{\nu} f_{\epsilon}\}$ is shown to be fundamental as $\epsilon \longrightarrow 0$ in the same way. From this and the completeness of the space $L_{p,\beta}(\Omega)$, it follows that $D^{\nu} f_{\epsilon}$ converges to some element F in $L_{p,\beta}(\Omega)$ as $\epsilon \longrightarrow 0$. By the definition of the generalized Sobolev derivative, for each fixed ϵ and for an arbitrary function $\psi \in C_0^{\infty}$ with compact support in Ω, the following equation holds:

$$\int_{\Omega} f_{\varepsilon} \frac{\partial^{\nu} \psi}{\partial x_1^{\alpha_1} \ldots \partial x_n^{\alpha_n}}\, dx = (-1)^{|\nu|} \int_{\Omega} \psi D^{\nu} f_{\varepsilon}\, dx.$$

Note that $f(x)$ is a locally integrable function and hence that $f_{\epsilon}(x) \longrightarrow f(x)$ in $L^{\mathrm{loc}}(\Omega)$, so that by passing to the limit as $\epsilon \longrightarrow 0$ we obtain

$$\int_{\Omega} f \frac{\partial^{\nu}\psi}{\partial x_1^{\alpha_1} \ldots \partial x_n^{\alpha_n}}\, dx = (-1)^{|\nu|} \int_{\Omega} \psi F\, dx,$$

from which it follows that the limit F of the sequence $\{D^{\nu} f_{\epsilon}\}$ is the generalized derivative $D^{\nu} f$ of the function $f(x)$.

Now suppose $\kappa > 0$ ($\kappa - \epsilon_1 = 0$, $\epsilon_1 > 0$). In this case the proof is carried out in the same way as in the previous case. Let us consider certain parts of the proof for the case $\kappa > 0$.

For $\alpha - \beta = k$ ($\gamma = 0$), to bound $M_i^{(1)}$ in the limiting case we simply use the Minkowski inequality and the bound for the convolution in place of the Zygmund-Calderón theorem about singular integrals. Later we will have an inequality analogous to (15) in the nonlimiting case:

$$M_i^{(2)} \leqslant \left\| \int_{E^n} \chi(y; R) \left(\sum_{i=1}^{n} |y_i|^{l_i} \right)^{-|\sigma|+\varepsilon_1} \left| 1 - \frac{\pi^{\beta}(x)}{\pi^{\alpha-k}(x+y)} \right| |F_i'(x+y)|\, dy \right\|_{L_p(E^n)} . \tag{22}$$

Since the integration with respect to y in (22) is actually carried out on the horn $R(l)$, it is clear that we only increase the right side of (20) if we write it in the form of (15). The subsequent bound is similar. We handle the case when $\alpha - \beta = k + \gamma$ ($0 < \gamma < 1$) in the same way. The theorem is proved.

Note that results analogous to Theorem 1 were obtained by S. V. Uspenskiĭ [7], and then complemented by T. A. Timan [8] (for $\beta > -1/p$) for a region with a sufficiently smooth boundary $\partial\Omega \subset C^{(l)}$. These results are contained in Theorem 1.

In the theorem just proved, as in the others that follow, we shall consider the case $\alpha \geqslant \beta$, which in our view is the most interesting, since the limiting case ($\kappa = 0$) follows from this. Indeed, by unweighted imbedding theorems we have

$$1 - \sum_{i=1}^{n} \frac{\nu_i}{l_i} \geqslant 0,$$

and therefore the case when κ is equal to zero can be obtained by considering the case when $\alpha - \beta \geqslant 0$.

THEOREM 2. *Let* $f \in W^1_{p,\alpha}(\Omega)$, $1 < p < \infty$, $\kappa = 1 - \Sigma_1^n \nu_i/l_i - (\alpha - \beta)/l_n > 0$, $\alpha \geqslant \beta - 1/p$, $r_i = l_i\kappa$, *and* $\nu = (\nu_1, \ldots, \nu_n)$, *where the* ν_i ($i = 1, \ldots, n$) *are integers. Then* $D^{\nu} f(x) \equiv f^{(\nu)}(x) \in L^{\mathbf{r}}_{p,\beta}(\Omega)$, *and the following bound holds*:

$$\| D^{\nu} f \|_{L^{\mathbf{r}}_{p,\beta(\Omega)}} \leqslant C \| f \|_{W^{1}_{p,\alpha(\Omega)}};$$

moreover, C does not depend on $f(x)$.

PROOF. From (8), apply the differential operators $D_j^{r_j}$ ($j = 1, \ldots, n$), take the $L_{p,\beta}$ norm, and compute the power of h to obtain

$$\| \pi^{\beta}(x) D_j^{r_j} f_{\varepsilon}^{(\nu)}(x) \|_{L_p(\Omega)} \leqslant C(r) \left\| \int_{E^n} \Phi_0^{(\widetilde{\nu})} \left(\frac{y}{r^{\sigma}} \right) \pi^{\beta}(x) F_0(x+y)\, dy \right\|_{L_p(\Omega)}$$

$$+ \sum_{i=1}^{n} \left\| \int_{\varepsilon}^{r} \int_{E^n} h^{-1-|\sigma|+\frac{1}{l_n}(\alpha-\beta)} \Phi_i^{(\widetilde{\nu})} \left(\frac{y}{h^{\sigma}} \right) \pi^{\beta}(x) F_i(x+y)\, dy dh \right\|_{L_p(\Omega)} = I + I_1,$$

where $\widetilde{\nu} = \bar{\nu} + r_j e_j$ and $\bar{\nu}$ and $F_i(x)$ $(i = 1, \ldots, n)$ have the same meaning as in the proof of Theorem 1. Note that here, as in the proof of Theorem 1, $\alpha - \beta$ is represented in the form $\alpha - \beta = k + \gamma$ $(k = 0, 1, 2, \ldots\ ;\ 0 \leqslant \gamma < 1)$, and k is fixed. We shall divide the rest of the proof into two cases: $\gamma = 0$ and $0 < \gamma < 1$, as was done in Theorem 1.

Let $\alpha - \beta = k$. Then

$$I_1 \leqslant \sum_{i=1}^{n} \left\| \int_{\varepsilon}^{r} \int_{E^n} h^{-1-|\sigma|} \Phi_i^{(\widetilde{\nu})}\left(\frac{y}{h^{\sigma}}\right) \pi^{\beta}(x)\, F_i(x+y)\, dy dh \right\|_{L_p(\Omega)}.$$

The right side of the last expression is bounded by $C\|f\|_{L^1_{p,\alpha}(\Omega)}$ because of (12). On the other hand, by (11)

$$I \leqslant C \| f \|_{L_{p,\alpha}(\Omega)}.$$

Thus for $\gamma = 0$, by collecting bounds we obtain

$$\| \pi^{\beta}(x)\, D_j^{r_j} f_{\varepsilon}^{(\nu)}(x) \|_{L_p(\Omega)} \leqslant C \| f \|_{W^1_{p,\alpha}(\Omega)},$$

from which we have

$$\| f_{\varepsilon}^{(\nu)} \|_{L^{\mathbf{r}}_{p,\beta}(\Omega)} = \sum_{j=1}^{n} \| \pi^{\beta}(x)\, D_j^{r_j} f_{\varepsilon}^{(\nu)}(x) \|_{L_p(\Omega)} \leqslant C \| f \|_{W^1_{p,\alpha}(\Omega)}. \tag{23}$$

For $0 < \gamma < 1$, (23) is established by using (20) and (21).

Furthermore, by an argument which is analogous to the corresponding argument from the proof of Theorem 1, we find that

$$\| D^{\nu} f \|_{L^{\mathbf{r}}_{p,\beta}(\Omega)} \leqslant C \| f \|_{W^1_{p,\alpha}(\Omega)},$$

which was to be shown.

The next result follows from Theorems 1 and 2.

COROLLARY 1. *Suppose that the conditions of Theorem 2 are fulfilled. Then* $D^{\nu} f(x) \in W^{\mathbf{r}}_{p,\beta}(\Omega)$ *and*

$$\| D^{\nu} f \|_{W^{\mathbf{r}}_{p,\beta}(\Omega)} \leqslant C \| f \|_{W^1_{p,\alpha}(\Omega)},$$

where C does not depend on $f(x)$.

Corollary 1 contains the corresponding result of Uspenskiĭ ([9], Theorem 1) proved for $\Omega = E^n_+$ and then transferred to the case of a region with a sufficiently smooth boundary.

Theorem 2 generalizes the results of A. D. Džabrailov ([10], Theorem 6) established for $\alpha \geqslant \beta \geqslant 0$ in the nonlimiting case, also for $\Omega = E^n_+$.

THEOREM 3. *Let* $f(x) \in W^1_{p,\alpha}(\Omega)$, $1 < p < q < \infty$, *and*

$$\varkappa_1 = 1 - \left(\frac{1}{p} - \frac{1}{q}\right) \sum_{i=1}^{n} \frac{1}{l_i} - \sum_{i=1}^{n} \frac{\nu_i}{l_i} - \frac{1}{l_n}(\alpha - \beta) \geqslant 0,$$

$$a \geqslant \beta - 1/q, \quad \alpha \neq k - 1/p \quad (k = 0, 1, 2, \ldots).$$

Then $D^{\nu} f(x) \in L_{q,\beta}(\Omega)$, *and*

$$\| D^{\nu} f \|_{L_{q,\beta}(\Omega)} \leqslant C \| f \|_{W^1_{p,\alpha}(\Omega)},$$

where the constant C does not depend on $f(x)$.

The proof of this theorem is quite similar to the proof of Theorem 1. Thus we shall not give the development and computations in detail, but only refer to the corresponding places in the proof of Theorem 1. In addition, note that some of the notation of Theorem 1 remains in effect in this case.

We shall prove the theorem initially for the case $\kappa_1 = 0$. The conditions of the theorem imply that $\alpha - \beta$ can be represented in the form $0 \leqslant \alpha - \beta < k + 1/s$, where $1/s = 1 - 1/p + 1/q$ $(k = 0, 1, \dots)$. From (8) we have (for fixed k)

$$\| D^{\nu} f_{\varepsilon} \|_{L_{q,\beta}(\Omega)} \leqslant \| D^{\nu} f_{r^{\sigma}} \|_{L_{q,\beta}(\Omega)} + \sum_{i=1}^{n} \Big\| \int_{\varepsilon}^{r} \int_{E^n} h^{-|\sigma|-(\sigma,\nu)-k\sigma_n} \Phi_i^{(\bar{\nu})} \Big(\frac{y}{h^{\sigma}}\Big) F_i(x+y)\, dy\, dh \Big\|_{L_{q,\beta}(\Omega)}.$$

Set

$$N = \sum_{i=1}^{n} \Big\| \int_{\varepsilon}^{r} \int_{E^n} h^{-|\sigma|-(\sigma,\nu)-k\sigma_n} \Phi_i^{(\bar{\nu})} \Big(\frac{y}{h^{\sigma}}\Big) F_i(x+y)\, dy dh \Big\|_{L_{q,\beta}(\Omega)}.$$

Then

$$N \leqslant \sum_{i=1}^{n} \Big\| \int_{\varepsilon}^{r} \int_{E^n} h^{-|\sigma|-(\sigma,\nu)-k\sigma_n} \Phi_i^{(\bar{\nu})} \Big(\frac{y}{h^{\sigma}}\Big) \frac{\pi^{\beta}(x)}{\pi^{\alpha-k}(x+y)} F_i'(x+y)\, dy dh \Big\|_{L_q(E^n)}. \tag{24}$$

Note that $\bar{\nu}$, F_i and F_i' have the same meanings as in Theorem 1.

Computing the powers of h (since $\kappa_1 = 0$) in (24), we obtain

$$N \leqslant \sum_{i=1}^{n} \Big\| \int_{\varepsilon}^{r} \int_{E^n} h^{\tau-1} \Phi_i^{(\bar{\nu})} \Big(\frac{y}{h^{\sigma}}\Big) \frac{\pi^{\beta}(x)}{\pi^{\alpha-k}(x+y)} F_i'(x+y)\, dy dh \Big\|_{L_q(E^n)}, \tag{25}$$

where

$$\tau = -|\sigma|(1 - 1/p + 1/q) - k\sigma_n + 1/l_n(\alpha - \beta).$$

Using (14), (16), (17) and the Fubini theorem, and interchanging y and x as was done in the proof of Theorem 1, we obtain from (24) that

$$N \leqslant \sum_{i=1}^{n} \Big\| \int_{E^n} \chi(\eta; E^n_+) \Big(\sum_{i=1}^{n} |\eta_i|^{l_i}\Big)^{\tau} \frac{\xi_n^{\beta}}{(\eta_n + \xi_n)^{\alpha-k}} |\bar{F}_i'(\eta + \xi)|\, d\eta \Big\|_{L_q(E^n)}, \tag{26}$$

where $\bar{F}_i'(\eta) = F_i'(\eta', \eta_n + \varphi(\eta'))$.

We shall use the following well-known one-dimensional inequality:

$$\| f * g \|_q \leqslant \| g \|_s \, \| f \|_p, \quad 1/p - 1/q = 1 - 1/s, \quad 1 \leqslant p \leqslant q \leqslant \infty,$$

which we apply to (26) first with respect to ξ_1, then with respect to ξ_2, and so forth on up to ξ_{n-1}. We obtain

$$N \leqslant \sum_{i=1}^{n} \left\| \int_0^\infty |\eta_n - \xi_n|^{\tau_1} \frac{\xi_n^\beta}{\eta_n^{\alpha-k}} \| \bar{F}_i'(\cdot, \eta_n) \|_{L_p(E^{n-1})} d\eta_n \right\|_{L_q(E')}, \tag{27}$$

where $\tau_1 = -(1 - 1/p + 1/q) + (\alpha - k) - \beta$.

Indeed, in our case

$$g(\eta_1, \ldots, \eta_n) = \left(\sum_{i=1}^{n} |\eta_i|^{l_i}\right)^\tau,$$

from which an estimate of the L_s norm with respect to η_1 of the kernel

$$g(\eta_1, \ldots, \eta_n) = \left(|\eta_1|^{l_1} + \sum_{i=2}^{n} |\eta_i|^{l_i}\right)^\tau = (|\eta_1|^{l_1} + a)^\tau$$

is obtained in the form

$$\left\{\int_0^\infty (|\eta_1|^{l_1} + a)^{\tau s} d\eta_1\right\}^{1/s} \leqslant \left\{\int_0^\infty (|\eta_1| + a^{1/l_1})^{\tau s l_1} d\eta_1\right\}^{1/s} \leqslant \left(\frac{1}{\tau s l_1 + 1}\right)^{1/s} a^{\tau + 1/s l_1},$$

since

$$\tau s l_1 + 1 = -\frac{l_1}{l_2} - \ldots \frac{l_1}{l_{n-1}} - \frac{l_1}{l_n}(1 + ks - s(\alpha - \beta)) < 0,$$

because $1 + ks - s(\alpha - \beta) > 0$, which follows from the condition $\alpha - \beta < k + 1/s$ of the theorem.

In an analogous manner, estimates for the L_s-norms with respect to $\eta_2, \ldots, \eta_{n-1}$ of the kernels are obtained. We rewrite (27) in the more convenient form

$$N \leqslant \sum_{i=1}^{n} \left\| \int_0^\infty \frac{\| \bar{F}_i'(\cdot, \eta_n) \|_{L_p(E^{n-1})} d\eta_n}{|\eta_n - \xi_n|^{1-1/p+1/q-(\alpha-k)+\beta} \xi_n^{-\beta} \eta_n^{\alpha-k}} \right\|_{L_q(E')}$$

From this and the weighted inequality

$$1 < p < q < \infty, \quad h < 1 - 1/p, \quad m < 1/q, \; h + m \geqslant 0,$$

$$\left\| \int_0^\infty \frac{f(x)\, dx}{|x - y|^{1-1/p+1/q-h-m} x^h y^m} \right\|_{L_q} \leqslant K \| f \|_{L_p}$$

of Hardy and Littlewood [11], and the conditions of the theorem, we obtain that

$$N \leqslant C \sum_{i=1}^{n} \| F_i' \|_{L_p(E^n)}.$$

By virtue of the Hardy inequality and (19), we obtain from the last bound that

$$N \leqslant C \sum_{i=1}^{n} \| \pi^\alpha D_i^{l_i} f \|_{L_p(E^n)} = C \| f \|_{L^1_{p,\alpha}(\Omega)}. \tag{28}$$

Note that for $k = 0$, by applying the Hardy-Littlewood inequality, we obtain directly that

$$N \leqslant C \sum_{i=1}^{n} \| D_i^{l_i} f \|_{L_{p,\alpha}(\Omega)} = C \| f \|_{L^1_{p,\alpha}(\Omega)}.$$

The bound

$$\| D^{\nu} f_{r\sigma} \|_{L_{q,\beta}(\Omega)} \leqslant C \| f \|_{L_{p,\alpha}(\Omega)} \tag{29}$$

is established similarly with the help of (20).

Thus we now have the bound

$$\| D^{\nu} f_{\varepsilon} \|_{L_{q,\beta}(\Omega)} \leqslant C \| f \|_{W^{1}_{p,\alpha}(\Omega)},$$

where C does not depend on $f(x)$ or ϵ.

From the last inequality and the fact that the generalized differential operator is closed, this inequality follows as $\epsilon \longrightarrow 0$:

$$\| D^{\nu} f \|_{L_{q,\beta}(\Omega)} \leqslant C \| f \|_{W^{1}_{p,\alpha}(\Omega)}.$$

The proof is analogous to the proof of the corresponding result in Theorem 1. In the case where $\kappa_1 = \epsilon > 0$, the differences from the above proof will be the same as in Theorem 1 for the cases $\kappa = 0$ and $\kappa > 0$. The theorem is proved.

THEOREM 4. *Let $f(x) \in W^{1}_{p,\alpha}(\Omega)$, $1 < p < q < \infty$, $r_i = l_i \kappa_1$, $\nu = (\nu_1, \ldots, \nu_n)$, where the ν_i $(i = 1, \ldots, n)$ are integers, and*

$$\varkappa_1 = 1 - \left(\frac{1}{p} - \frac{1}{q}\right) \sum_{i=1}^{n} \frac{1}{l_i} - \sum_{i=1}^{n} \frac{\nu_i}{l_i} - \frac{1}{l_n}(\alpha - \beta) > 0,$$

$$\alpha \geqslant \beta > -\frac{1}{q}, \quad \alpha \neq k - \frac{1}{p} \quad (k = 0, 1, 2, \ldots).$$

Then $D^{\nu} f(x) \equiv f^{(\nu)}(x) \in L^{\mathbf{r}}_{q,\beta}(\Omega)$, and

$$\| D^{\nu} f \|_{L^{\mathbf{r}}_{q,\beta}(\Omega)} \leqslant C \| f \|_{W^{1}_{p,\alpha}(\Omega)},$$

where the constant C does not depend on $f(x)$.

PROOF. As was done in Theorem 3, fix k and represent $0 \leqslant \alpha - \beta < k + 1/s$. We differentiate (8), then evaluate the $L_{q,\beta}$ norms in the notation introduced in Theorem 3 to obtain that

$$\begin{aligned} \| \pi^{\beta}(x) D_j^{r_j} f_{\varepsilon}^{(\nu)}(x) \|_{L_q(\Omega)} &\leqslant \| \pi^{\beta}(x) D_j^{r_j} f_{r\sigma}^{(\nu)}(x) \|_{L_q(\Omega)} \\ &+ \sum_{i=1}^{n} \Bigg\| \int_{\varepsilon}^{r} \int_{E^n} h^{-|\sigma| - (\sigma,\nu) - k\sigma_n - r_j\sigma_j} \Phi_i^{(\widetilde{\nu})}\left(\frac{y}{h^{\sigma}}\right) \\ &\times \frac{\pi^{\beta}(x)}{\pi^{\alpha-k}(x+y)} F_i'(x+y)\, dy dh \Bigg\|_{L_q(E^n)} = I + I_1. \end{aligned}$$

Here $\widetilde{\nu} = \bar{\nu} + r_j e_j$.

Consider I_1. By computing the powers of h, we obtain

$$I_1 \leqslant \sum_{i=1}^{n} \Bigg\| \int_{\varepsilon}^{r} \int_{E^n} h^{\mu_1} \Phi_i^{(\widetilde{\nu})}\left(\frac{y}{h^{\sigma}}\right) \frac{\pi^{\beta}(x)}{\pi^{\alpha-k}(x+y)} F_i'(x+y)\, dy dh \Bigg\|_{L_q(E^n)}, \tag{30}$$

where

$$\mu_1 = -|\sigma|(1 - 1/p + 1/q) - k\sigma_n + 1/l_n(\alpha - \beta).$$

Since the right side of (30) is similar to the right side of (25), we have by (28) that

$$I_1 \leqslant C\|f\|_{L^1_{p,\alpha}(\Omega)}.$$

Moreover, with the help of (29) it is easy to obtain the bound $I \leqslant C\|f\|_{L_{p,\alpha}(\Omega)}$. Collecting bounds, we obtain that $\|\pi^\beta D_j^{r_j} f_\epsilon^{(\nu)}\|_{L_q(\Omega)} \leqslant C\|f\|_{W^1_{p,\alpha}(\Omega)}$. Therefore

$$\|D^\nu f_\varepsilon\|_{L^{\mathbf{r}}_{q,\beta}(\Omega)} = \sum_{j=1}^{n} \|\pi^\beta D_j^{r_j} f_\varepsilon^{(\nu)}\|_{L_q(\Omega)} \leqslant C\|f\|_{W^1_{p,\alpha}(\Omega)}.$$

In order to complete the proof of the theorem, we use the fact that the generalized differential operator is closed (as $\epsilon \longrightarrow 0$) to obtain

$$\|D^\nu f\|_{L^{\mathbf{r}}_{q,\beta}(\Omega)} \leqslant C\|f\|_{W^1_{p,\alpha}(\Omega)}.$$

The proof of this is similar to the proof of the corresponding assertion in Theorem 1.

Theorems 3 and 4 imply the following assertion:

COROLLARY 2. *Suppose that the conditions of Theorem* 4 *are fulfilled. Then* $D^\nu f \in W^{\mathbf{r}}_{q,\beta}(\Omega)$, *and the following bound holds*:

$$\|D^\nu f\|_{W^{\mathbf{r}}_{q,\beta}(\Omega)} \leqslant C\|f\|_{W^1_{p,\alpha}(\Omega)},$$

where C does not depend on $f(x)$.

It is the author's pleasure to thank O. V. Besov for his constant help in this research.

BIBLIOGRAPHY

1. S. L. Sobolev, *Applications of functional analysis in mathematical physics*, Izdat. Leningrad. Gos. Univ., Leningrad, 1950; English transl., Transl. Math. Monographs, vol. 7, Amer. Math. Soc., Providence, R. I., 1963. MR **14**, 565; **29** #2624.

2. V. P. Il'in, *The properties of some classes of differentiable functions of several variables defined in an n-dimensional region,* Trudy Mat. Inst. Steklov. **66** (1962), 227–363; English transl., Amer. Math. Soc. Transl. (2) **81** (1969), 91–256. MR **27** #3750.

3. O. V. Besov and V. P. Il'in, *Natural extension of the class of regions in imbedding theorems*, Mat. Sb. **75** (117) (1968), 483–495 = Math. USSR Sb. **4** (1968), 445–456. MR **37** #1987.

4. S. M. Nikol'skiĭ, *Approximation of functions of several variables and embedding theorems*, "Nauka", Moscow, 1969; English transl., Die Grundlehren der math. Wissenschaften, Band 205, Springer-Verlag, Berlin and New York, 1974. MR **46** #9714.

5. O. V. Besov, *The continuation of functions in* L^l_p *and* W^l_p, Trudy Mat. Inst. Steklov. **89** (1967), 5–17 = Proc. Steklov Inst. Math. 89 (1967), 1–15. MR **35** #5920.

6. V. P. Il'in, *Integral representations of differentiable functions and their applications to questions of continuation of functions of classes* $W_p^{(l)}(G)$, Sibirsk, Mat. Ž. **8** (1967), 573–586 = Siberian Math. J. **8** (1967), 421–432. MR **36** #5683.

7. S. V. Uspenskiĭ, *Mixed derivatives of functions summable with a weight,* Differencial'nye Uravnenija **3** (1967), 139–154 = Differential Equations 3 (1967), 70–77. MR **35** #5928.

8. T. A. Timan, *Existence conditions and estimates of transforms of Calderón-Zygmund type in weighted* L_p*-spaces,* Trudy Mat. Inst. Steklov. **105** (1969), 213–229 = Proc. Steklov Inst. Math. **105** (1969), 261–280. MR **41** #4317.

9. S. V. Uspenskiĭ, *Imbedding theorems for weighted classes,* Trudy Mat. Inst. Steklov. **60** (1961), 282–303; English transl., Amer. Math. Soc. Transl. (2) **87** (1970), 121–145. MR **25** #440; **41** #3202.

10. A. D. Džabrailov, *On the direct and inverse imbedding theorems of weight spaces,* Dokl. Akad. Nauk SSSR **164** (1965), 24–27 = Soviet Math. Dokl. **6** (1965), 1142–1146. MR **32** #8147.

11. G. H. Hardy, J. E. Littlewood and G. Pólya, *Inequalities,* Cambridge Univ. Press, New York, 1934.

Translated by W. L. BYNUM

Trudy Mat. Inst. Steklov.
131 (1974)

Proc. Steklov Inst. Math.
131 (1974)

SOME INEQUALITIES FOR FRACTIONAL SEMINORMS

UDC 517.518.22

Ju. V. KUZNECOV

ABSTRACT. For $1 < p < \infty$ the function spaces $w_p^r(a, b)$, $1/p < r < 1$, as well as the spaces $b_p^r(a, b)$ and $B_p^r(a, b)$, $1/p < r \leqslant 1$, are considered; they consist of all functions $f(x)$ satisfying

$$\|f\|_{w_p^r(a, b)} \equiv \left(\int_a^b\int_a^b \frac{|f(x)-f(y)|^p}{|x-y|^{1+pr}}\,dx\,dy\right)^{1/p} < \infty,$$

$$\|f\|_{b_p^r(a, b)} \equiv \left(\int_a^b\int_a^b \frac{\left|f(x)-2f\left(\frac{x+y}{2}\right)+f(y)\right|^p}{|x-y|^{1+pr}}\,dx\,dy\right)^{1/p} < \infty,$$

$$\|f\|_{B_p^r(a, b)} \equiv \|f\|_{L_p(a, b)} + \|f\|_{b_p^r(a,b)} < \infty,$$

respectively. For any pair of intersecting intervals (a, b) and (c, d) the following results are proved.

THEOREM 1. *If $f \in w_p^r(a, b)$ and $f \in w_p^r(c, d)$, then*

$$\|f\|_{w_p^r(a, d)} \leqslant \frac{5}{r-1/p}\left(\|f\|_{w_p^r(a, b)} + \|f\|_{w_p^r(c, d)}\right).$$

THEOREM 2. *If $f \in B_p^r(a, b)$ and $f \in B_p^r(c, d)$, then*

$$\|f\|_{b_p^r(a, d)} \leqslant c_1\left(\|f\|_{b_p^r(a, b)} + \|f\|_{b_p^r(c, d)}\right) + c_2\left(\frac{\|f\|_{L_p(a, b)}}{(b-a)^r} + \frac{\|f\|_{L_p(c, d)}}{(d-c)^r}\right),$$

where $c_1 = c_1(p, r)$, $c_2 = c_2(p, r)$.

The case of functions of several variables is also considered.

Bibliography: 5 items.

Let $f(x)$ be a measurable function defined on the interval (a, b), $-\infty \leqslant a < b \leqslant +\infty$, and let, as usual,

$$\|f\|_{w_p^r(a, b)} = \left(\int_a^b\int_a^b \frac{|f(x)-f(y)|^p}{|x-y|^{1+pr}}\,dxdy\right)^{1/p},$$

$$\|f\|_{W_p^r(a,b)} = \|f\|_{L_p(a,b)} + \|f\|_{w_p^r(a, b)},$$

where $1 \leqslant p < \infty$ and $0 < r < 1$, and

$$\|f\|_{b_p^r(a, b)} = \left(\int_a^b\int_a^b \frac{\left|f(x)-2f\left(\frac{x+y}{2}\right)+f(y)\right|^p}{|x-y|^{1+pr}}\,dxdy\right)^{1/p},$$

$$\|f\|_{B_p^r(a, b)} = \|f\|_{L_p(a,b)} + \|f\|_{b_p^r(a,b)},$$

where $1 \leqslant p < \infty$ and $0 < r \leqslant 1$.

AMS (MOS) subject classifications (1970). Primary 46E35, 26A86.

Further, let two intersecting intervals (a, b) and (c, d) be given, $-\infty \leqslant a \leqslant c < b \leqslant d \leqslant +\infty$, and let

$$m = \max\left\{\frac{b-c}{b-a};\ \frac{b-c}{d-c}\right\} > 0.$$

V. I. Burenkov [1] has proved that if $f \in w_p^r(a, b)$ and $f \in w_p^r(c, d)$, then $f \in w_p^r(a, d)$ and

$$\|f\|_{w_p^r(a,d)} \leqslant C(m)\left(\|f\|_{w_p^r(a,b)} + \|f\|_{w_p^r(c,d)}\right). \tag{1}$$

Quite similarly, if $f \in b_p^r(a, b)$ and $f \in b_p^r(c, d)$, then $f \in b_p^r(a, d)$ and

$$\|f\|_{b_p^r(a,d)} \leqslant C(m)\left(\|f\|_{b_p^r(a,b)} + \|f\|_{b_p^r(c,d)}\right). \tag{2}$$

It is of interest to study the behavior of the constant $C(m)$ as $m \to 0$ for various relations between r and p. In [1] it is shown that for $r < 1/p$ we have $C(m) \to \infty$ as $m \to 0$. It can be shown that the same result holds for $r = 1/p$.

In the present paper the case $r > 1/p$ is considered. For the spaces w_p^r we prove

THEOREM 1. *Let $f \in w_p^r(a, b)$ and $f \in w_p^r(c, d)$, where $1 < p < \infty$ and $1/p < r < 1$. Then*

$$\|f\|_{w_p^r(a,d)} \leqslant \frac{5}{r - 1/p}\left(\|f\|_{w_p^r(a,b)} + \|f\|_{w_p^r(c,d)}\right). \tag{3}$$

Thus for the spaces w_p^r with $r > 1/p$ the constant $C(m)$ does not depend on m, and in (3) it is possible to take the limit as $m \to 0$.

For the spaces b_p^r the analog of this statement turns out to be false. Indeed, consider the function

$$f(x) = \begin{cases} 0, & x < 0, \\ x, & x \geqslant 0 \end{cases}$$

and set $a = -1$, $0 < b < 1$, $c = 0$ and $d = 1$, so that

$$m = \max\left\{\frac{b-c}{b-a};\ \frac{b-c}{d-c}\right\} = b > 0.$$

It is not difficult to verify that

$$\|f\|_{b_p^r(c,d)} = 0, \qquad \|f\|_{b_p^r(a,b)} \to 0$$

as $m = b \to 0$, and consequently the sum of the seminorms on the right-hand side of (2) tends to zero as $m \to 0$. At the same time,

$$\|f\|_{b_p^r(a,d)} = \text{const} \neq 0,$$

Thus $C(m) \to \infty$ as $m \to 0$, i.e. the constant $C(m)$ appearing in (2) essentially depends on m.

Nevertheless, we have

THEOREM 2. *Let $f \in B_p^r(a, b)$ and $f \in B_p^r(c, d)$, where $1 < p < \infty$ and $1/p < r \leqslant 1$. Then*

$$\|f\|_{b_p^r(a,d)} \leqslant C_1\left(\|f\|_{b_p^r(a,b)} + \|f\|_{b_p^r(c,d)}\right) + C_2\left(\frac{\|f\|_{L_p(a,b)}}{(b-a)^r} + \frac{\|f\|_{L_p(c,d)}}{(d-c)^r}\right), \tag{4}$$

where

$$C_1 = \frac{242}{(r-1/p)(1-r+1/p)}; \qquad C_2 = \frac{288}{1-r+1/p}.$$

Having considered the one-dimensional case, we now turn our attention to the case of an arbitrary n. Let $x = (x_1, \ldots, x_n)$, $p = (p_1, \ldots, p_n)$, $1 \leqslant p_i < \infty$, $r = (r_1, \ldots, r_n)$, $0 < r_i < \infty$, and $\alpha_i = r_i - [r_i] > 0$, and let Ω be a region in E_n. We will study the spaces $w_p^r(\Omega)$ (the definition of these spaces appears in [1], p. 30=36).* Then, as a corollary of Theorem 1, the following assertion can be obtained:

THEOREM 3. *Let Ω_1 and Ω_2 be arbitrary regions in E_n. Let $f \in w_p^r(\Omega_1)$ and $f \in w_p^r(\Omega_2)$, where $1 < p_i < \infty$ and $1/p_i < \alpha_i < 1$. Then $f \in w_p^r(\Omega_1 \cup \Omega_2)$ and*

$$\|f\|_{w_p^r(\Omega_1 \cup \Omega_2)} \leqslant C\left(\|f\|_{w_p^r(\Omega_1)} + \|f\|_{w_p^r(\Omega_2)}\right),$$

where

$$C = \max_i \frac{5}{\alpha_i - 1/p_i}.$$

Theorem 3 says that, in the case $1/p_i < \alpha_i < 1$, the spaces w_p^r are additive with respect to any pair of regions $\Omega_1, \Omega_2 \subset E_n$.

For proving Theorem 3 it is sufficient to repeat, with insignificant changes, the evaluations performed in the proof of Theorem 1 in [1]; thus we content ourselves here with the formulation of this theorem. On the other hand, for proving Theorems 1 and 2 we will need some auxiliary assertions, to which we now turn.

REMARK 1. It is well known that if $f \in W_p^r(a, b)$, $1/p < r < 1$, $-\infty \leqslant a < b \leqslant +\infty$, then there exists a function $\tilde{f} \in C[a, b]$ such that $\tilde{f} \sim f$ and

$$\|\tilde{f}\|_{C[a,b]} \leqslant K \|f\|_{W_p^r(a,b)}. \tag{5}$$

So we may assume from the beginning that $f(x)$ is continuous on $[a, b]$. A similar statement holds for $f \in B_p^r(a, b)$, $1/p < r \leqslant 1$, $-\infty \leqslant a < b \leqslant +\infty$.

REMARK 2. If $f \in w_p^r(a, b)$, $-\infty < a < b \leqslant +\infty$, then it is not difficult to verify that $f \in L_p(a, b)$ and, consequently, $f \in W_p^r(a, b)$. Therefore, on account of Remark 1, we may assume that any function $f \in w_p^r(a, b)$, $1/p < r < 1$, $-\infty \leqslant a < b \leqslant +\infty$, is continuous on $[a, b]$.

In the statements of Lemmas 1–3 as given below it is assumed that the function $f(x)$ is already replaced by a continuous function equivalent to it, and so it is possible to speak of the value of this function at a point.

LEMMA 1. *If $f \in w_p^r(a, b)$, $1/p < r < 1$, $-\infty < a < b \leqslant +\infty$, then*

$$\left(\int_a^b \frac{|f(x) - f(a)|^p}{(x-a)^{pr}}\, dx\right)^{1/p} \leqslant \frac{2}{r - 1/p} \|f\|_{w_p^r(a,b)}. \tag{6}$$

Inequality (6) without any estimate of the constant has been proved by G. N. Jakovlev (see [2] for $a = 0$, $b = +\infty$, and [3], Theorem 2.3, for $-\infty < a < b < +\infty$).

LEMMA 2. *If $f \in B_p^r(a, b)$, $1/p < r \leqslant 1$, $-\infty < a < b \leqslant +\infty$, then*

**Editor's note.* In the first line of p. 36 of the English translation of [1], read w in place of W.

$$\left(\int_a^b \frac{\left|f(x) - 2f\left(\frac{x+a}{2}\right) + f(a)\right|^p}{(x-a)^{pr}}\,dx\right)^{1/p} \leqslant \frac{10}{r - 1/p}\|f\|_{b_p^r(a,\,b)}. \tag{7}$$

In the case $a = 0$, $b = +\infty$ inequality (7) has been proved (with another constant) by V. A. Solonnikov [4]. The proof presented below is valid for the general case and yields a better constant.

LEMMA 3. *If* $f \in B_p^r(a, b)$, $1/p < r \leqslant 1$, $-\infty < a < b \leqslant +\infty$, *then*

$$\left(\int_a^b \frac{|f(x) - f(a)|^p}{(x-a)^{pr}}\,dx\right)^{1/p} \leqslant C_1\|f\|_{b_p^r(a,\,b)} + C_2\frac{\|f\|_{L_p(a,\,b)}}{(b-a)^r}, \tag{8}$$

where

$$C_1 = \frac{40}{(r - 1/p)(1 - r + 1/p)}; \quad C_2 = \frac{24}{1 - r + 1/p}.$$

If $a = 0$ and $b = \infty$, then inequality (8) takes the form

$$\left(\int_0^\infty \frac{|f(x) - f(0)|^p}{x^{pr}}\,dx\right)^{1/p} \leqslant C_1\|f\|_{b_p^r(0,\infty)}. \tag{8'}$$

Inequality (8′) (with another constant) was proved in [4].

Now let us turn to the proof of Lemmas 1–3. First suppose that $b < +\infty$ and the function $f(x)$ is continuously differentiable on the segment $[a, b]$. Then, obviously, the integrals standing on the left-hand side of the inequalities (6)–(8) are finite.

Let us prove (6). Since

$$f(x) - f(a) = \frac{1}{x-a}\int_a^x [f(x) - f(a)]\,d\xi,$$

it follows that

$$\frac{|f(x) - f(a)|}{(x-a)^r} = \frac{1}{(x-a)^{r+1}}\left|\int_a^x [f(x) - f(a)]\,d\xi\right|.$$

Thus

$$I \equiv \left(\int_a^b \frac{|f(x) - f(a)|^p}{(x-a)^{pr}}\,dx\right)^{1/p} = \left(\int_a^b \frac{dx}{(x-a)^{pr+p}}\left|\int_a^x [f(x) - f(a)]\,d\xi\right|^p\right)^{1/p}$$
$$\leqslant \left(\int_a^b \frac{dx}{(x-a)^{pr+p}}\left|\int_a^x [f(x) - f(\xi)]\,d\xi\right|^p\right)^{1/p}$$
$$+ \left(\int_a^b \frac{dx}{(x-a)^{pr+p}}\left|\int_a^x [f(\xi) - f(a)]\,d\xi\right|^p\right)^{1/p} \equiv I_1 + I_2,$$

i.e.

$$I \leqslant I_1 + I_2. \tag{9}$$

For estimating the integral I_1, note that

$$\left|\int_a^x [f(x)-f(\xi)]\,d\xi\right|^p \leqslant \int_a^x |f(x)-f(\xi)|^p\,d\xi\cdot(x-a)^{p-1}$$

and consequently

$$I_1 \leqslant \left(\int_a^b \frac{dx}{(x-a)^{1+pr}} \int_a^x |f(x)-f(\xi)|^p\,d\xi\right)^{1/p} = \left(\int_a^b d\xi \int_\xi^b \frac{|f(x)-f(\xi)|^p}{(x-a)^{1+pr}}\,dx\right)^{1/p}.$$

Taking into account that

$$x-a \geqslant x-\xi, \qquad \frac{1}{x-a} \leqslant \frac{1}{x-\xi} = \frac{1}{|x-\xi|},$$

we obtain

$$I_1 \leqslant \|f\|_{w_p^r(a,\,b)}. \tag{10}$$

Let us estimate I_2:

$$I_2 = \left(\int_a^b dx \left|\int_a^x \frac{f(\xi)-f(a)}{(x-a)^{r+1}}\,d\xi\right|^p\right)^{1/p} = \left(\int_a^b dx \left|\int_0^1 \frac{f[(1-t)a+tx]-f(a)}{(x-a)^r}\,dt\right|^p\right)^{1/p}$$

$$\leqslant \int_0^1 dt \left(\int_a^b \frac{|f[(1-t)a+tx]-f(a)|^p}{(x-a)^{pr}}\,dx\right)^{1/p} = \int_0^1 t^{r-1/p}\,dt \left(\int_a^{(1-t)a+tb} \frac{|f(\xi)-f(a)|^p}{(\xi-a)^{pr}}\,d\xi\right)^{1/p}$$

$$\leqslant \int_0^1 t^{r-1/p}\,dt \left(\int_a^b \frac{|f(\xi)-f(a)|^p}{(\xi-a)^{pr}}\,d\xi\right)^{1/p} = \frac{1}{r-1/p+1}\,I.$$

Thus

$$I_2 \leqslant \frac{1}{r-1/p+1}\,I. \tag{11}$$

In view of (9)–(11) we find that

$$I \leqslant \|f\|_{w_p^r(a,\,b)} + \frac{1}{r-1/p+1}\,I,$$

whence

$$I \leqslant \frac{r-1/p+1}{r-1/p}\,\|f\|_{w_p^r(a,\,b)} \leqslant \frac{2}{r-1/p}\,\|f\|_{w_p^r(a,\,b)}.$$

Inequality (6) is proved.

Let us prove (7). We will use the following short notation:

$$f(x)-2f\left(\frac{x+y}{2}\right)+f(y) \equiv \Delta^2 f(x,\,y).$$

Since

$$\Delta^2 f(x,\,a) = \frac{1}{x-a}\int_a^x \Delta^2 f(x,\,a)\,d\xi,$$

$$\frac{|\Delta^2 f(x,\,a)|}{(x-a)^r} = \frac{1}{(x-a)^{r+1}}\left|\int_a^x \Delta^2 f(x,\,a)\,d\xi\right|,$$

we have

$$I \equiv \left(\int_a^b \frac{|\Delta^2 f(x, a)|^p}{(x-a)^{pr}}\, dx\right)^{1/p} = \left(\int_a^b \frac{dx}{(x-a)^{pr+p}} \left|\int_a^x \Delta^2 f(x, a)\, d\xi\right|^p\right)^{1/p}$$

$$\leq \left(\int_a^b \frac{dx}{(x-a)^{pr+p}} \left|\int_a^x [\Delta^2 f(x, a) - \Delta^2 f(\xi, a)]\, d\xi\right|^p\right)^{1/p}$$

$$+ \left(\int_a^b \frac{dx}{(x-a)^{pr+p}} \left|\int_a^x \Delta^2 f(\xi, a)\, d\xi\right|^p\right)^{1/p} \equiv I_1 + I_2,$$

or

$$I \leq I_1 + I_2. \tag{12}$$

Estimating I_1 in the same way as it was done in the proof of (6), we can show that

$$I_1 \leq \left(\int_a^b d\xi \int_\xi^b \frac{|\Delta^2 f(x, a) - \Delta^2 f(\xi, a)|^p}{(x-a)^{1+pr}}\, dx\right)^{1/p}.$$

We now use the identity

$$\Delta^2 f(x, a) - \Delta^2 f(\xi, a) = \Delta^2 f(x, \xi) + 2\Delta^2 f\left(\frac{x+\xi}{2}; \frac{a+\xi}{2}\right) - 2\Delta^2 f\left(\frac{x+a}{2}; \xi\right)$$

which is easy to verify, and the Minkowski inequality. As a result we obtain

$$I_1 \leq \left(\int_a^b d\xi \int_\xi^b \frac{|\Delta^2 f(x, \xi)|^p}{(x-a)^{1+pr}}\, dx\right)^{1/p} + 2\left(\int_a^b d\xi \int_\xi^b \frac{\left|\Delta^2 f\left(\frac{x+\xi}{2}; \frac{a+\xi}{2}\right)\right|^p}{(x-a)^{1+pr}}\, dx\right)^{1/p}$$

$$+ 2\left(\int_a^b d\xi \int_\xi^b \frac{\left|\Delta^2 f\left(\frac{x+a}{2}; \xi\right)\right|^p}{(x-a)^{1+pr}}\, dx\right)^{1/p} \equiv i_1 + 2i_2 + 2i_3.$$

Here

$$i_1 \leq \|f\|_{b_p^r(a, b)}, \qquad i_2 = 2^{1/p}\left(\int_a^b d\xi \int_\xi^{\frac{\xi+b}{2}} \frac{\left|\Delta^2 f\left(\eta, \frac{a+\xi}{2}\right)\right|^p}{(2\eta - \xi - a)^{1+pr}}\, d\eta\right)^{1/p}$$

$$\leq 2^{1/p}\left(\int_a^b d\xi \int_\xi^b \frac{\left|\Delta^2 f\left(\eta, \frac{a+\xi}{2}\right)\right|^p}{(2\eta - \xi - a)^{1+pr}}\, d\eta\right)^{1/p} = 2^{1/p}\left(\int_a^b d\eta \int_a^\eta \frac{\left|\Delta^2 f\left(\eta, \frac{a+\xi}{2}\right)\right|^p}{(2\eta - \xi - a)^{1+pr}}\, d\xi\right)^{1/p}$$

$$= \left(\frac{1}{2}\right)^{r-1/p}\left(\int_a^b d\eta \int_a^{\frac{a+\eta}{2}} \frac{|\Delta^2 f(\eta, \zeta)|^p}{(\eta - \zeta)^{1+pr}}\, d\zeta\right)^{1/p} \leq \|f\|_{b_p^r(a, b)},$$

$$i_3 = \left(\frac{1}{2}\right)^r \left(\int_a^b d\xi \int_{\frac{a+\xi}{2}}^{\frac{a+b}{2}} \frac{|\Delta^2 f(\eta, \xi)|^p}{(\eta - a)^{1+pr}}\, d\eta\right)^{1/p}.$$

Since

$$a \leq \xi \leq b, \qquad a \leq \frac{a+\xi}{2} \leq \eta \leq \frac{a+b}{2},$$

we have

$$\eta - \xi = (\eta - a) - (\xi - a) \leqslant \eta - a,$$

and

$$\eta - \xi = -(\eta - a) + 2\eta - (a + \xi) \geqslant -(\eta - a)$$

$$|\eta - \xi| \leqslant \eta - a, \qquad \frac{1}{\eta - a} \leqslant \frac{1}{|\eta - \xi|};$$

thus, in view of the relation $(1/2)^r \leqslant 1$, we find

$$i_3 \leqslant \left(\int_a^b d\xi \int_{\frac{a+\xi}{2}}^{\frac{a+b}{2}} \frac{|\Delta^2 f(\eta, \xi)|^p}{|\eta - \xi|^{1+pr}} d\eta\right)^{1/p} \leqslant \|f\|_{b_p^r(a, b)}.$$

Combination of the above estimates yields

$$I_1 \leqslant 5 \| f \|_{b_p^r(1, b)}. \tag{13}$$

The integral I_2 can be estimated in the same manner as the corresponding integral in deriving (6); as a result we obtain

$$I_2 \leqslant \frac{1}{r - 1/p + 1} I. \tag{14}$$

Now from (12)–(14) it follows that

$$I \leqslant 5 \|f\|_{b_p^r(a, b)} + \frac{1}{r - 1/p + 1} I.$$

Thus

$$I \leqslant \frac{5(r - 1/p + 1)}{r - 1/p} \|f\|_{b_p^r(a, b)} \leqslant \frac{10}{r - 1/p} \|f\|_{b_p^r(a, b)},$$

and (7) is proved.

Let us prove (8). We have

$$I \equiv \left(\int_a^b \frac{|f(x) - f(a)|^p}{(x - a)^{pr}} dx\right)^{1/p} \leqslant \left(\int_a^{\frac{a+b}{2}} \frac{|f(x) - f(a)|^p}{(x - a)^{pr}} dx\right)^{1/p}$$

$$+ \left(\int_{\frac{a+b}{2}}^b \frac{|f(x) - f(a)|^p}{(x - a)^{pr}} dx\right)^{1/p} \equiv I_1 + I_2.$$

In order to estimate I_1, we make the substitution $x = (\xi + a)/2$ in it:

$$I_1 = \left(\frac{1}{2}\right)^{-r+1/p} \left(\int_a^b \frac{\left|f\left(\frac{\xi + a}{2}\right) - f(a)\right|^p}{(\xi - a)^{pr}} d\xi\right)^{1/p}.$$

Next we use the identity

$$f\left(\frac{\xi + a}{2}\right) - f(a) = \frac{1}{2}\left[(f(\xi) - f(a)) - \Delta^2 f(\xi, a)\right]$$

and the Minkowski inequality. Then

$$I_1 \leqslant \left(\frac{1}{2}\right)^{1-r+1/p} \left[\left(\int_a^b \frac{|f(\xi) - f(a)|^p}{(\xi - a)^{pr}} d\xi\right)^{1/p}\right.$$

$$\left. + \left(\int_a^b \frac{|\Delta^2 f(\xi, a)|^p}{(\xi - a)^{pr}} d\xi\right)^{1/p}\right] \equiv \left(\frac{1}{2}\right)^{1-r+1/p} (I + i_1).$$

Taking into account that

$$\left(\frac{1}{2}\right)^{1-r+1/p} < 1 - \frac{1-r+1/p}{2} \equiv 1-\alpha < 1,$$

and also the fact that, by (7),

$$i_1 \leqslant \frac{10}{r-1/p} \| f \|_{b_p^r(a,\,b)},$$

we obtain

$$I_1 \leqslant (1-\alpha) I + (1-\alpha) \frac{10}{r-1/p} \| f \|_{b_p^r(a,\,b)},$$

or

$$I_1 \leqslant (1-\alpha) I + \frac{10}{r-1/p} \| f \|_{b_p^r(a,\,b)}. \tag{15}$$

For estimating I_2 we use the identity

$$f(x) - f(a) = 2\left[f(x) - f\left(\frac{x+a}{2}\right)\right] - \Delta^2 f(x,\,a)$$

and the Minkowski inequality. As a result, we find

$$I_2 \leqslant 2\left(\int\limits_{\frac{a+b}{2}}^{b} \frac{\left|f(x) - f\left(\frac{x+a}{2}\right)\right|^p}{(x-a)^{pr}}\,dx\right)^{1/p} + \left(\int\limits_{\frac{a+b}{2}}^{b} \frac{|\Delta^2 f(x,\,p)|^p}{(x-a)^{pr}}\,dx\right)^{1/p} \equiv 2i_2 + i_3.$$

Since here

$$\begin{aligned} i_2 &\leqslant \left(\frac{2}{b-a}\right)^r \left(\int\limits_{\frac{a+b}{2}}^{b} \left|f(x) - f\left(\frac{x+a}{2}\right)\right|^p dx\right)^{1/p} \\ &\leqslant \left(\frac{2}{b-a}\right)^r \left[\left(\int\limits_a^b |f(x)|^p\,dx\right)^{1/p} + \left(\int\limits_a^b \left|f\left(\frac{x+a}{2}\right)\right|^p dx\right)^{1/p}\right] \\ &= \left(\frac{2}{b-a}\right)^r \left[\left(\int\limits_a^b |f(x)|^p\,dx\right)^{1/p} + 2^{1/p}\left(\int\limits_a^{\frac{a+b}{2}} |f(\xi)|^p\,d\xi\right)^{1/p}\right] \leqslant \frac{6}{(b-a)^r} \| f \|_{L_p(a,\,b)} \end{aligned}$$

and, moreover,

$$i_3 \leqslant \frac{10}{r-1/p} \| f \|_{b_p^r(a,\,b)},$$

it follows that

$$I_2 \leqslant \frac{12}{(b-a)^r} \| f \|_{L_p(a,\,b)} + \frac{10}{r-1/p} \| f \|_{b_p^r(a,\,b)}.$$

Hence and from (15) we obtain

$$I \leqslant (1-\alpha) I + \frac{20}{r-1/p} \| f \|_{b_p^r(a,\,b)} + \frac{12}{(b-a)^r} \| f \|_{L_p(a,\,b)},$$

$$I \leqslant \frac{20}{\alpha(r-1/p)} \| f \|_{b_p^r(a,\,b)} + \frac{12}{\alpha(b-a)^r} \| f \|_{L_p(a,\,b)},$$

and, since

$$\frac{1}{\alpha} = \frac{2}{1 - r + 1/p},$$

we obtain (8).

So if $b < +\infty$ and $f(x)$ is continuously differentiable on $[a, b]$, then the inequalities (6)–(8) are satisfied.

Now let $f(x)$ be any function belonging to $w_p^r(a, b)$, $b < +\infty$. Then, by Remark 2, $f \in W_p^r(a, b)$ and, as mentioned before, it may be considered continuous on $[a, b]$. Since the set of all continuously differentiable functions is everywhere dense in $W_p^r(a, b)$, it contains a sequence $\{\varphi_n\}$ such that, for $n \to \infty$,

$$\| \varphi_n - f \|_{W_p^r(a, b)} \to 0$$

and consequently

$$\| \varphi_n - f \|_{w_p^r(a, b)} \to 0.$$

As the functions φ_n are continuously differentiable, for every n we have, by what has already been proved,

$$I_n \equiv \left(\int_a^b \frac{|\varphi_n(x) - \varphi_n(a)|^p}{(x-a)^{pr}} dx \right)^{1/p} \leqslant \frac{2}{r - 1/p} \| \varphi_n \|_{w_p^r(a, b)}. \tag{16}$$

Then, for $m, n \to \infty$,

$$|I_n - I_m| \leqslant \left(\int_a^b \frac{|(\varphi_n - \varphi_m)(x) - (\varphi_n - \varphi_m)(a)|^p}{(x-a)^{pr}} dx \right)^{1/p}$$
$$\leqslant \frac{2}{r - 1/p} \| \varphi_n - \varphi_m \|_{w_p^r(a, b)} \to 0,$$

i.e. there exists

$$\lim_{n\to\infty} I_n = \lim_{n\to\infty} \left(\int_a^b \frac{|\varphi_n(x) - \varphi_n(a)|^p}{(x-a)^{pr}} dx \right)^{1/p}.$$

Further, since (cf. (5))

$$\| \varphi_n - f \|_{C[a, b]} \leqslant K \| \varphi_n - f \|_{W_p^r(a, b)} \to 0,$$

for any $x \in [a, b]$ we have

$$\lim_{n\to\infty} \varphi_n(x) = f(x).$$

But then everywhere on $[a, b]$, except at the point $x = a$,

$$\lim_{n\to\infty} \frac{|\varphi_n(x) - \varphi_n(a)|^p}{(x-a)^{pr}} = \frac{|f(x) - f(a)|^p}{(x-a)^{pr}},$$

and, applying a corollary of Fatou's theorem ([5], Chapter VI, Corollary of Theorem 9),

$$\left(\int_a^b \frac{|f(x) - f(a)|^p}{(x-a)^{pr}} dx \right)^{1/p} \leqslant \lim_{n\to\infty} \left(\int_a^b \frac{|\varphi_n(x) - \varphi_n(a)|^p}{(x-a)^{pr}} dx \right)^{1/p}.$$

Consequently, passing to the limit as $n \to \infty$ in (16), we obtain the inequality (6) for any function $f \in w_p^r(a, b)$, where $b < +\infty$. But if $f \in w_p^r(a, +\infty)$, then it is not difficult to obtain (6) by taking the limit as $b \to +\infty$.

The deduction of (7) and (8) can be completed in exactly the same way.

Let us prove Theorem 1. If in the course of evaluations an integrand remains unchanged, we denote it by the letter Λ. We have

$$\|f\|_{w_p^r(a,\,d)} = \left(\int_a^d\int_a^d \frac{|f(x)-f(y)|^p}{|x-y|^{1+pr}}\,dx\,dy\right)^{1/p}$$

$$\leqslant \left(\int_a^b dx\int_a^b \Lambda\,dy\right)^{1/p} + \left(\int_c^d dx\int_c^d \Lambda\,dy\right)^{1/p} + \left(\int_a^c dx\int_b^d \Lambda\,dy\right)^{1/p} + \left(\int_b^d dx\int_a^c \Lambda\,dy\right)^{1/p}.$$

As

$$\left(\int_a^b dx\int_a^b \Lambda\,dy\right)^{1/p} = \|f\|_{w_p^r(a,\,b)},\qquad \left(\int_c^d dx\int_c^d \Lambda\,dy\right)^{1/p} = \|f\|_{w_p^r(c,\,d)},$$

therefore, introducing the notation

$$\left(\int_a^c dx\int_b^d \Lambda\,dy\right)^{1/p} = \left(\int_b^d dx\int_a^c \Lambda\,dy\right)^{1/p} \equiv I,$$

we obtain

$$\|f\|_{w_p^r(a,\,d)} \leqslant \|f\|_{w_p^r(a,\,b)} + \|f\|_{w_p^r(c,\,d)} + 2I. \tag{17}$$

Let $z = (b+c)/2$. Then

$$I = \left(\int_a^c dx\int_b^d \Lambda\,dy\right)^{1/p} \leqslant \left(\int_a^z dx\int_z^d \Lambda\,dy\right)^{1/p}$$

$$\leqslant \left(\int_a^z dx\int_z^d \frac{|f(x)-f(z)|^p}{|x-y|^{1+pr}}\,dy\right)^{1/p} + \left(\int_a^z dx\int_z^d \frac{|f(z)-f(y)|^p}{|x-y|^{1+pr}}\,dy\right)^{1/p},$$

or

$$I \leqslant I_1 + I_2. \tag{18}$$

Since

$$I_1 = \left(\int_a^z |f(x)-f(z)|^p\,dx\int_z^d \frac{dy}{(y-x)^{1+pr}}\right)^{1/p}$$

$$\leqslant \left(\frac{1}{pr}\right)^{1/p}\left(\int_a^z \frac{|f(x)-f(z)|^p}{(z-x)^{pr}}\,dx\right)^{1/p} \leqslant \left(\int_a^z \frac{|f(x)-f(z)|^p}{(z-x)^{pr}}\,dx\right)^{1/p},$$

on applying Lemma 1 we find

$$I_1 \leqslant \frac{2}{r-1/p}\|f\|_{w_p^r(a,\,z)} \leqslant \frac{2}{r-1/p}\|f\|_{w_p^r(a,\,b)}. \tag{19}$$

The integral I_2 can be estimated in just the same way:

$$I_2 \leqslant \frac{2}{r-1/p}\|f\|_{w_p^r(c,\,d)}. \tag{20}$$

Substitution of (19) and (20) in (18) yields

$$I \leqslant \frac{2}{r-1/p}\left(\|f\|_{w_p^r(a,\,b)} + \|f\|_{w_p^r(c,\,d)}\right),$$

Finally, substituting this estimate in (17), we obtain

$$\|f\|_{w_p^r(a,d)} \leqslant \frac{r-1/p+4}{r-1/p}(\|f\|_{w_p^r(a,b)} + \|f\|_{w_p^r(c,d)}) \leqslant \frac{5}{r-1/p}(\|f\|_{w_p^r(a,b)} + \|f\|_{w_p^r(c,d)}).$$

Theorem 1 is proved.

Let us prove Theorem 2. In the same manner as in the proof of Theorem 1, one can show that

$$\|f\|_{b_p^r(a,d)} \leqslant \|f\|_{b_p^r(a,b)} + \|f\|_{b_p^r(c,d)} + 2I, \tag{21}$$

where

$$I \equiv \left(\int_a^c dx \int_b^d \frac{|\Delta^2 f(x,y)|^p}{|x-y|^{1+pr}}\, dy\right)^{1/p} \leqslant \left(\int_a^z dx \int_z^d \frac{|\Delta^2 f(x,y)|^p}{|x-y|^{1+pr}}\, dy\right)^{1/p},$$

the notation $z = (b+c)/2$ being used as before. Since

$$\Delta^2 f(x,y) = [f(x) - f(z)] + [f(y) - f(z)] + 2\left[f(z) - f\left(\frac{x+y}{2}\right)\right],$$

on account of the Minkowski inequality we have

$$I \leqslant \left(\int_a^z dx \int_z^d \frac{|f(x)-f(z)|^p}{|x-y|^{1+pr}}\, dy\right)^{1/p} + \left(\int_a^z dx \int_z^d \frac{|f(y)-f(z)|^p}{|x-y|^{1+pr}}\, dy\right)^{1/p}$$
$$+ 2\left(\int_a^z dx \int_z^d \frac{\left|f(z) - f\left(\frac{x+y}{2}\right)\right|^p}{|x-y|^{1+pr}}\, dy\right)^{1/p} \equiv I_1 + I_2 + 2I_3,$$

or

$$I \leqslant I_1 + I_2 + 2I_3. \tag{22}$$

In the same way as in the proof of Theorem 1, we obtain

$$I_1 \leqslant \left(\int_a^z \frac{|f(z)-f(x)|^p}{(z-x)^{pr}}\, dx\right)^{1/p} \equiv J_1. \tag{23}$$

Similarly,

$$I_2 \leqslant \left(\int_z^d \frac{|f(x)-f(z)|^p}{(x-z)^{pr}}\, dx\right)^{1/p} \equiv J_2.$$

Furthermore,

$$I_3 = \left(\frac{1}{2}\right)^r \left(\int_a^z dx \int_{\frac{x+z}{2}}^{\frac{x+d}{2}} \frac{|f(z)-f(\eta)|^p}{(\eta-x)^{1+pr}}\, d\eta\right)^{1/p}$$
$$\leqslant \left(\int_a^z dx \int_{\frac{x+z}{2}}^{d} \Lambda\, d\eta\right)^{1/p} \leqslant \left(\int_a^z dx \int_{\frac{x+z}{2}}^{z} \Lambda\, d\eta\right)^{1/p} + \left(\int_a^z dx \int_z^d \Lambda\, d\eta\right)^{1/p} \equiv i_1 + i_2. \tag{24}$$

Since

$$i_1 = \left(\int_{\frac{a+z}{2}}^{z} |f(z)-f(\eta)|^p\, d\eta \int_a^{2\eta-z} \frac{dx}{(\eta-x)^{1+pr}}\right)^{1/p}$$
$$\leqslant \left(\frac{1}{pr}\right)^{1/p} \left(\int_{\frac{a+z}{2}}^{z} \frac{|f(z)-f(\eta)|^p}{(z-\eta)^{pr}}\, d\eta\right)^{1/p} \leqslant \int_a^z \frac{|f(z)-f(\eta)|^p}{(z-\eta)^{pr}}\, d\eta\Big)^{1/p} = J_1$$

and, in exactly the same way, $i_2 \leqslant J_2$, it follows that

$$I_2 \leqslant J_1 + J_2. \tag{25}$$

Substituting (23)–(25) in (22), we have $I \leqslant 3(J_1 + J_2)$. Applying Lemma 3, we find

$$I \leqslant 3M_1 (\|f\|_{b^r_p(a,z)} + \|f\|_{b^r_p(z,d)}) + 3M_2 \left(\frac{\|f\|_{L_p(a,z)}}{(z-a)^r} + \frac{\|f\|_{L_p(z,d)}}{(d-z)^r} \right),$$

where

$$M_1 = \frac{40}{(r-1/p)(1-r+1/p)}, \qquad M_2 = \frac{24}{1-r+1/p}.$$

Since $z = (b+c)/2$ we have

$$\|f\|_{b^r_p(a,z)} \leqslant \|f\|_{b^r_p(a,b)}, \qquad \|f\|_{b^r_p(z,d)} \leqslant \|f\|_{b^r_p(c,d)},$$
$$\|f\|_{L_p(a,z)} \leqslant \|f\|_{L_p(a,b)}, \qquad \|f\|_{L_p(z,d)} \leqslant \|f\|_{L_p(c,d)}.$$

Moreover,

$$z - a \geqslant \frac{b-a}{2}; \qquad \frac{1}{(z-a)^r} \leqslant \frac{2^r}{(b-a)^r} \leqslant \frac{2}{(b-a)^r};$$
$$d - z \geqslant \frac{d-c}{2}; \qquad \frac{1}{(d-z)^r} \leqslant \frac{2^r}{(d-c)^r} \leqslant \frac{2}{(d-c)^r}.$$

Thus

$$I \leqslant 3M_1 (\|f\|_{b^r_p(a,b)} + \|f\|_{b^r_p(c,d)}) + 6M_2 \left(\frac{\|f\|_{L_p(a,b)}}{(b-a)^r} + \frac{\|f\|_{L_p(c,d)}}{(d-c)^r} \right).$$

Substituting the latter estimate in (21), we obtain

$$\|f\|_{b^r_p(a,d)} \leqslant (1+6M_1)(\|f\|_{b^r_p(a,b)} + \|f\|_{b^r_p(c,d)}) + 12M_2 \left(\frac{\|f\|_{L_p(a,b)}}{(b-a)^r} + \frac{\|f\|_{L_p(c,d)}}{(d-c)^r} \right),$$

which implies (4), since

$$1 + 6M_1 \leqslant \frac{242}{(r-1/p)(1-r+1/p)}, \qquad 12M_2 = \frac{288}{1-r+1/p}.$$

BIBLIOGRAPHY

1. V. I. Burenkov, *On the additivity of the spaces W^r_p and B^r_p and on imbedding theorems for regions of a general kind,* Trudy Mat. Inst. Steklov. **105** (1969), 30–45 = Proc. Steklov. Inst. Math. **105** (1969), 35–53. MR **41** #2374.

2. G. N. Jakovlev, *Boundary properties of functions of class $W^{(l)}_p$ on regions with angular points,* Dokl. Akad. Nauk SSSR **140** (1961), 73–76 = Soviet Math. Dokl. **2** (1961), 1177–1180. MR **25** #448.

3. ———, *Traces of functions in the space W^l_p on piecewise smooth surfaces,* Mat. Sb. **74 (116)** (1967), 526–543 = Math. USSR Sb. **3** (1967), 481–498. MR **36** #3120.

4. V. A. Solonnikov, *A priori estimates for second-order parabolic equations,* Trudy Mat. Inst. Steklov. **70** (1964), 133–212; English transl., Amer. Math. Soc. Transl. (2) **65** (1967), 51–137. MR **28** #5267.

5. I. P. Natanson, *Theory of functions of a real variable,* 2nd rev. ed., GITTL, Moscow, 1957; English transl., Ungar, New York, 1961. MR **26** #6309.

Translated by J. BOGNÁR

Trudy Mat. Inst. Steklov.
131 (1974)

Proc. Steklov Inst. Math.
131 (1974)

PROPERTIES OF FUNCTIONS IN THE SPACES $\Lambda^r_{p,\theta}$

UDC 517.518.22

P. I. LIZORKIN

ABSTRACT. In this paper we introduce new function spaces $\Lambda^r_{p,\theta}(R^n)$, constructed by means of sectioning the Fourier transforms of the functions being considered into "packets". For $\theta = 2$ the spaces $\Lambda^0_{p,\theta}(R^n)$ coincide to within equivalence of norms with the L_p-spaces of functions whose pth powers are integrable over R^n, $1 < p < \infty$. When the index θ increases, the space $\Lambda^0_{p,\theta}(R^n)$ becomes larger. We consider theorems on multipliers of Fourier integrals for functions in $\Lambda^0_{p,\theta}$. We prove an imbedding theorem involving a variation in the index p and a theorem on boundary values. We compare the spaces $\Lambda^r_{p,\theta}$ with other spaces. We construct various examples.

Bibliography: 10 items.

Introduction

Application of the Fourier transform to the study of various families of function spaces has been very fruitful. Recently the frequently encountered function scales (L^r_p, H^r_p, $B^r_{p,\theta}$) have been described from the point of view of Fourier transforms (see [1] and [2]). This approach makes it easy to obtain the corresponding imbedding theorems in a unified fashion.

In this work we introduce and investigate a new class of function spaces. This class arises naturally in the study of the Lebesgue spaces L_p from the point of view of Fourier transforms by means of the Littlewood-Paley theorem on sections.

Let R^n be an n-dimensional euclidean space and let $x = (x_1, \dots, x_n)$ be a point in this space; let $|x| = (\Sigma_1^n x_i^2)^{1/2}$. We write $R^1 = R$. As usual, $L_p(R^n) = L_p$ is the space of measurable functions $f(x)$, $x \in R^n$, whose pth powers are integrable over R^n, $1 \leqslant p \leqslant \infty$, with the norm

$$\|f\|_p = \begin{cases} \left(\int\limits_{R^n} |f(x)|^p\right)^{1/p}, & 1 \leqslant p < \infty; \\ \operatorname*{ess\,sup}\limits_{x \in R^n} |f(x)|, & p = \infty. \end{cases}$$

Suppose for simplicity that $n = 1$ and $1 < p \leqslant 2$ (for the general case see §1). Then the Fourier transform $\tilde{f}$ of $f \in L_p(R)$ exists:

$$\tilde{f}(\lambda) = \frac{1}{\sqrt{2\pi}} \int\limits_{-\infty}^{\infty} f(x)\, e^{-i\lambda x}\, dx.$$

Here $\tilde{f} \in L_{p'}(R^n)$, $1/p + 1/p' = 1$, and the following functions are well defined:

AMS (MOS) subject classifications (1970). Primary 42A68, 46E35; Secondary 46E30, 42A18.

$$f_m(x) = \frac{1}{\sqrt{2\pi}} \int\limits_{2^m<|\lambda|<2^{m+1}} \tilde{f}(\lambda) e^{i\lambda x} d\lambda, \qquad m = 0, \pm 1, \pm 2, \ldots .$$

The Littlewood-Paley theorem tells us that in this case we have the equivalence relation

$$\left\| \left\{ \sum_{m=-\infty}^{\infty} |f_m(x)|^2 \right\}^{1/2} \right\|_p \sim \|f\|_p;$$

that is, there exist constants A and B $(0 < A \leqslant B < \infty)$ such that

$$A\|f\|_p \leqslant \left\| \left\{ \sum_{-\infty}^{\infty} |f_m(x)|^2 \right\}^{1/2} \right\|_p \leqslant B\|f\|_p.$$

This equivalence relation carries over to the n-dimensional case for arbitrary p, $1 < p < \infty$ (see §1).

Thus the space L_p can be regarded as the set of functions for which the following norm is finite:

$$\left\| \left\{ \sum |f_m(x)|^2 \right\}^{1/2} \right\|_p .$$

Later (see the end of §1 and §2) we will consider spaces with the norm

$$\left\| \left\{ \sum |f_m(x)|^\theta \right\}^{1/\theta} \right\|_p, \qquad 1 \leqslant \theta \leqslant \infty.$$

For $\theta = 2$ we obtain the space $L_{p,2}$ which coincides (to within equivalence of norms) with the space L_p, and for $\theta < 2$ is smaller and for $\theta > 2$ is larger than L_p. Thus the index θ provides for an additional gradation of properties.

The isomorphic image of $L_{p,\theta}$ under the Bessel integration operator I_r is called $L^r_{p,\theta}$. In other words, $L^r_{p,\theta}$ consists of functions $f \in L_{p,\theta}$ whose derivatives of order r belong to $L_{p,\theta}$. For the spaces $L^r_{p,\theta}(R^n)$ we prove an imbedding theorem and a theorem on boundary values (see §4). The problem of characterizing the trace (in particular, the continuation problem) remains unsolved.

In §3 we present auxiliary results on multipliers, which are of interest in themselves.

We note that the spaces $L^r_{p,\theta}$ can be characterized by means of difference norms (without using the Fourier transform).

In the Appendix we present some auxiliary assertions and examples.

§1. Various forms of the Littlewood-Paley theorem on sections, and the spaces $\Lambda_{p,\theta}$

We consider a partition of the euclidean space R^n of points $\lambda = (\lambda_1, \ldots, \lambda_n)$ into "packets" and "corridors".

Let $\mathbf{m} = (m_1, \ldots, m_n)$ be a set of integers. We let

$$Q_{2\mathbf{m}} = \{\lambda \in R^n;\ 2^{m_j} \leqslant |\lambda_j| < 2^{m_j+1}, \quad j = 1, \ldots, n\}.$$

Thus the packet $Q_{2\mathbf{m}}$ consists of the parallelipiped $\{\lambda \in R^n,\ 2^{m_j} < \lambda_j \leqslant 2^{m_j+1}, j = 1, \ldots, n\}$ and its mirror images with respect to all of the coordinate hyperplanes. Clearly the packets $Q_{2\mathbf{m}}$ do not intersect and they form a partition $T = \{Q_{2\mathbf{m}}\}$ of the space

$$(R \setminus \{0\})^n = (R \setminus \{0\}) \times (R \setminus \{0\}) \ldots (R \setminus \{0\})$$

$$(R\setminus\{0\})^n = \bigcup_{\mathbf{m}} Q_{2^{\mathbf{m}}},$$

where the union is taken over all $\mathbf{m}$ ($m_j = 0, \pm 1, \pm 2, \ldots; j = 1, \ldots, n$).

Let m be an integer. We set

$$D_m = \{\lambda \in R^n;\ |\lambda_j| < 2^m,\quad j = 1, \ldots, n\},\qquad \Gamma_m = D_m \setminus D_{m-1}.$$

Thus the set Γ_m is a corridor obtained by removing the cube D_{m-1} from the larger cube D_m. It is clear that the corridors Γ_m do not intersect and the collection $T = \{\Gamma_m\}$ forms a covering of $R^n\setminus\{\theta\}$, $\theta = (0, \ldots, 0) \in R^n$, $R^n - \{\theta\} = \bigcup_m \Gamma_m$.

Along with the partitions T and T we will also consider partitions $\overset{+}{T}$ and $\overset{+}{T}$. The partition $\overset{+}{T}$ is obtained from T by discarding the hyperplanes which correspond to negative m_j. That is, we let

$$Q^*_{2^{\mathbf{m}}} = \{\lambda;\ \zeta(m_j) \leqslant |\lambda_j| < 2^{m_j},\quad j = 1, \ldots, n\},$$

where $\zeta(m_j) = 2^{m_j - 1}$ if $m_j \geqslant 1$ and $\zeta(m_j) = 0$ if $m_j = 0$. Then $\overset{+}{T} = \{Q^*_{2^{\mathbf{m}}}\}_{\mathbf{m} \geqslant 0}$, where the notation $\mathbf{m} \geqslant 0$ means that $m_j \geqslant 0$, $j = 1, \ldots, n$.

Analogously we let

$$\Gamma^*_m = \{\lambda;\ \eta(m) \leqslant \max_i |\lambda_j| < 2^m\},$$

where $\eta(m) = 2^{m-1}$ if $m \geqslant 1$ and $\eta(m) = 0$ if $m = 0$. The partition $\overset{+}{T}$ is the union of the Γ^*_m, $m \geqslant 0$.

Let $S(R^n) = S$ be the Schwartz space of test functions on R^n. Recall that a function $\varphi(x) \in S$ is infinitely differentiable and decreases at infinity together with its derivatives

$$D^l\varphi = \frac{\partial^l \varphi}{\partial x_1^{l_1} \ldots \partial x_n^{l_n}},\qquad l = l_1 + \ldots + l_n,$$

faster than any power of $|x|$. Let

$$F[\varphi](\lambda) = \widetilde{\varphi}(\lambda)$$

denote the Fourier transform of $\varphi(x)$:

$$\widetilde{\varphi}(\lambda) = (2\pi)^{-n/2} \int_{R^n} \varphi(x)\, e^{-i\lambda x}\, dx.$$

It is known that $\widetilde{\varphi} \in S$. The inverse Fourier transform of a function $\psi(x) \in S$ will be denoted by $F^{-1}[\psi](\lambda) \equiv \hat{\psi}(x)$. Thus

$$F^{-1}[\widetilde{\varphi}] = \hat{\widetilde{\varphi}} = \varphi(x) = (2\pi)^{-n/2} \int_{R_n} \widetilde{\varphi}(\lambda)\, e^{i\lambda x}\, d\lambda.$$

We construct a "section" of a function $\varphi \in S$ corresponding to the partitions T, $\overset{+}{T}$, T and $\overset{+}{T}$ respectively. Let $\chi_\Omega(\lambda)$ denote the characteristic function of a set Ω. Let

$$\varphi_{2^m}(x)=\frac{1}{(2\pi)^{n/2}}\int_{R^n}\chi_{Q_{2^m}}(\lambda)\widetilde{\varphi}(\lambda)e^{i\lambda x}\,d\lambda=\frac{1}{(2\pi)^{n/2}}\int_{Q_{2^m}}\widetilde{\varphi}(\lambda)e^{i\lambda x}\,d\lambda,$$

$$\varphi^*_{2^m}(x)=\frac{1}{(2\pi)^{n/2}}\int_{R^n}\chi_{Q^*_{2^m}}(\lambda)\widetilde{\varphi}(\lambda)e^{i\lambda x}\,d\lambda=\frac{1}{(2\pi)^{n/2}}\int_{Q^*_{2^m}}\widetilde{\varphi}(\lambda)e^{i\lambda x}\,d\lambda,$$

$$\varphi_m(x)=\frac{1}{(2\pi)^{n/2}}\int_{R^n}\chi_{\Gamma_m}(\lambda)\widetilde{\varphi}(\lambda)e^{i\lambda x}\,d\lambda=\frac{1}{(2\pi)^{n/2}}\int_{\Gamma_m}\widetilde{\varphi}(\lambda)e^{i\lambda x}\,d\lambda,$$

$$\varphi^*_m(x)=\frac{1}{(2\pi)^{n/2}}\int_{R^n}\chi_{\Gamma^*_m}(\lambda)\widetilde{\varphi}(\lambda)e^{i\lambda x}\,d\lambda=\frac{1}{(2\pi)^{n/2}}\int_{\Gamma^*_m}\widetilde{\varphi}(\lambda)e^{i\lambda x}\,d\lambda.$$

We will call the functions φ_{2^m} and $\varphi^*_{2^m}$ *packets*, and the functions φ_m and φ^*_m *corridors*. It is easy to show that

$$\varphi(x)=\sum_m\varphi_{2^m}(x),\qquad \varphi(x)=\sum_{m\geqslant 0}\varphi^*_{2^m}(x),$$

$$\varphi(x)=\sum_{-\infty}^{\infty}\varphi_m(x),\qquad \varphi(x)=\sum_0^{\infty}\varphi^*_m(x),$$

in the sense of pointwise convergence (the sums in the first two formulas are performed with respect to the ordering by corridors described in part I of the Appendix). The above representations of φ will be called *sections* of φ (respectively T, $\overset{+}{T}$, T and $\overset{+}{T}$-sections of φ).(1)

The general form of the Littlewood-Paley theorem on sections asserts that for $1<p<\infty$ we have

$$\Big\|\Big\{\sum_m|\varphi_{2^m}(x)|^2\Big\}^{1/2}\Big\|_p\sim\|\varphi\|_p, \tag{I}$$

$$\Big\|\Big\{\sum_{m\geqslant 0}|\varphi^*_{2^m}(x)|^2\Big\}^{1/2}\Big\|_p\sim\|\varphi\|_p, \tag{II}$$

$$\Big\|\Big\{\sum_{-\infty}^{\infty}|\varphi_m(x)|^2\Big\}^{1/2}\Big\|_p\sim\|\varphi\|_p, \tag{III}$$

$$\Big\|\Big\{|S_\varphi(x)|^2+\sum_1^{\infty}|\varphi^*_m(x)|^2\Big\}^{1/2}\Big\|_p\sim\|\varphi\|_p. \tag{IV}$$

Since S is dense in L_p, $1<p<\infty$, these relations are valid on all of L_p for $1<p<\infty$. Here the packets and corridors can be interpreted as singular convolutions.

Relation (I) was obtained by J. T. Schwartz for $n=1$ in [3] and by the author for arbitrary n in [4] (see also [5]); relation (III) was proved by Krée in [6]; relation (II) was proved in [7]; relation (IV) follows easily from (III). For suppose that (III) is true. Then we have the chain of equivalence relations

$$\|\varphi\|_p=\Big\|S_\varphi(x)+\sum_1^{\infty}\varphi^*_m(x)\Big\|_p\sim\|S_\varphi(x)\|_p+\Big\|\sum_1^{\infty}\varphi^*_m(x)\Big\|_p\sim$$

(using (III) to justify the second term)

---

(1) It is convenient to speak of "sectioning φ into packets" or "sectioning φ into corridors". We let $S_\varphi(x)$ denote $\varphi^*_0(x)$.

$$\sim \| S_\varphi(x) \|_p + \left\| \left\{ \sum_1^\infty |\varphi_m^*(x)|^2 \right\}^{1/2} \right\|_p \sim | S_\varphi(x) | + \left\{ \sum_1^\infty |\varphi_m^*(x)|^2 \right\}^{1/2} \Bigg\|_p$$

$$\sim \left\| \left\{ |S_\varphi(x)|^2 + \sum_1^\infty |\varphi_m^*(x)|^2 \right\}^{1/2} \right\|_p .$$

Thus if we take the completion of S with respect to any of the norms on the left sides of the relations (I)–(IV) we obtain the space L_p, $1 < p < \infty$.

On the set S we now introduce the more general norms

$$|\varphi, L_{p,\theta}| = \left\| \left\{ \sum_{m} |\varphi_{2^m}(x)|^\theta \right\}^{1/\theta} \right\|_p,$$

$$|\varphi, \overset{+}{L}_{p,\theta}| = \left\| \left\{ \sum_{m \geqslant 0} |\varphi_{2^m}^*(x)|^\theta \right\}^{1/\theta} \right\|_p,$$

$$|\varphi, \mathscr{L}_{p,\theta}| = \left\| \left\{ \sum_{-\infty}^{\infty} |\varphi_m(x)|^\theta \right\}^{1/\theta} \right\|_p,$$

$$|\varphi, \overset{+}{\mathscr{L}}_{p,\theta}| = \left\| \left\{ |S_\varphi(x)|^\theta + \sum_0^\infty |\varphi_m^*(x)|^\theta \right\}^{1/\theta} \right\|_p .$$

We can see that these norms are finite for $\varphi \in S$, $1 < p < \infty$, $1 \leqslant \theta \leqslant \infty$.(2) The completion of S with respect to these norms will be denoted by $L_{p,\theta}$, $\overset{+}{L}_{p,\theta}$, $\mathscr{L}_{p,\theta}$ and $\overset{+}{\mathscr{L}}_{p,\theta}$ respectively.

We can see that each of the above spaces for $\theta = 2$ and $1 < p < \infty$ coincides with L_p. For $\theta \neq 2$, as we will see later, the spaces $L_{p,\theta}$, $\overset{+}{L}_{p,\theta}$, $\mathscr{L}_{p,\theta}$ and $\overset{+}{\mathscr{L}}_{p,\theta}$ are distinct and do not equal L_p. For the sake of brevity we will let $\Lambda_{p,\theta}$ denote any of the spaces $L_{p,\theta}$, $\overset{+}{L}_{p,\theta}$, $\mathscr{L}_{p,\theta}$ or $\overset{+}{\mathscr{L}}_{p,\theta}$.

We note that in the one-dimensional case ($n = 1$) the space $L_{p,\theta}$ coincides with $\mathscr{L}_{p,\theta}$, and $\overset{+}{L}_{p,\theta}$ coincides with $\overset{+}{\mathscr{L}}_{p,\theta}$.

§2. Basic properties of functions in $\Lambda_{p,\theta}$. The spaces $\Lambda_{p,\theta}^{r}$

It is natural to expect that the space $\Lambda_{p,\theta}$ is not very different from L_p when θ is near 2. We will not present a detailed discussion of the "deviation" of $\Lambda_{p,\theta}$ from L_p. It follows immediately from our definitions and well-known summation inequalities that for $1 \leqslant \theta < 2$ and $1/\theta + 1/\theta' = 1$ we have the imbeddings

$$\Lambda_{p,1} \to \Lambda_{p,\theta} \to \Lambda_{p,2} \sim L_p \to \Lambda_{p,\theta'} \to \Lambda_{p,\infty}.\text{(3)} \tag{1}$$

For example, for $\mathscr{L}_{p,\theta}$ we have

$$\left\| \left\{ \sum_{-\infty}^{\infty} |\varphi_m(x)|^2 \right\}^{1/2} \right\|_p \leqslant \left\| \left\{ \sum_{-\infty}^{\infty} |\varphi_m(x)|^\theta \right\}^{1/\theta} \right\|_p \leqslant \left\| \sum_{-\infty}^{\infty} |\varphi_m(x)| \right\|_p ,$$

which gives us the imbeddings listed on the left side of (1). Thus for $1 \leqslant \theta < 2$ we have $\mathscr{L}_{p,\theta} \to L_p$; the imbedding is strict (see II in the Appendix).

What are the elements of $\Lambda_{p,\theta}$ which are adjoined to S when we take its completion

---

(2) See I in the Appendix.

(3) Thus the space $\Lambda_{p,\theta}$ grows monotonically when θ increases.

for $\theta > 2$? Let f be such an element. It is generated by some convergent (in the norm of $\Lambda_{p,\theta}$) sequence of functions $\varphi^k \in S$. From the structure of the norm it is clear that the mth components φ_m^k of these functions (to fix ideas, we consider $L_{p,\theta}$) will converge in the norm of L_p to some (entire) function f_m whose spectrum is concentrated on Γ_m.

In fact, convergence of the sequence $\{\varphi^k\}$ means that

$$\left\|\left\{\sum |\varphi_m^k - \varphi_m^l|^\theta\right\}^{1/\theta}\right\|_p < \varepsilon \quad \text{for} \quad k, l > N(\varepsilon).$$

Consequently

$$\|\varphi_m^k - \varphi_m^l\|_p < \varepsilon \quad \text{for} \quad k, l > N(\varepsilon).$$

Since L_p is complete, the φ_m^k converge in the norm of L_p to some function $f_m \in L_p$. Clearly the support of the Fourier transform of f_m (in the sense of generalized functions) is concentrated on Γ_m. Clearly convergence of φ_m^k to f_m as $k \to \infty$ implies that

$$\langle \varphi_m^k, \widetilde{\psi}\rangle \to \langle f_m, \widetilde{\psi}\rangle, \quad \psi \in S.$$

If we use the definition of the Fourier transform in S' we find that

$$\langle \widetilde{\varphi}_m^k, \psi\rangle \to \langle \widetilde{f}_m, \psi\rangle.$$

Therefore if the function ψ vanishes outside $\overline{\Gamma}_m$, then $\langle \widetilde{f}_m, \psi\rangle = 0$. This means that the support of $\widetilde{f}_m$ is concentrated on Γ_m.

It is also clear that

$$\lim_{k\to\infty}\left\|\left\{\sum_{-\infty}^{\infty} |\varphi_m^k|^\theta\right\}^{1/\theta}\right\|_p = \left\|\left\{\sum_{m=-\infty}^{\infty} |f_m|^\theta\right\}^{1/\theta}\right\|_p;$$

and also

$$\lim_{k\to\infty}\left\|\left\{\sum_{-\infty}^{\infty} |f_m - \varphi_m^k|^\theta\right\}^{1/\theta}\right\|_p = 0. \tag{2}$$

The space $L_{p,\theta}$ can be regarded as a subspace of the space $L_p(l_\theta)$ of functions $\mathbf{f}(x)$ with values in the space of (two-sided) sequences whose θ-powers are summable. This sequence space is complete. Therefore a sequence $\varphi^k = \{\varphi_m^k\}_{m=-\infty}^{\infty}$ which converges in the norm of this space has a limit $\{f_m\}_{-\infty}^{\infty}$:

$$\{f_m\}_{-\infty}^{\infty} \overset{\text{def}}{=} \mathbf{f} \in L_p(l_\theta).$$

We have shown that the limit function $\mathbf{f}$ has a special structure; namely, each of its components f_m has a spectrum which is concentrated on Γ_m. This property characterizes $L_{p,\theta}$. Therefore we write φ^k in the form φ^k, and consequently we can write $\mathbf{f}$ in the form f. Then equation (2) can be rewritten in the form

$$\lim_{k\to\infty} \varphi^k \overset{\mathscr{L}_{p,\theta}}{=} f;$$

that is, f is the limit of the φ^k in the sense of norm convergence. Thus we have described the elements which are adjoined to S when we take its completion.

From what we have shown, it follows that the series Σf_m converges to f in the norm of $L_{p,\theta}$. Thus for $\varphi \in S$ we have the convergence (see I in the Appendix)

$$\sum_{-M}^{N} \varphi_m \xrightarrow{\mathscr{L}_{p,\theta}} \varphi \quad \text{as} \quad M, N \to \infty.$$

Now we will show that the series $\Sigma_{-\infty}^{\infty} f_m(x)$ converges to f in S'; that is, as $M \longrightarrow \infty$ and $N \longrightarrow \infty$ we have

$$\left\langle \sum_{-M}^{N} f_m, \varphi \right\rangle \to \langle f, \varphi \rangle \qquad \forall \varphi \in S.$$

In fact, the finite sum $\Sigma_{-M}^{N} f_m(x)$ is an element of S' since(4)

$$\begin{aligned} \left| \left\langle \sum_{-M}^{N} f_m, \varphi \right\rangle \right| &= \left| \int_{R^n} \left(\sum_{-M}^{N} f_m \right) \varphi(x)\, dx \right| \leqslant \sum_{-M}^{N} \left| \int_{R^n} f_m \varphi\, dx \right| \\ &= \sum_{-M}^{N} \left| \int_{R^n} f_m \varphi_m\, dx \right| \leqslant \int_{R^n} \sum_{-M}^{N} |f_m \varphi_m|\, dx \\ &\leqslant \int_{R^n} \left(\sum_{-M}^{N} |f_m|^{\theta} \right)^{1/\theta} \left(\sum_{-M}^{N} |\varphi_m|^{\theta'} \right)^{1/\theta'} dx \\ &\leqslant \left\| \left\{ \sum_{-\infty}^{\infty} |f_m|^{\theta} \right\}^{1/\theta} \right\|_p \left\| \left\{ \sum_{-\infty}^{\infty} |\varphi_m|^{\theta'} \right\}^{1/\theta'} \right\|_{p'}. \end{aligned} \tag{3}$$

It now follows that the quantity $|\varphi, L_{p',\theta'}|$ is majorized by a seminorm of φ in S of finite order (see (A.I.4) in the Appendix). Thus we have shown that $\Sigma_{-M}^{N} f_m(x) \in S'$. Since (3) is valid for arbitrary M and N, we have $\Sigma_{-\infty}^{\infty} f_m(x) \in S'$. Clearly $\Sigma_{-\infty}^{\infty} f_m(x) \overset{S'}{=} f$.

Thus an element $f \in L_{p,\theta}$ is a generalized function in S' which can be written in the form

$$f = \sum f_m, \tag{4}$$

where each of the terms f_m is an entrie function with a spectrum which is concentrated on Γ_m such that

$$\left\| \left\{ \sum |f_m(x)|^{\theta} \right\}^{1/\theta} \right\|_p < \infty.$$

The inadequate feature of this characterization is the fact that we cannot construct the expansion (4) for an arbitrary function $f \in S'$. Therefore it is desirable to work with a space of generalized functions which is smaller than S' and in which the construction of the functions f_m from f is well defined. In the $\overset{+}{L}_{p,\theta}$ and $\overset{+}{L}_{p,\theta}$ cases we can use the Schwartz space D'_{L_p} found in [8] (considered also by Nikol'skiĭ in [7], where it was called the space S'_p of generalized functions which are regular in the sense of L_p). We will return to this question later when we consider spaces of generalized functions, where it is natural to consider the spaces $L_{p,\theta}$ and $L_{p,\theta}$.

We will find the general form of a linear functional on $\Lambda_{p,\theta}$ and we will show that the dual space of $\Lambda_{p,\theta}$ is the space $\Lambda_{p',\theta'}$. To fix ideas, we will work with the space $L_{p,\theta}$.

---

(4) The function f_m is orthogonal to $\varphi_{m'}$ if $m' \neq m$. We can see this by virtue of the fact that $\varphi_k^m \xrightarrow{L_p} f_m$.

LEMMA. *Each continuous linear functional on* $L_{p,\theta}$, $1 < p < \infty$, $1 \leqslant \theta < \infty$, *can be written in the form*

$$\sum_{-\infty}^{\infty} (f_m, u_m), \qquad f = \sum_{-\infty}^{\infty} f_m \in \mathscr{L}_{p,\theta},$$

where $u = \Sigma_{-\infty}^{\infty} u_m$ *is an element of* $L_{p',\theta'}$.

PROOF. Suppose that we are given a continuous linear functional U on $L_{p,\theta}$ with norm $\|U\|$:

$$| Uf | \leqslant \| U \| \, |f, \mathscr{L}_{p,\theta}|.$$

We can regard $L_{p,\theta}$ as a closed subspace of the space $L_p(l_\theta)$ of vector-valued functions $\mathbf{f}(x)$, $x \in R^n$, with values in l_θ and with a finite norm $\{\int |\mathbf{f}(x), l_\theta|^p \, dx\}^{1/p}$. By the Hahn-Banach theorem the functional U can be extended from the subspace $L_{p,\theta}$ to the whole space $L_p(l_\theta)$ with preservation of the norm. We will denote the extended functional by the same symbol U, and thus

$$| U\mathbf{f} | \leqslant \| U \| \, | \mathbf{f}, \; L_p(l_\theta) |.$$

But it is known that each functional on $L_p(l_\theta)$ can be written in the form

$$((\mathbf{f}, \mathbf{g})) = \int_{R^n} (\mathbf{f}(x), \mathbf{g}(x)) \, dx = \int_{R^n} \sum_{-\infty}^{\infty} f_m(x) \, g_m(x) \, dx,$$

where $\mathbf{g}(x)$ is some element in $L_{p'}(l_{\theta'})$. Thus for some $\mathbf{u}(x) \in L_{p'}(l_{\theta'})$ we have

$$U\mathbf{f} = \int_{R^n} \sum f_m(x) \, u_m(x) \, dx.$$

We note that the interchange of integration and summation is valid since

$$\left| \sum f_m(x) \, u_m(x) \right| \leqslant \sum | f_m(x) \, u_m(x) | \leqslant \left(\sum | f_m(x) |^\theta \right)^{1/\theta} \left(\sum | u_m(x) |^{\theta'} \right)^{1/\theta'}.$$

If we write $U\mathbf{f}$ in the form

$$U\mathbf{f} = \sum \int_{R^n} f_m(x) \, u_m(x) \, dx, \tag{5}$$

we see that its value does not change if the transform of the Fourier function changes outside Γ_m.

We now show that formula (5) implies that the norm of the extended functional U equals $\| \{ \Sigma |u_m(x)|^{\theta'} \}^{1/\theta'} \|_{p'}$. We can show that the smallest value of this norm is obtained in the case where the spectrum of u_m is concentrated on Γ_m. Since extending U did not change its norm, we see that $\widetilde{u}_m$ equals zero outside $\overline{\Gamma}_m$, and thus an element $\mathbf{u} \in L_{p'}(l_{\theta'})$ can be regarded as a function $u = \Sigma u_m \in L_{p',\theta'}$. Consequently

$$Uf = \int_{R^n} \sum f_m(x) \, u_m(x) \, dx, \qquad f \in \mathscr{L}_{p,\theta},$$

where u is a completely determined element of $L_{p',\theta'}$.

Conversely, an expression of the form

$$\int_{R^n} \sum f_m(x)\, u_m(x)\, dx,$$

where $u = \Sigma u_m \in L_{p',\theta'}$, clearly determines a linear functional on $L_{p,\theta}$. This follows from the estimates

$$\Big| \int_{R^n} \sum f_m(x)\, u_m(x)\, dx \Big| \leqslant \int_{R^n} \sum | f_m(x)\, u_m(x) |\, dx$$

$$\leqslant \int_{R^n} \Big\{ \sum | f_m(x) |^{\theta} \Big\}^{1/\theta} \Big\{ \sum | u_m(x) |^{\theta'} \Big\}^{1/\theta'} dx \leqslant c\, | f, \mathcal{L}_{p,\theta} |,$$

$$c = \Big\| \Big\{ \sum | u_m(x) |^{\theta'} \Big\}^{1/\theta'} \Big\|_{p'}.$$

The lemma is proved.

As a corollary we obtain the following assertion.

COROLLARY. *The dual space of* $L_{p,\theta}(1 < p < \infty;\ 1 \leqslant \theta < \infty)$ *can be identified with the space* $L_{p',\theta'}$. *For* $1 < p < \infty$ *and* $1 < \theta < \infty$ *the space* $L_{p,\theta}$ *is reflexive.*

We note that if for $\theta < 2$ the space $L_{p,\theta}$ is defined as the set of functions in L_p such that $\|\{\Sigma |f_m|^{\theta}\}^{1/\theta}\|_p < \infty$, then for $\theta > 2$ it can be defined as the set of continuous linear functionals on $L_{p',\theta'}$.

Finally, we consider the relationship between $\overset{+}{L}_{p,\theta}$ and the Besov spaces $B_{p,\theta}$. Such a space is defined as the collection of functions $f = \Sigma f^*_m$ in S' (5) such that

$$\Big\{ \| S_f(x) \|_p^{\theta} + \sum_{m=1}^{\infty} \| f^*_m \|_p^{\theta} \Big\}^{1/\theta} < \infty.$$

For $1 \leqslant p \leqslant \theta$ we have

$$\Big\{ \sum \Big(\int_{R^n} | f^*_m(x) |^p\, dx \Big)^{\theta/p} \Big\}^{1/\theta} \leqslant \Big\{ \int_{R^n} \Big(\sum | f^*_m |^{\theta} \Big)^{p/\theta} dx \Big\}^{1/p},$$

and for $1 \leqslant \theta \leqslant p$ the direction of the inequality is reversed. Thus we have the imbeddings

$$B_{p,\theta} \to \overset{+}{\mathcal{L}}_{p,\theta}, \qquad 1 < \theta \leqslant p;$$

$$\overset{+}{\mathcal{L}}_{p,\theta} \to B_{p,\theta}, \qquad 1 \leqslant p \leqslant \theta.$$

In particular, for $p = \theta$ the spaces $\overset{+}{L}_{p,\theta}$ and $B_{p,\theta}$ coincide.

Now we are in a position to define the spaces $\Lambda^r_{p,\theta}$, where the index r is concerned with differentiation properties. On functions in S' the Bessel integration operator I_r is well defined (see [8]). On Fourier transforms this operator reduces to multiplication by the function $(1 + |\lambda|^2)^{-r/2}$, $r > 0$ (for $r < 0$ the corresponding operator will be called the Bessel differentiation operator). The function $(1 + |\lambda|^2)^{-r/2}$, $r > 0$, has as a Fourier pre-image the Bessel-Macdonald kernel

$$G_r(x) = \frac{1}{(2\pi)^{n/2}} \int_{R^n} \frac{e^{ix\lambda}\, d\lambda}{(1 + |\lambda|^2)^{r/2}} = 2^{\frac{2-r}{2}} \Gamma^{-1}\Big(\frac{r}{2}\Big) |x|^{-\frac{n-r}{2}} K_{\frac{n-r}{2}}(|x|),$$

(5) More precisely, $f \in S'_p$ (see [7]).

where $K_\nu(t)$ is the Macdonald function. The kernel $G_r(x)$ has been thoroughly investigated (see [7] or [8] and their bibliographies). The function $G_r(x)$ is integrable and satisfies the estimate

$$G_r(x) \leqslant \frac{c}{|x|^{n-r}}, \qquad 0<r<n,$$

which we will use later.

We note that on functions in L_p the operator I_r is a classical convolution

$$\mathcal{I}_r f = \frac{1}{(2\pi)^{n/2}} \int_{R^n} G_r(x-y) f(y)\, dy.$$

We have shown that $\Lambda_{p,\theta}$ is a part of S'. Therefore on functions in $\Lambda_{p,\theta}$ the operation I_r is defined. The isomorphic image of the space $\Lambda_{p,\theta}$ under the Bessel operator I_r is called $\Lambda^r_{p,\theta}$, $-\infty<r<\infty$, or, in symbols,

$$\Lambda^r_{p,\theta} \overset{\text{def}}{=} \mathcal{I}_r \Lambda_{p,\theta}.$$

By this definition, $f \in \Lambda^r_{p,\theta}$ if and only if $f = I_r g$, $g \in \Lambda_{p,\theta}$, where

$$|f, \Lambda^r_{p,\theta}| \overset{\text{def}}{=} |g, \Lambda_{p,\theta}|.$$

Hereafter we will concentrate on the study of the spaces $\Lambda^r_{p,\theta}$. In §3 we obtain new equivalent expressions for the norm $|f, \Lambda^r_{p,\theta}|$, and in §4 we establish a series of imbedding theorems for $\Lambda^r_{p,\theta}$.

§3. Theorems on multipliers and some corollaries

A function $\Phi(\lambda)$, $\lambda \in R^n$, is said to be a multiplier in $F\Lambda_{p,\theta}$ if $f \in \Lambda_{p,\theta}$ implies that $\Phi(\lambda)\widetilde{f}(\lambda)$ is the Fourier transform of some function $g \in \Lambda_{p,\theta}$ ($\widetilde{g}(\lambda) = \Phi(\lambda)\widetilde{f}(\lambda)$), where we have the estimate

$$|g, \Lambda_{p,\theta}| \leqslant c\, |f, \Lambda_{p,\theta}|,$$

with a constant c which does not depend on f. If a function is a multiplier in $F\Lambda_{p,\theta}$, then we will write $\Phi \in M(\Lambda_{p,\theta})$. In [9] the author proved various theorems on multipliers in $FL_{p,\theta}$. These theorems easily carry over to the spaces $L_{p,\theta}$ (with the appropriate changes). For the sake of completeness we present these theorems here with short proofs. Then we show how to obtain the corresponding theorems for the spaces $\overset{+}{L}_{p,\theta}$ and $\overset{+}{L}_{p,\theta}$.

THEOREM 1. *Suppose the function* $\Phi(\lambda)$ *can be written in the form*

$$\Phi(\lambda) = \int_{-\infty}^{\lambda_1} \cdots \int_{-\infty}^{\lambda_n} d\mu_m \tag{6}$$

in each of the corridors Γ_m, *where the* μ_m *are finite measures which are collectively bounded*:

$$\int_{R^n} |d\mu_m| \leqslant M.$$

Then $\Phi(\lambda)$ *is a multiplier in* $FL_{p,\theta}$, *and its norm is majorized by* M:

$$|\overline{\Phi(\lambda)\tilde{\varphi}(\lambda)}, \mathscr{L}_{p,\theta}| \leqslant cM|\varphi, \mathscr{L}_{p,\theta}|. \tag{7}$$

PROOF. First of all we note that for $\theta = 2$ this theorem coincides with a theorem of Krée in [6] on the spaces $L_p \sim L_{p,2}$.

We will use the method employed by the author in [9] for the case of the spaces $L_{p,\theta}$. We will proceed on the basis of the following Assertion A (see [9], Theorem 5 for $p = q$).

Let $1 < p < \infty$ and $1 < \theta < \infty$, and suppose that we are given a sequence of functions

$$\Phi_m(\lambda) = \int_{-\infty}^{\lambda_1} \dots \int_{-\infty}^{\lambda_n} d\mu_m,$$

where the μ_m are finite measures which are uniformly of bounded total variation:

$$\operatorname{var} \mu_m = \int_{R^n} |d\mu_m| \leqslant M.$$

Suppose the transformation $\mathscr{K}$ is defined on functions in $S \cap L_p(l_\theta^+)$ by the equation

$$\mathscr{K}\mathfrak{f} = \{\overline{\Phi_m(\lambda)\tilde{f}_m(\lambda)}\}_1^\infty.$$

Then $\mathscr{K}$ can be extended to a continuous transformation from $L_p(l_\theta^+)$ into $L_p(l_\theta^+)$. Here

$$|\mathscr{K}\mathfrak{f}, L_p(l_\theta^+)| \leqslant cM|\mathfrak{f}, L_p(l_\theta^+)|,$$

where the constant c does not depend on f or Φ.

We recall that here S is the family of vector-valued functions $\varphi(x)$, $x \in R^n$, with infinitely differentiable components φ_k, $k = 1, 2, \dots$, and $L_p(l_\theta^+)$ is the space of vector-valued functions

$$\mathfrak{f}(x) = (f_1(x), f_2(x), \dots, f_n(x), \dots)$$

with a finite norm

$$|\mathfrak{f}, L_p(l_\theta^+)| = \left\| \left\{ \sum_1^\infty |f_k(x)|^\theta \right\}^{1/\theta} \right\|_p.$$

It is clear that Assertion A remains true for functions with values in the space l_θ of two-sided sequences whose θ-powers are summable (in [9] it was applied to sequences with $2n$ inputs).

To a function $f \in L_{p,\theta}(R^n)$ we associate the vector-valued function $\mathfrak{f} = \{f_m\}_{-\infty}^\infty$. By virtue of the definition of the space $L_{p,\theta}(R^n)$ we have

$$\mathfrak{f} \in L_p(l_\theta), \qquad |f, \mathscr{L}_{p,\theta}| = \|\mathfrak{f}\|_{L_p(l_\theta)}.$$

Values of the function $\Phi(\lambda)$ in the domain Γ_m, defined outside Γ_m by (6), will be denoted by Φ_m. Then the operator $\mathscr{K}$ can be interpreted as an operator in $L_p(l_\theta)$ which acts on a vector-valued function $\mathfrak{f} \in L(l_\theta)$ according to the formula

$$\mathscr{K}\mathfrak{f} = \{\overline{\Phi_m \tilde{f}_m}\} = \{g_m\} = \mathfrak{g}.$$

Then by Assertion A (whose conditions are clearly satisfied) we have

$$\|\mathfrak{g}\|_{L_p(l_\theta)} \leqslant cM\|\mathfrak{f}\|_{L_p(l_\theta)}. \tag{8}$$

With the vector-valued function $\mathbf{g} \in L_p(l_\theta)$ we can associate a vector-valued function $g \in \mathcal{L}_{p,\theta}$ by setting $\widetilde{g}(\lambda) = \widetilde{g}_m(\lambda)$ for $\lambda \in \Gamma_m$. Since $|g, \mathcal{L}_{p,\theta}| = \|\mathbf{g}\|_{L_p(l_\theta)}$, by means of (8) we obtain (7). The theorem is proved.

We can see that other theorems on multipliers for $L_{p,\theta}$ which were proved in [9] carry over to $\mathcal{L}_{p,\theta}$ in a similar fashion. We will limit ourselves to presenting one more such result.

THEOREM 2. *Suppose that the function $\Phi(\lambda)$ is continuous outside all of the coordinate hyperplanes (that is, for $|\lambda_j| > 0, j = 1, \ldots, n$) and has continuous derivatives there*:

$$\frac{\partial^k \Phi}{\partial \lambda_1^{k_1} \ldots \partial \lambda_n^{k_n}}, \qquad 0 \leqslant k_1 + \ldots + k_n = k \leqslant n, \qquad k_j = 0, 1,$$

where

$$|\lambda|^{k_1 + \ldots + k_n} \left| \frac{\partial^k \Phi}{\partial \lambda_1^{k_1} \ldots \partial \lambda_n^{k_p}} \right| \leqslant M. \tag{9}$$

Then Φ is a multiplier in $F\mathcal{L}_{p,\theta}$, and the norm of the corresponding multiplier transformation

$$T : f \to \widehat{\Phi \widetilde{f}}$$

is majorized by M; that is,

$$|\widehat{T\widetilde{f}}, \mathcal{L}_{p,\theta}| \leqslant cM |f, \mathcal{L}_{p,\theta}|.$$

The proof is very similar to the proof for the spaces $L_{p,\theta}$ given in [9], but of course it depends on Theorem 1 above. We will be brief. For each m we can construct from the function Φ a function Φ_m by means of the formula

$$\Phi_m(\lambda) = \begin{cases} \Phi(\lambda), & \lambda \in \Gamma_m; \\ 0, & \lambda \notin \Gamma_m. \end{cases}$$

We take the (generalized) derivative of the function Φ_m:

$$\frac{\partial^n \Phi}{\partial \lambda_1 \ldots \partial \lambda_n}.$$

This derivative is a measure distributed inside Γ_m with density $\partial^n \Phi / \partial \lambda_1 \cdots \partial \lambda_n$, along the $(n-1)$-dimensional faces $\lambda_j = \text{const}$ with density $\partial^{n-1}\Phi/\partial\lambda_1 \cdots \partial\lambda_n$, and so forth (so that at zero-dimensional faces, that is, vertices of Γ_m, we obtain δ-functions multiplied by the corresponding values of the function). The mass (total variation) of such a measure turns out to be uniformly bounded with respect to m (some multiple of M) by virtue of (9), and we can use Theorem 1. The theorem is proved.

Theorems on multipliers for the spaces $\overset{+}{\mathcal{L}}_{p,\theta}$ are obtained as corollaries of the corresponding theorems for the spaces $\mathcal{L}_{p,\theta}$ and the following obvious equivalence relation:

$$|f, \overset{+}{\mathcal{L}}_{p,\theta}| = \left\| \left\{ |S_f(x)|^\theta + \sum_{m>1} |\overset{*}{f}_m(x)|^\theta \right\}^{1/\theta} \right\|_p$$

$$\sim \|S_f(x)\|_p + \left\| \left\{ \sum_{m \geqslant 1} |\overset{*}{f}_m(x)|^\theta \right\}^{1/\theta} \right\|_p. \tag{10}$$

Suppose that a function Φ is a multiplier in $\mathcal{L}_{p,\theta}$ and also in $\mathcal{L}_{p,2} = L_p$. We let T denote the corresponding multiplier transformation

$$T\colon \mathscr{L}_{p,\theta} \ni f \to \overline{\Phi \tilde{f}} \in \mathscr{L}_{p,\theta}.$$

This transformation is defined and bounded on $\overset{+}{L}_{p,\theta}$. In fact, a function f in $\overset{+}{L}_{p,\theta}$ can be written in the form

$$f = S_f(x) + h(x), \qquad h(x) = f(x) - S_f(x).$$

Clearly $S_f \in L_p$ (see [10]), and $h \in L_{p,\theta}$. Therefore

$$Tf = TS_f + Th,$$

$$|Tf, \overset{+}{\mathscr{L}}_{p,\theta}| \leqslant |TS_f, \overset{+}{\mathscr{L}}_{p,\theta}| + |Th, \overset{+}{\mathscr{L}}_{p,\theta}|.$$

Since a multiplier transformation does not enlarge the support of the spectrum, by (10) we have

$$|TS_f, \overset{+}{\mathscr{L}}_{p,\theta}| \sim \|TS_f\|_p.$$

It is also clear that

$$|Th, \overset{+}{\mathscr{L}}_{p,\theta}| = \Big\| \Big\{ \sum_{m \geqslant 0} |(Tf)^*_m|^\theta \Big\}^{1/\theta} \Big\|_p = \Big\| \Big\{ \sum_{-\infty}^{\infty} |(Tf)_m|^\theta \Big\}^{1/\theta} \Big\|_p = |Th, \mathscr{L}_{p,\theta}|.$$

If we continue the preceding inequality, noting that T is bounded in L_p and $L_{p,\theta}$, we obtain (in the last step we again use (10))

$$\begin{aligned} |Th, \overset{+}{\mathscr{L}}_{p,\theta}| &\leqslant c\,(\|S_f\|_p + |h, \mathscr{L}_{p,\theta}|) \\ &= c\Big(\|S_f\|_p + \Big\|\Big\{\sum |f^*_m|^\theta\Big\}^{1/\theta}\Big\|_p\Big) \sim |f, \overset{+}{\mathscr{L}}_{p,\theta}|. \end{aligned} \tag{11}$$

Thus the transformation T is bounded in $\overset{+}{L}_{p,\theta}$, and consequently Theorems 1 and 2 are valid not only in $L_{p,\theta}$ but also in $\overset{+}{L}_{p,\theta}$.

On the basis of Theorem 1 we can renorm the spaces $\overset{+}{L}{}^r_{p,\theta}$.

THEOREM 3. *On functions in* $\overset{+}{L}{}^r_{p,\theta}$ *we have the equivalence relation*

$$|f, \overset{+}{\mathscr{L}}{}^r_{p,\theta}| \sim \Big\| \Big\{ |S_f(x)|^\theta| + \sum_{m \geqslant 1} 2^{m\theta r} |f^*_m(x)|^\theta \Big\}^{1/\theta} \Big\|_p. \tag{12}$$

PROOF. If $f \in \overset{+}{L}{}^r_{p,\theta}$, then

$$f = G_r * g, \qquad g \in \overset{+}{\mathscr{L}}_{p,\theta},$$

$$|f, \overset{+}{\mathscr{L}}{}^r_{p,\theta}| = |g, \overset{+}{\mathscr{L}}_{p,\theta}|.$$

If we take Fourier transforms, we obtain

$$\tilde{f} = \tilde{G}_r \tilde{g};$$

that is,

$$\tilde{g} = (1 + |\lambda|^2)^{r/2} \tilde{f} = \frac{(1 + |\lambda|^2)^{r/2}}{2^{mr}} 2^{mr} \tilde{f}(\lambda) \quad \text{for} \quad \lambda \in \Gamma_m.$$

The factor

$$\Phi(\lambda) = \frac{(1 + |\lambda|^2)^{r/2}}{2^{mr}} \quad \text{for} \quad \lambda \in \Gamma_m$$

is a multiplier from $L_{p,\theta}$ into $L_{p,\theta}$. In fact, if we define $\Phi(\lambda)$ to be zero outside Γ_m^*, we obtain a function $\Phi_m(\lambda)$ which can be represented in Γ_m^* by means of the formula

$$\Phi_m(\lambda)=\int_{-\infty}^{\lambda_1}\cdots\int_{-\infty}^{\lambda_n} d\mu_m,$$

where μ_m is the generalized derivative $\partial^n\Phi_m/\partial\lambda_1\cdots\partial\lambda_n$. By means of easy calculations we can see that the total variation of this measure is uniformly bounded with respect to m (see the concluding arguments in Theorem 2). Thus

$$\left\|\left\{\sum_0^\infty |g_m^*(x)|^\theta\right\}^{1/\theta}\right\|_p \leqslant c\left\|\left\{\sum |2^{mr}f_m^*(x)|^\theta\right\}^{1/\theta}\right\|_p. \tag{13}$$

Since (by virtue of similar considerations) the function

$$\frac{2^{mr}}{(1+|\lambda|^2)^{r/2}} \quad \text{for} \quad \lambda\in\Gamma_m$$

is also a multiplier from $L_{p,\theta}$ into $L_{p,\theta}$, the estimate which is the converse of (13) is also valid. Therefore the equivalence relation (11) implies (12). Theorem 3 is proved.

We mention the special case of Theorem 3 with $\theta=2$. Here the space $\overset{+}{L}{}^r_{p,2}$ to within equivalence of norms coincides with the space L_p^r of Bessel potentials, and Theorem 3 asserts equivalence of the quantities

$$|f, L_p^r|\sim\left\|\left\{|S_f(x)|^2+\sum_1^\infty 2^{2mr}|f_m^*|^2\right\}^{1/2}\right\|_p.$$

This characterization of L_p^r can be used in finding estimates for singular numbers of the imbedding operator in spaces of Bessel potentials.

We note that on functions in $L^r_{p,\theta}$ the corresponding equivalence relation is of the form (for arbitrary r)

$$|f, \mathscr{L}^r_{p,\theta}|\sim\left\|\left\{\sum_{-\infty}^\infty (1+2^m)^{\theta r}|f_m(x)|^\theta\right\}^{1/\theta}\right\|_p. \tag{14}$$

As a corollary of Theorem 3 we obtain the imbedding

$$\overset{+}{\mathscr{L}}{}^r_{p,\theta}\to L_p, \qquad r>0,\ \forall\theta. \tag{15}$$

First of all we note that the operator I_r has the group property (see [8])

$$\mathscr{I}_\rho(\mathscr{I}_r f)=\mathscr{I}_{r+\rho}f.$$

From this fact and the definition of the norm in $L_{p,\theta}$ we obtain

$$|\mathscr{I}_\rho g, \overset{+}{\mathscr{L}}{}^r_{p,\theta}|=|g, \overset{+}{\mathscr{L}}{}^{r-\rho}_{p,\theta}|.$$

If $f\in\overset{+}{L}{}^r_{p,\theta}$, then $f=I_r g$, $g\in\overset{+}{L}_{p,\theta}$. We have

$$f=\mathscr{I}_r g, \qquad g\in\overset{+}{\mathscr{L}}_{p,\theta}.$$

$$\|f\|_p = \|\mathcal{I}_r g\|_p \sim |\mathcal{I}_r g, \overset{+}{\mathcal{L}}_{p,2}| = |g, \mathcal{L}^{-r}_{p,2}|$$
$$\sim \left\| \left\{ |S_g(x)|^2 + \sum_1^\infty 2^{-2mr} |\overset{*}{g}_m(x)|^2 \right\}^{1/2} \right\|_p$$
$$\leqslant \left\| |S_g(x)| + \sum_1^\infty 2^{-mr} |\overset{*}{g}_m(x)| \right\} \Big\|_p$$
$$\sim \left\| \left\{ |S_g(x)|^\theta + \left(\sum_1^\infty 2^{-mr} |\overset{*}{g}_m(x)| \right)^\theta \right\}^{1/\theta} \right\|_p$$
$$\leqslant c_\theta \left\| \left\{ |S_g(x)|^\theta + \sum_1^\infty |\overset{*}{g}_m(x)|^\theta \right\}^{1/\theta} \right\|_p = c_0 |f, \overset{+}{\mathcal{L}}{}^r_{p,\theta}|.$$

Here we have used the Hölder inequality

$$\left(\sum_1^\infty 2^{-mr} |g_m(x)| \right)^\theta \leqslant \sum_1^\infty |\overset{*}{g}_m(x)|^\theta \left(\sum_1^\infty 2^{-mr\theta'} \right)^{\frac{\theta}{\theta'}}.$$

Thus the imbedding (15) is proved.

§4. Imbedding theorems for the spaces $\Lambda^r_{p,\theta}$

Imbedding theorems for the spaces $\Lambda^r_{p,\theta}$ are easily obtained by means of estimates on integrals of potential type for vector-valued functions. These estimates could be obtained directly by the methods used in [10]. For the sake of brevity we will make use of reductions to the scalar case, also considered in [10]. We will prove the fundamental lemma in a more general form than we need in the context of this paper (since the generality does not require any additional effort).

Let B be a Banach space with norm $\|\cdot\|$, and let B^* be its dual space with norm $\|\cdot\|_{B^*}$. A strongly measurable B-valued function $f(x) = f(x_1, \ldots, x_n)$ defined on an n-dimensional space R^n belongs, by definition, to the space $L_{\mathbf{p}}(B, R^n) = L_{\mathbf{p}}(B)$, $\mathbf{p} = (p_i, \ldots, p_n)$, $1 \leqslant p_j \leqslant \infty$, $j = 1, \ldots, n$, if the following norm is finite:

$$\|f\|_{L_{\mathbf{p}}(B)} = \left\{ \int_{-\infty}^\infty dx_1 \left[\int_{-\infty}^\infty dx_2 \ldots \left(\int_{-\infty}^\infty \|f(x)\|_B^{p_n} dx_n \right)^{\frac{p_{n-1}}{p_n}} \ldots \right]^{\frac{p_1}{p_2}} \right\}^{\frac{1}{p_1}}$$

(if p_j is infinite we use the ess sup). For $p_1 = \cdots = p_n = p$, the space $L_{\mathbf{p}}(B)$ coincides with the space $L_p(B)$ of functions $f(x)$ with a finite norm

$$\|f\|_{L_p(B)} = \left\{ \int_{R^n} \|f(x)\|_B^p dx \right\}^{1/p}.$$

LEMMA. *Let*

$$h(x) \in L_{\mathbf{p}}(B), \quad g(x) \in L_{\mathbf{q}'}(B^*),$$

$$\lambda = \sum \varkappa_j \left(\frac{1}{q_j} + \frac{1}{p_j'} \right), \quad \frac{1}{p_j} + \frac{1}{p_j'} = 1, \quad \frac{1}{q_j} + \frac{1}{q_j'} = 1, \quad 1 < p_j < q_j \leqslant \infty$$

(here $q_j = \infty$ is allowed only for $j > 1$). Then the integral

$$I = \int_{R^n}\int_{R^n} \frac{\langle g(x), h(y)\rangle}{\| x - y\|^{\lambda}}\, dx\, dy, \tag{16}$$

where

$$\| x\| = \sqrt{\sum_1^n | x_j |^{2/\varkappa_j}}, \qquad \varkappa_j > 0,$$

and $\langle g, h\rangle$ *denotes the value of a functional* $g \in B^*$ *on an element* $h \in B$, *satisfies the estimate*

$$I \leqslant k^n \| h \|_{L_{\mathbf{p}}(B)} \| g \|_{L_{\mathbf{p}'}(B^*)}, \tag{17}$$

where k is the constant in the Hardy-Littlewood inequality.

PROOF. We use the inequality

$$\left| \int_{R^n}\int_{R^n} \frac{\langle g(x), h(y)\rangle}{\| x - y\|^{\lambda}}\, dx dy \right| \leqslant \int_{R^n}\int_{R^n} \frac{\| g(x)\|_{B^*} \| h(y)\|_B}{\| x - y\|^{\lambda}}\, dx\, dy.$$

We apply Lemma 3 in [10] to the right side of this inequality and we obtain (17). The lemma is proved.

In this work we are interested mainly in the case

$$p_1 = p_2 = \ldots = p_n = p, \quad q_1 = q_2 = \ldots = q_n = q, \quad 1 < p < q < \infty,$$

$$\varkappa_1 = \varkappa_2 = \ldots = \varkappa_n = 1.$$

In this case the quantity $\| x - y \|$ reduces to the euclidean distance

$$| x - y | = \sqrt{\sum_{j=1}^n (x_j - y_j)^2}, \quad \lambda = n(1/q + 1/p').$$

Inequality (17) can be rewritten in the form

$$\left| \int_{R^n}\int_{R^n} \frac{\langle g(x), h(y)\rangle}{| x - y |^{n(1/q + 1/p')}}\, dx dy \right| \leqslant k^n \| h \|_{L_p(B)} \| g \|_{L_{q'}(B^*)} .$$

From the preceding lemma we immediately obtain the following result.

THEOREM 4. *Let* $h(x) \in L_p(B)$, $1 < p < \infty$, *and*

$$f(x) = \int \frac{h(y)\, dy}{| x - y |^{n-r}}, \qquad 0 < r < n. \tag{18}$$

Then $f(x) \in L_q(B)$, $1/q - 1/p = r/n$, $1 < p < q < \infty$, *and we have the estimate*

$$\| f \|_{L_q(B)} \leqslant c \| h \|_{L_p(B)}, \tag{19}$$

where the constant c does not depend on h.

PROOF. Let q be defined by the equation $1/q - 1/p = r/n$; that is, $n - r = n(1/q + 1/p')$. For an arbitrary function $g \in L_{q'}(B^*)$ we have

$$\left|\int_{R^n} \langle g(x), f(x)\rangle\, dx\right| \leqslant \int_{R^n} |\langle g(x), h(x)\rangle|\, dx$$
$$\leqslant \int_{R^n} \|g(x)\|_{B^*} \|f(x)\|_B\, dx = \int_{R^n} \|g(x)\|_{B^*} \left\|\int_{R^n} \frac{h(y)\,dy}{|x-y|^{n-r}}\right\|_B dx$$
$$\leqslant \int_{R^n}\int_{R^n} \frac{\|g(x)\|_{B^*}\|h(y)\|_B}{|x-y|^{n(1/q+1/p')}}\, dxdy \leqslant k^n \|h\|_{L_p(B)} \|g\|_{L_{q'}(B^*)}.$$

Then by a known result in functional analysis we obtain (19).

From Lemma 1 we can also obtain an estimate for the norm of a function in $L_q(R^m)$. But this estimate can also be obtained directly from a theorem of Il'in after taking the B-norm of both sides of (18):

$$\|f(x)\|_B = \left\|\int_{R^n} \frac{h(y)\,dy}{|x-y|^{n-r}}\right\|_B \leqslant \int_{R^n} \frac{\|h(y)\|_B}{|x-y|^{n-r}}\, dy.$$

By the theorem of Il'in the right side, as a function, is majorized in the norm of $L_q(R^m)$, $r - n/p = -m/q$, by the quantity

$$\left\{\int_{R^n} \|h(y)\|_B^p dy\right\}^{1/p}.$$

From these relations we obtain the estimate

$$\left\{\int_{R^m} \|f(x)\|_B^q dx\right\}^{1/q} \leqslant c\left\{\int_{R^n} \|h(y)\|_B^p dy\right\}^{1/p}, \qquad r = \frac{n}{p} - \frac{m}{q}. \tag{20}$$

Now we are in a position to obtain imbedding theorems for the spaces $\Lambda^r_{p,\theta}$.

THEOREM 5 ("imbedding of different metrics"). *Let $1 < p < q < \infty$, $1 \leqslant \theta \leqslant \infty$, $r - n/p = \rho - n/q$, $\rho \geqslant 0$. If $f(x) \in L^r_{p,q}(R^n)$, then $f(x) \in \Lambda^\rho_{q,\theta}(R^n)$, and we have the estimate*

$$|f, \Lambda^\rho_{q,\theta}| \leqslant c\,|f, \Lambda^r_{p,\theta}|,$$

where the constant c does not depend on f. Thus we have the imbedding

$$\Lambda^r_{p,\theta}(R^n) \to \Lambda^\rho_{q,\theta}(R^n), \quad r - \frac{n}{p} = \rho - \frac{n}{q}.$$

PROOF. To fix ideas, let $f \in L^r_{p,\theta}$; then

$$f(x) = G_r * g, \qquad g \in \mathscr{L}_{p,\theta}(R^n).$$

It is clear that this convolution can be written in the form

$$f_m(x) = \int_{R^n} G_r(x-y)\, g_m(y)\, dy.$$

If we take the l_θ-norm of both sides and use estimates for the kernel G_r, we obtain

$$\left(\sum_{-\infty}^{\infty} |f_m(x)|^\theta\right)^{1/\theta} \leqslant \int_{R^n} G_r(x-y)\left\{\sum |g_m(y)|^\theta\right\}^{1/\theta} dy \leqslant c\int_{R^n} \frac{\left(\sum |g_m(y)|^\theta\right)^{1/\theta}}{|x-y|^{n-r}}\, dy. \tag{21}$$

Then, as we proved above, we have

$$\left\| \left\{ \sum | f_m(x) |^\theta \right\}^{1/\theta} \right\|_q \leqslant c \left\| \left\{ \sum | g_m(y) |^\theta \right\}^{1/\theta} \right\|_p, \qquad r - \frac{n}{p} = -\frac{n}{q};$$

or in other words

$$| f, \mathscr{L}_{q,\theta} | \leqslant c \, | g, \mathscr{L}_{p,\theta} |; \tag{22}$$

that is,

$$\mathscr{L}^r_{p,\theta}(R^n) \to \mathscr{L}_{q,\theta}(R^n).$$

In the general case, by using the group property of G_r we can show that

$$f(x) = G_\rho * (G_{r-\rho} * g). \tag{23}$$

From what we have shown, if $g \in L_{p,\theta}$, then

$$h = G_{r-\rho} * g \in \mathscr{L}_{q,\theta}, \qquad r - \rho - \frac{n}{p} = -\frac{n}{q}$$

and thus (23) can be written in the form

$$f(x) = G_\rho * h, \qquad h \in \mathscr{L}_{q,\theta}. \tag{24}$$

This means that

$$f \in \mathscr{L}^\rho_{q,\theta}, \qquad r - \frac{n}{p} = \rho - \frac{n}{q}.$$

Also, from (22) and (24) we have

$$| f, \mathscr{L}^\rho_{q,\theta} | = | h, \mathscr{L}_{q,\theta} | \leqslant c \, | g, \mathscr{L}_{p,\theta} |$$

and finally we obtain

$$| f, \mathscr{L}^\rho_{q,\theta} | \leqslant c \, | f, \mathscr{L}^r_{p,\theta} |.$$

The theorem is proved.

We note that Theorem 5 can be proved in a more general form:

$$f \in \Lambda^r_{p,\theta}(R^n) \Rightarrow \frac{\partial^\lambda f}{\partial x_1^{\lambda_1} \ldots \partial x_n^{\lambda_n}} \in \Lambda^\rho_{q,\theta}(R^m), \qquad r - \frac{n}{p} = \rho + \lambda - \frac{n}{q}.$$

Here we should note that each differentiation reduces the differential index of the class by one. This fact can be verified by means of a theorem on multipliers (where we restrict the values of θ to be strictly between 1 and ∞).

THEOREM 6 ("imbedding of different dimensions"). *Let $1 < p < q < \infty$, $1 \leqslant \theta \leqslant \infty$, $r - n/p = \rho - m/q$ and $\rho \geqslant 0$. If $f(x_1, \ldots, x_n) \in \Lambda^r_{p,\theta}(R^n)$, then for arbitrary fixed values of $x_{m+1}, \ldots, x_n$ the function $f(x_1, \ldots, x_m, x^0_{m+1}, \ldots, x^0_n)$ belongs to the space $\Lambda^\rho_{q,\theta}(R^n)$, and we have the estimate*

$$| f, \Lambda^\rho_{q,\theta}(R^m) | \leqslant c \, | f, \Lambda^r_{p,\theta}(R^n) |,$$

where the constant c does not depend on f. Thus we have the imbedding

$$\Lambda^r_{p,\theta}(R^n) \to \Lambda^\rho_{q,\theta}(R^m), \qquad r - \frac{n}{p} = \rho - \frac{m}{q}.$$

PROOF. The imbedding

$$\mathcal{L}^r_{p,\theta}(R^n) \to \mathcal{L}_{q,\theta}(R^m), \qquad r - \frac{n}{p} = -\frac{m}{q}$$

follows from (21). In fact, by the Il'in theorem we see from (21) that

$$\left\| \left\{ \sum_{-\infty}^{\infty} | f_m(x) |^\theta \right\}^{1/\theta} \right\|_{L_q(R^m)} \leqslant c \left\| \left\{ \sum_{-\infty}^{\infty} | g_m(y) |^\theta \right\}^{1/\theta} \right\|_p, \qquad r - \frac{n}{p} = -\frac{m}{q}.$$

For the proof of the general imbedding $L^r_{p,\theta}(R^n) \to L^\rho_{q,\theta}(R^n)$ we proceed as follows. The Fourier transform of the function

$$f = G_r * g$$

can be written, because of the convolution theorem, in the following form $(\lambda = (\lambda_1, \ldots, \lambda_m))$:(6)

$$\tilde{f} = \tilde{G}_r \tilde{g} = (1 + |\lambda|^2)^{-r/2} \tilde{g} = (1 + |\lambda'|^2)^{-\rho/2} (1 + |\lambda|^2)^{-\frac{r-\rho}{2}} \frac{(1 + |\lambda'|^2)^{\rho/2}}{(1 + |\lambda|^2)^{\rho/2}} \tilde{g},$$

where

$$|\lambda'| = \sqrt{\lambda_1^2 + \ldots + \lambda_m^2}, \quad 0 < \rho < r.$$

The function $\tilde{g}(\lambda)$ is the Fourier transform of $g \in L_{p,\theta}(R^n)$. By a theorem on multipliers the function

$$\tilde{h} = \frac{(1 + |\lambda'|^2)^{\rho/2}}{(1 + |\lambda|^2)^{r/2}} \tilde{g}$$

is also the Fourier transform of some function $h \in L_{p,\theta}(R^n)$, where

$$| h, \mathcal{L}_{p,\theta} | \leqslant c \, | g, \mathcal{L}_{p,\theta} |. \tag{25}$$

Thus

$$\tilde{f} = (1 + |\lambda'|^2)^{-\frac{\rho}{2}} (1 + |\lambda|^2)^{\frac{r-\rho}{2}} \tilde{h}, \qquad h \in \mathcal{L}_{p,0}(R^n),$$

which in the basic space can be written in the form(7)

$$f = \int_{R^n} G^m_\rho (x' - y') \, dy' \int_{R^n} G^n_{r-\rho} (y - t) \, h(t) \, dt = f(x_1, \ldots, x_m, y_{m+1} \ldots y_n). \tag{26}$$

By the preceding theorem the function

$$\int_{R^n} G^n_{r-\rho} (x - t) \, h(t) \, dt = \Phi(x_1, \ldots, x_n)$$

belongs to $L_{q,\theta}(R^m)$ if $r - \rho - n/p = -m/q$, and

$$| \Phi(\cdot, y''), \mathcal{L}_{q,\theta}(R^m) | \leqslant c \, | h, \mathcal{L}_{p,\theta}(R^n) |, \tag{27}$$

where c does not depend on h or $y'' = (y_{m+1}, \ldots, y_n)$. By virtue of (26) we can write

---

(6) This equation can also be written for the components $\tilde{f}_m = \tilde{G}_r \tilde{g}_m$, and so forth.

(7) In the general case this notation should be interpreted as a convolution (in the theory of generalized functions).

$$f = \int_{R^m} G_\rho^m (x' - y') \Phi (y', y'') \, dy' = G_\rho^m * \Phi (\cdot, y'').$$

Thus for each fixed y'' the function f is an element of $L^\rho_{q',\theta}(R^m)$, and (by virtue of (25) and (27))

$$| f, \mathscr{L}^\rho_{q,\theta} (R^m) | \overset{\text{def}}{=} | \Phi (\cdot, y''), \mathscr{L}_{q,\theta} (R^m) | \leqslant c \, | h, \mathscr{L}_{p,\theta} | \leqslant c_1 | g, \mathscr{L}_{p,\theta} | = c_1 | f, \mathscr{L}^r_{p,\theta} |.$$

Therefore $L^r_{p,\theta}(R^n) \longrightarrow L^\rho_{q,\theta}$ for $r - n/p = \rho - n/q$. The theorem is proved.

Appendix

I. *Finiteness of the norm* $\|\varphi, \Lambda_{p,\theta}\|$, $\varphi \in S$. *Convergence of the series* Σf_m *and* Σf_{2m}. For the sake of simplicity we will restrict our calculations to the case $n = 1$ for the space $L_{p,\theta}$. Without loss of generality we can assume that $\varphi \in S$ and $\tilde{\varphi}(\lambda)$ equals zero for negative λ. We have

$$| \varphi, \mathscr{L}_{p,\theta} | = \left\{ \int_{-\infty}^{\infty} \left(\sum_{-\infty}^{\infty} | \varphi_m (x) |^\theta \right)^{p/\theta} \right\}^{1/p} \leqslant \left\{ \int_{-\infty}^{\infty} \left(\sum | \varphi_m (x) | \right)^p dx \right\}^{1/p}$$

$$\leqslant \left\{ \int_{|x|<1} \left(\sum_{-\infty}^{0} | \varphi_m (x) | \right)^p dx \right\}^{1/p} + \left\{ \int_{|x|>1} \left(\sum_{-\infty}^{0} | \varphi_m (x) | \right)^p dx \right\}^{1/p}$$

$$+ \left\{ \int_{|x|<1} \left(\sum_{1}^{\infty} | \varphi_m (x) | \right)^p dx \right\}^{1/p} + \left\{ \int_{|x|>1} \left(\sum_{1}^{\infty} | \varphi_m (x) | \right)^p dx \right\}^{1/p} = \mathcal{I}_1 + \mathcal{I}_2 + \mathcal{I}_3 + \mathcal{I}_4.$$

To get an estimate for I_1 we use the inequality

$$| \varphi_m (x) | = \frac{1}{\sqrt{2\pi}} \left| \int_{2^m}^{2^{m+1}} \tilde{\varphi} (\lambda) \, e^{i\lambda x} d\lambda \right| \leqslant \frac{c_0}{\sqrt{2\pi}} 2^m, \qquad c_0 = \max_\lambda | \tilde{\varphi} (\lambda) |.$$

We have

$$\mathcal{I}_1 \leqslant \frac{c_0}{\sqrt{2\pi}} \left\{ \int_{-1}^{1} \left(\sum_{-\infty}^{0} 2^m \right)^p dx \right\}^{1/p} = \frac{c_0}{\sqrt{2\pi}} 2^{1+1/p}. \tag{A.I.1}$$

To get an estimate for I_2 we use integration by parts and write φ_m in the form

$$\sqrt{2\pi} \, \varphi_m (x) = \int_{2^m}^{2^{m+1}} \tilde{\varphi} (\lambda) \, e^{i\lambda x} d\lambda = \frac{[\tilde{\varphi} (2^{m+1}) - \tilde{\varphi} (2^m)]}{2^m} \, 2^m \, \frac{e^{i2^{m+1}}}{ix}$$

$$+ \tilde{\varphi} (2^m) \, e^{i2^m x} \, \frac{[e^{i2^m x} - 1]}{ix} - \frac{1}{ix} \int_{2^m}^{2^{m+1}} \tilde{\varphi}' (\lambda) \, e^{i\lambda x} d\lambda.$$

By the mean-value theorem we have

$$\left| \frac{\tilde{\varphi} (2^{m+1}) - \tilde{\varphi} (2^m)}{2^m} \right| \leqslant | \tilde{\varphi}' (\xi) | \leqslant \max_\lambda | \tilde{\varphi}' (\lambda) | = c_1.$$

Consequently

$$\sqrt{2\pi} \sum_{-\infty}^{-M} | \varphi_m (x) | \leqslant \frac{c_1}{|x|} \sum_{-\infty}^{-M} 2^m + \sum_{-\infty}^{-M} | \tilde{\varphi} (2^m) | \frac{| e^{i2^m x} - 1 |}{|x|} + \frac{1}{|x|} \int_0^{2^{-M+1}} | \tilde{\varphi}'(\lambda) | \, d\lambda. \tag{A.I.2}$$

Therefore we have

$$\sqrt{2\pi}\,\mathcal{I}_2 \leqslant 3c_1\left(\frac{2}{p-1}\right)^{1/p} + \sum_{-\infty}^{0} |\widetilde{\varphi}(2^m)| \left\{2\int_0^\infty \frac{|e^{i2^m x}-1|^p}{|x|^p}\,dx\right\}^{1/p}.$$

Since

$$\int^\infty \frac{e^{i2^m x}-1|}{|x|^p}\,dx < \int_0^\infty \frac{|e^{i2^m x}-1|^p}{x^{p-1}}\frac{dx}{x} = 2^{m(p-1)}\int_0^\infty \frac{|e^{iu}-1|^p}{u^p}\,du = ce^{m(p-1)},$$

we have

$$\sqrt{2\pi}\,\mathcal{I}_2 \leqslant 3c_1\left(\frac{2}{p-1}\right)^{1/p} + c_0 c\sum_{-\infty}^{0} 2^{m(p-1)} = \text{const},\quad c_0 = \max_\lambda |\widetilde{\varphi}(\lambda)|.$$

To get an estimate for I_3 we note that $\widetilde{\varphi} \in S$ and therefore $|\widetilde{\varphi}(\lambda)| \leqslant a/\lambda^2$. It follows that

$$\mathcal{I}_3 = \frac{1}{\sqrt{2\pi}}\left\{\int_{-1}^{1}\left(\sum_1^\infty \int_{2^m}^{2^{m+1}} |\widetilde{\varphi}(\lambda)|\,d\lambda\right)^p dx\right\}^{1/p} \leqslant \frac{2^{1/p}}{\sqrt{2\pi}}\,a\sum_1^\infty \int_{2^m}^{2^{m+1}} \frac{d\lambda}{\lambda^2} = \frac{a2^{1/p}}{\sqrt{2\pi}}\sum_1^\infty 2^{-m}.$$

To get an estimate for I_4 we use the fact that

$$|\widetilde{\varphi}(\lambda)| < \frac{b}{|\lambda|},\quad |\widetilde{\varphi'}(\lambda)| < \frac{d}{\lambda^2}.$$

We have

$$\sqrt{2\pi}\,\mathcal{I}_4 \leqslant \left\{\int_{|x|>1}\left(\sum_1^\infty\left|\frac{\widetilde{\varphi}(2^{m+1})\,e^{i2^m x} - \widetilde{\varphi}(2^m)\,e^{i2^m x}}{ix} - \frac{1}{ix}\int_{2^m}^{2^{m+1}} \widetilde{\varphi}' e^{i\lambda x}\,d\lambda\right|\right)^p dx\right\}^{1/p}$$

$$\leqslant \left\{\int_{|x|>1} \frac{1}{|x|^p}\left[\sum_1^\infty (|\widetilde{\varphi}(2^{m+1})| + |\widetilde{\varphi}(2^m)|\right]^p dx\right\}^{1/p}$$

$$+ \left\{\int_{|x|>1} \frac{1}{|x|^p}\left(\sum_1^\infty \int_{2^m}^{2^{m+1}} |\widetilde{\varphi}'(\lambda)|\,d\lambda\right)^p dx\right\}^{1/p} = \text{const}.$$

Thus we have shown that $|\varphi, L_{p,\theta}| < \infty$ for $\varphi \in S$, $1 < p < \infty$, $1 \leqslant \theta \leqslant \infty$. We note that for $\theta \geqslant 2$ there is a shorter proof, using monotonicity of the l_θ-norm and the Littlewood-Paley theorem on sections:

$$\left\{\int_{-\infty}^\infty \left(\sum |\varphi_m(x)|^\theta\right)^{\frac{1}{\theta}p} dx\right\}^{1/p} \leqslant \left\{\int_{-\infty}^\infty \left(\sum |\varphi_m(x)|^2\right)^{p/2} dx\right\}^{1/p} \leqslant B\|\varphi\|_p.$$

We should mention two important facts.

1. The calculations can be performed in such a way that we obtain the estimate

$$|\varphi, \mathcal{L}_{p,\theta}| \leqslant c \sup_{\substack{x\\ k,\,|l|\leqslant 2}} (1+|\lambda|^2)^k D^l\widetilde{\varphi}(\lambda). \tag{A.I.3}$$

Since the Fourier transformation is an isomorphism of the countably-normed space S onto itself, the seminorm on the right side of (A.I.3) can be estimated in terms of a seminorm of finite order N of the function φ; that is,

$$|\varphi, \mathcal{L}_{p,\theta}| \leqslant c \sup_{\substack{x\\ k, |l|\leqslant N}} (1+|x|^2)^{k/2} D^l \varphi(x). \tag{A.I.4}$$

2. Essentially the same calculations enable us to prove norm-convergence of the series $\Sigma\varphi_m$ to φ. Here we should note that the quantity

$$\left|\varphi - \sum_{-M}^{N} \varphi_m, \mathcal{L}_{p,\theta}\right| = \left\| \left\{ \sum_{-\infty}^{-(M+1)} |\varphi_m|^\theta + \sum_{N+1}^{\infty} |\varphi_m|^\theta \right\}^{1/\theta} \right\|_p$$

can be made arbitrarily small for large M and N. As before, we obtain four terms I'_1, I'_2, I'_3, I'_4, this time containing the above truncated sums. Here I'_1 is small since for large M the sum $\Sigma_{-\infty}^{-(M+1)} 2^m$ is small (see (A.I.1)). To get an estimate for I'_2 we use (A.I.2). The term I'_3 is small since $\Sigma_{N+1}^{\infty} 2^{-m}$ is small for large N, and so forth.

We note that pointwise convergence of the series $\Sigma\varphi_m$ to φ follows from the general theory of the Fourier integral; for an integrable function at points of locally bounded variation we have an inversion formula in the sense of principal values.

All of the preceding facts carry over to the spaces $L_{p,\theta}(R^n)$ and $\overset{+}{L}_{p,\theta}(R^n)$; naturally the calculations become more complicated. When we consider the spaces $L_{p,\theta}$ and $\overset{+}{L}_{p,\theta}$ we also have to settle the question of the order of summation of the multiple series.

On the set of vectors $\mathbf{m} = (m_1, \ldots, m_n)$ with nonnegative integer components m_j ($m_j = 0, 1, 2, \ldots; j = 1, \ldots, n$) we introduce an ordering as follows. Two such vectors $\mathbf{m}^1$ and $\mathbf{m}^2$ are said to be equal if all of their components are equal. We will write $\mathbf{m}^1 < \mathbf{m}^2$ if $\max_j m_j^1 < \max_i m_i^2$, or if $\max_j m_j^1 = \max_i m_i^2$ and the first nonequal components are related by the symbol $<$. For example, if $n = 3$ this rule gives us the distribution

000
001, 010, 011, 100, 101, 110, 111
002, 012, 020, 021, 022, 102, 112, 120, 121, 122
200, 201, 202, 210, 211, 212, 220, 221, 222.

We will call this ordering an *ordering by corridors.*

A series $\Sigma_{\mathbf{m}}\varphi_{2\mathbf{m}}$ will be written as a sum of 2^n terms:

$$\sum_{\mathbf{m}} \varphi_{2\mathbf{m}} = \sum_{\sigma} \sum_{\mathbf{m}\in\sigma} \varphi_{2\mathbf{m}},$$

where the inner sum is taken over those vectors $\mathbf{m}$ which have the same signature (the signature is the distribution of signs of the components of the vector), and the outer sum is taken over all signatures. If the signature contains a minus sign in some position, to the inner sum we assign the terms corresponding to those $\mathbf{m}$ which have a zero in this position. Thus σ is that n-ant(8) of R^n which is half-closed in a natural way (by attaching zero components to the ones with negative values). In the case of an arbitrary (but identical!) signature for $\mathbf{m}^1$ and $\mathbf{m}^2$, we will say that $\mathbf{m}^1 \leqslant \mathbf{m}^2$ if

$$(|m_1{}^1|, |m_2{}^1|, \ldots, |m_n{}^1|) \leqslant (|m_1{}^2|, |m_2{}^2|, \ldots, m_n{}^2).$$

Vectors with distinct signatures are not comparable. We will interpret convergence of the series $\Sigma_{\mathbf{m}}\varphi_{2\mathbf{m}}$ in terms of convergence of the collection of series $\Sigma_{\mathbf{m}\in\sigma}\varphi_{2\mathbf{m}}$. In each such

(8) For $n = 2$ this means quadrant; for $n = 3$ it means octant; and so forth.

series the vectors $\mathbf{m}$ can be enumerated by natural numbers $N_{\mathbf{m}}$. We set

$$S_k^\sigma = \sum_{\substack{\mathbf{m}\in\sigma \\ N_{\mathbf{m}}\leqslant k}} \varphi_{2\mathbf{m}}.$$

Here $\lim_{k\to\infty} S_k^\sigma$ means the sum of the series $\Sigma_{\mathbf{m}\in\sigma}\varphi_{2\mathbf{m}} = S^\sigma$. Then

$$\sum \varphi_{2\mathbf{m}} \overset{\text{def}}{=} \sum_\sigma S^\sigma.$$

With these conventions we can establish finiteness of the norms $|\varphi, L_{p,\theta}|$ and $|\varphi, \overset{+}{L}_{p,\theta}|$, $\varphi \in S$, and convergence of the series $\Sigma_{\mathbf{m}}\varphi_{2\mathbf{m}}$ pointwise and with respect to the norms in $L_{p,\theta}$ and $\overset{+}{L}_{p,\theta}$. Because the situation is clear enough, we will not dwell on it. This method of summing multiple series will be called *summation by ordered corridors.*

II. *Examples of strictness of the imbedding* (1). Suppose that f is a function such that

$$\tilde{f} = \begin{cases} c_m, & 2^m - 1 < \lambda < 2^m, \quad m = 2, 3, \dots; \\ 0 & \text{for other } \lambda. \end{cases}$$

Then

$$f_{m-1}(x) = \frac{c_m}{\sqrt{2\pi}} \int_{2^m-1}^{2^m} e^{i\lambda x} d\lambda = \frac{c_m}{\sqrt{2\pi}} e^{i(2^m-1/2)x} \frac{\sin x/2}{x/2}$$

and consequently

$$\left(\sum | f_m(x)|^\theta\right)^{1/\theta} = \frac{1}{\sqrt{2\pi}} \left|\frac{\sin x/2}{x/2}\right| \left(\sum | c_m|^\theta\right)^{1/\theta}.$$

Thus if the series $\Sigma|c_m|^\theta$ converges, then $f \in \Lambda_{p,\theta}$.

Membership of f in $L_p \sim L_{p,2}$ means convergence of the series $\Sigma|c_m|^2$. We also know that we have the imbedding (1):

$$\Lambda_{p,\theta} \to L_p \text{ for } 1 \leqslant \theta < 2.$$

If we choose c_m such that the series $\Sigma|c_m|^\theta$ diverges and the series $\Sigma|c_m|^2$ still converges, we see that the imbedding is strict. If $\theta > 2$ we take

$$c_m = \frac{1}{(m \ln^2 m)^{1/\theta}}.$$

Then the series $\Sigma|c_m|^\theta$ converges but $\Sigma|c_m|^2$ diverges. Consequently $f \in \Lambda_{p,\theta}$, but $f \notin \Lambda_{p,2} \sim L_p$, and thus the imbedding $L_p \longrightarrow L_{p,\theta}$ is strict.

BIBLIOGRAPHY

1. P. I. Lizorkin, *Generalized Liouville differentiation and the method of multipliers in the theory of imbeddings of classes of differentiable functions,* Trudy Mat. Inst. Steklov. **105** (1969), 89–167 = Proc. Steklov Inst. Math. **105** (1969), 105–202. MR **41** #7419.

2. ———, *Generalized Hölder spaces* $B^{(r)}_{p,\theta}$ *and their relation with the Sobolev spaces* $L^{(r)}_p$, Sibirsk. Mat. Ž. 9 (1968), 1127–1152 = Siberian Math. J. 9 (1968), 837–858. MR **38** #3726.

3. J. T. Schwartz, *A remark on inequalities of Calderón-Zygmund type for vector-valued functions,* Comm. Pure Appl. Math. **14** (1961), 785–799. MR **26** #597.

4. P. I. Lizorkin, (L_p, L_q)*-multipliers of Fourier integrals,* Dokl. Akad. Nauk SSSR **152** (1963), 808–811 = Soviet Math. Dokl. 4 (1963), 1420–1424. MR **27** #4016.

5. W. Littman, C. McCarthy and N. Rivière, *L^p-multiplier theorems,* Studia Math. **30** (1968), 193–217. MR **37** #6681.

6. P. Krée, *Sur les multiplicateurs dans FL^p*, C. R. Acad. Sci. Paris **260** (1965), 4400–4403. MR **31** #2634.

7. S. M. Nikol'skiĭ, *Approximation of functions of several variables and embedding theorems,* "Nauka", Moscow, 1969; English transl., Die Grundlehren der math. Wissenschaften, Band 205, Springer-Verlag, Berlin and New York, 1974. MR **46** #9714.

8. L. Schwartz, *Théorie des distributions.* Tomes I, II, Actualités Sci. Indust., nos. 1091, 1122, Hermann, Paris, 1950, 1951. MR **12**, 31, 833.

9. P. I. Lizorkin, *On multipliers of Fourier integrals in the spaces $L_{p,\theta}$*, Trudy Mat. Inst. Steklov. **89** (1967), 231–248 = Proc. Steklov Inst. Math. **89** (1967), 269–290. MR **36** #608.

10. ———, *Multipliers of Fourier integrals and bounds of convolutions in spaces with mixed norm. Applications,* Izv. Akad. Nauk SSSR Ser. Mat. **34** (1970), 218–247 = Math. USSR Izv. **4** (1970), 225–254. MR **41** #7420.

Translated by F. A. CEZUS

Trudy Mat. Inst. Steklov.
131 (1974)

Proc. Steklov Inst. Math.
131 (1974)

THE BEHAVIOR AT INFINITY OF FUNCTIONS WITH DIFFERENTIAL-DIFFERENCE PROPERTIES GIVEN IN L_p

UDC 517.518.22

Ju. S. NIKOL'SKIĬ

ABSTRACT. Functions of several real variables with differential-difference properties given in L_p are considered. It is shown that either they or their derivatives "look like a polynomial" in a well-defined sense as $|x| \longrightarrow \infty$, and those which "look like zero" are approximable with arbitrary precision by infinitely differentiable functions with compact support.

Bibliography: 9 items.

We shall consider the space $b^{\mathbf{l}(\mathbf{r})}_{p,\beta}$ of functions $f(x)$, $x \in E_n$, which have generalized x_i-derivatives of order s_i, $i = 1, \ldots, n$, with finite seminorm

$$\|f\|_{b^{\mathbf{l}(\mathbf{r})}_{p,\beta}} = \sum_{i=1}^{n} \left\{ \int_0^\infty t^{-1-p(l_i - s_i)} dt \int_{E_n} |[\rho(x) + t^{1/\alpha_i}]^\beta \Delta_i^{\sigma_i}(t)\, D_i^{s_i} f(x)|^p dx \right\}^{1/p}, \ 1 \leqslant p < \infty, \quad (1)$$

$$\|f\|_{b^{\mathbf{l}(\mathbf{r})}_{\infty,\beta}} = \sum_{i=1}^{n} \operatorname*{ess\,sup}_{x \in E_n,\, t>0} |t^{-l_i+s_i}[\rho(x) + t^{1/\alpha_i}]^\beta \Delta_i^{\sigma_i}(t)\, D_i^{s_i} f(x)|, \quad (2)$$

where

$$\mathbf{l} = (l_1, \ldots, l_n), \quad \mathbf{r} = (r_1, \ldots, r_n), \ r_i = \sigma_i + s_i > l_i > s_i \geqslant 0,$$

$$\rho(x) = \sum_{i=1}^{n} |x_i|^{1/\alpha_i}, \qquad \alpha_i > 0,$$

$$\Delta_i^\sigma(t)\,\varphi(x) = \sum_{j=0}^{\sigma} (-1)^{\sigma-j} C_\sigma^j \varphi(x_1, \ldots, x_{i-1}, x_i + jt, \ x_{i+1}, \ldots, x_n).$$

We shall denote by $B^{\mathbf{l}(\mathbf{r})}_{p,\beta}$ the space of functions with the norm

$$\|f\|_{B^{\mathbf{l}(\mathbf{r})}_{p,\beta}} = \|f\|_{L_p(1<|x|<2)} + \|f\|_{b^{\mathbf{l}(\mathbf{r})}_{p,\beta}}. \quad (3)$$

In this paper it is shown that functions $f \in b^{\mathbf{l}(\mathbf{r})}_{p,\beta}$ or their derivatives "look like a polynomial" in a well-defined sense as $|x| \longrightarrow \infty$, and those which "look like zero" are approximated with an arbitrary precision by infinitely differentiable functions with respect to the norm (3). These results are a generalization of results of Uspenskiĭ [1], Sobolev [2], and Besov [3] on the behavior of differentiable functions from the whole spaces L_p^1 at infinity, and also of the corresponding results of the author [4] for the weight spaces $L^1_{p,\beta}$ to the fractional spaces $b^{\mathbf{l}(\mathbf{r})}_{p,\beta}$.

The following notation will be used: for an arbitrary vector $\alpha = (\alpha_1, \ldots, \alpha_n)$, $\alpha_i > 0$,

AMS (MOS) subject classifications (1970). Primary 46E30, 41A30; Secondary 26A24, 26A69.

$i = 1, \dots, n$, let us put $|\alpha| = \alpha_1 + \cdots + \alpha_n$, $\alpha! = \alpha_1! \cdots \alpha_n!$, if the $\alpha_i > 0$ are integers, $\alpha - 1 = (\alpha_1 - 1, \dots, \alpha_n - 1)$, $y = (y_1, \dots, y_n)$ is a point of Euclidean space E_n, $h > 0$, $y/h^\alpha = (y_1/h^{\alpha_1}, \dots, y_n/h^{\alpha_n})$, $\mathbf{k} = (k_1, \dots, k_n)$, k_i are nonnegative integers, $D_i^{k_i} = \partial^{k_i}/\partial y_i^{k_i}$, $D_y^k = D_1^{k_1} \cdots D_n^{k_n}$, $\mathbf{k}\alpha = k_1\alpha_1 + \cdots + k_n\alpha_n$, $\mathbf{k} < \mathbf{r}$, if $k_i < r_i$, $i = 1, \dots, n$, $1/p + 1/p' = 1$ and $\varphi^{(i)}(x, t) = t^{-1/p - l_i + s_i} \Delta_i^{\sigma_i}(t) D_i^{s_i} f(x)$. If the region Ω of integration is not indicated under the integral sign, this means that $\Omega = E_n$. For simplicity of writing, we shall write $A \ll B$ instead of the inequality $A \leqslant cB$, where $A = A(f)$ and $B = B(t)$ are certain expressions containing a function f of the class being considered.

Let us define the weight spaces $L_{p,\beta}$ and $L_{p,\beta,\alpha_i}(E^+_{n+1})$, $i = 1, \dots, n$, $1 \leqslant p \leqslant \infty$, of the functions $f(x)$, $x \in E_n$, and $F(x, t)$, $(x, t) \in E^+_{n+1} = \{(x, t) \colon t > 0\}$, respectively, with finite norms

$$\| f \|_{p,\beta} = \| [\rho(x)]^\beta f(x) \|_p, \quad \| F \|_{p,\beta,\alpha_i} = \| [\rho(x) + t^{1/\alpha_i}]^\beta F(x, t) \|_{L_p(E^+_{n+1})}.$$

Everywhere in the sequel we shall assume that $\alpha_i = 1/l_i$, $i = 1, \dots, n$, if it is not explicitly mentioned.

For any function $f \in b^{\mathbf{l}(\mathbf{r})}_{p,\beta}$ and for any vector $\mathbf{k}$ one has the integral representation of V. P. Il'in [5]

$$\begin{aligned} D^{\mathbf{k}} f_\varepsilon(x) - D^{\mathbf{k}} f_H(x) = D^{\mathbf{k}} f_{\varepsilon, H}(x) &= (-1)^{|\mathbf{k}|} \int_\varepsilon^H \sum_{i=1}^n h^{-\delta_i - \mathbf{k}\alpha}\, dh \\ &\times \iint T_i^{(\mathbf{k})}\left(\frac{y}{h^\alpha}, \frac{t}{h^{\alpha_i}}\right) \Delta_i^{\sigma_i}\left(\frac{t}{\sigma_i}\right) D_i^{s_i} f(x + y)\, dt\, dy, \end{aligned} \tag{4}$$

where $0 < \epsilon < H < \infty$, and if $\mathbf{k}\alpha < 1$, then $\epsilon \geqslant 0$, $D^{\mathbf{k}} f_0(x) = D^{\mathbf{k}} f(x)$ and for $\epsilon = 0$ the first equality in (4) is understood in the sense of "almost everywhere", $\delta_i = 1 + |\alpha| - s_i\alpha_i + \alpha_i$, and

$$D^{\mathbf{k}} f_\varepsilon(x) = (-1)^{|\mathbf{k}|} \varepsilon^{-|\alpha| - \mathbf{k}\alpha} \int \Omega^{(\mathbf{k})}\left(\frac{y}{\varepsilon^\alpha}\right) f(x + y)\, dy.$$

Here $T_i(y, t)$ and $\Omega(y)$ are certain infinitely differentiable functions with compact support respectively defined in E_{n+1} and E_n whose supports are respectively concentrated in the regions $0 < y_j < 1$, $j = 1, \dots, n$, $0 < t < 1$, $0 < t + y_i < 1$ and $0 < y_j < 1$, $j = 1, \dots, n$.

In connection with this representation let us study properties of the integral

$$I_{\varepsilon, H}(x) = \int_\varepsilon^H h^{-\delta_i - \sigma}\, dh \iint t^{1/p + l_i - s_i}\, T\left(\frac{y}{h^\alpha}, \frac{t}{h^{\alpha_i}}\right) F(x + y, t)\, dt\, dy, \quad l_i > s_i \geqslant 0,$$

for $F \in L_{p,\beta,\alpha_i}(E^+_{n+1})$ and a piecewise continuous finite function $T(y, t)$ with support concentrated in the region $0 \leqslant y_j < 1$, $j = 1, \dots, n$, $0 \leqslant t < 1$, $0 < t + y_i < 1$.

LEMMA 1. *Let $1 \leqslant p \leqslant q \leqslant \infty$, $\beta < (|\alpha| + \alpha_i)/p'$, $0 \leqslant \mu < |\alpha|/q + \beta$ (for $p = 1$ or $q = \infty$ one can respectively put $\beta = 0$ or $\mu = \beta$), and let $F \in L_{p,\beta,\alpha_i}(E^+_{n+1})$. Then*

$$\left\| h^{-\delta_i} \iint t^{1/p + l_i - s_i}\, T\left(\frac{y}{h^{\alpha_i}}, \frac{t}{h^{\alpha_i}}\right) F(x + y, t)\, dt\, dy \right\|_{q, \beta - \mu} \leqslant c h^{-|\alpha|(1/p - 1/q) - \mu} \| F \|_{p,\beta,\alpha_i}, \tag{5}$$

where the constant c does not depend on F or h.

PROOF. Put $\varphi(x, t) = [\rho(x) + t^{1/\alpha_i}]^\beta F(x, t)$. Since the Minkowski inequality

$$\rho(x+y) \leqslant c_0(\rho(x)+\rho(y)) \tag{6}$$

is satisfied for the function $\rho(x)$, it is easy to show that for $\rho(x) < 2c_0 a$, $\rho(y) < a$, $t < a^{\alpha_i}$, $a > 0$ (c_0 is the constant from (6)), we have

$$c_1\rho(x) \leqslant \rho(x+y) + t^{1/\alpha_i} \leqslant c_2\rho(x), \tag{7}$$

where the positive constants c_1 and c_2 do not depend on a.

Using properties of the kernel $T(y, t)$ and the inequalities (7) for $a = nh$, applying the Hölder inequality for the three functions 1, $[\varphi(x+y)t^{1/p+l_i-s_i}]^{q-p/q}$, and $[\varphi(x+y)t^{1/p+l_i-s_i}]^{p/q}$ with indices $p/(p-1)$, $pq/(q-p)$, and q (in the cases $p=1$, $p=q$ and $q=\infty$ the ordinary Hölder inequality suffices), we obtain, after simple computations,

$$\begin{aligned}
&\left\| h^{-\delta_i} \iint t^{1/p+l_i-s_i} T\left(\frac{y}{h^{\alpha}}, \frac{t}{h^{\alpha_i}}\right) F(x+y, t)\,dt\,dy \right\|_{L_{q,\beta-\mu}(\rho(x) > 2c_0 nh)} \\
&\leqslant h^{-\delta_i-\mu}\left(\int\limits_{0<y_j<h^{\alpha_j}} dy \int\limits_0^{h^{\alpha_i}} dt\right)^{\frac{p-1}{p}} \left(\iint\limits_0^{h^{\alpha_i}} t^{1+p(l_i-s_i)} |\varphi|^p\, dt\, dy\right)^{\frac{q-p}{pq}} \\
&\times \left(\int dx \int\limits_{0<y_j<h^{\alpha_j}} dy \int\limits_0^{h^{\alpha_i}} t^{1+p(l_i-s_i)} |\varphi(x+y,t)|^p\, dt\right)^{\frac{1}{q}} \leqslant h^{-|\alpha|\left(\frac{1}{p}-\frac{1}{q}\right)-\mu} \|F\|_{p,\beta,\alpha_i}.
\end{aligned} \tag{8}$$

Let us apply the Hölder inequality, noting that one can determine a positive number d, not depending on h, such that the set $\{(y, t)\colon y_j - x_j < h^{\alpha_j}, j = 1, \ldots, n, 0 < t < h^{\alpha_i}, \rho(x) < 2c_0 nh\} \subset \{(y, t)\colon \rho(y) + t^{1/\alpha_i} < dh\}$. In two integrals let us pass to generalized spherical coordinates (r, ξ) and (ρ, η), respectively, according to the formulas $y_j = r^{\alpha_j}\xi_j$, $j = 1, \ldots, n$, $t = r^{\alpha_i}\xi_{n+1}$ (here $r = \rho(y) + t^{1/\alpha_i}$, $\xi = (\xi_1, \ldots, \xi_{n+1})$ is the point of intersection of the unit r-sphere ($r(\xi) = 1$) with the trajectory $(u^{\alpha_1}\xi_1, \ldots, u^{\alpha_n}\xi_n, u^{\alpha_i}\xi_{n+1})$, $0 < u < \infty$) passing through the point (y, t), and $x_i = \rho^{\alpha_i}\eta_i$, $i = 1, \ldots, n$ (here $\rho = \rho(x)$ and $\eta = (\eta_1, \ldots, \eta_n)$ is the point of intersection of the unit ρ-sphere with the trajectory $(u^{\alpha_1}\eta_1, \ldots, u^{\alpha_n}\eta_n)$, $0 < u < \infty$) [6]. As a result we obtain

$$\begin{aligned}
&h^{-\delta_i} \left\| \iint t^{\frac{1}{p}+l_i-s_i} T\left(\frac{y}{h^{\alpha}}, \frac{t}{h^{\alpha_i}}\right) F(x+y, t)\,dt\,dy \right\|_{L_{q,\beta-\mu}(\rho(x) < 2c_0 nh)} \\
&\leqslant h^{-\delta_i + \frac{\alpha_i}{p} + \alpha_i(l_i-s_i)} \left(\int\limits_0^{dh} r^{-\beta p'+|\alpha|+\alpha_i-1}\,dr\right)^{1/p'} \left(\int\limits_0^{2c_0 nh} \rho^{\beta q-\mu q+|\alpha|-1} d\rho\right)^{1/q} \|F\|_{p,\beta,\alpha_i} \\
&\leqslant h^{-|\alpha|\left(\frac{1}{p}-\frac{1}{q}\right)-\mu} \|F\|_{p,\beta,\alpha_i}.
\end{aligned} \tag{9}$$

For $p = 1$ or $q = \infty$ the first integral in the middle part of the latter inequalities must be replaced by $\sup_{r<dh} r^{-\beta}$, and the second integral by $\sup_{\rho<2c_0 nh}\rho^{\beta-\mu}$.

Using (8) and (9), we obtain the required estimate (5).

LEMMA 2. *Suppose that the conditions of Lemma* 1 *are satisfied and* $\kappa = 1 - |\alpha|(1/p - 1/q) - \mu - \sigma < 0$. *Then for* $0 < \epsilon < H \leqslant \infty$

$$\| I_{\varepsilon, H}(x) \|_{q,\beta-\mu} \leqslant c\varepsilon^{\varkappa} \| F \|_{p,\beta,\alpha_i},$$

where the constant c *does not depend on* ϵ, H, *or* F.

The proof immediately follows from the generalized Minkowski inequality and Lemma 1.

LEMMA 3. *Let* $1 \leqslant p \leqslant q < \infty$, $\alpha = (\alpha_1, \ldots, \alpha_n)$, $\alpha_i > 0$, $\bar{x} = (x_1, \ldots, x_m, 0, \ldots, 0)$, $1 \leqslant m \leqslant n$, $|\bar{\alpha}| = \alpha_1 + \cdots + \alpha_m$, $\beta < |\alpha|/p'$, $\gamma < |\bar{\alpha}|/q$, $\lambda = |\bar{\alpha}|/p' + |\alpha|/q$, $f \in L_p(E_n)$,

$$I(\bar{x}) = \int \frac{f(y)\,dy}{[\rho(\bar{x})]^{\gamma}\,[\rho(y)]^{\beta}\,[\rho(\bar{x}-y)]^{\lambda-\beta-\gamma}},$$

where $\gamma + \beta > 0$, *and* $\gamma + \beta \geqslant 0$ *for* $1 < p < q < \infty$ ($\gamma = \beta = 0$ *for* $p = 1, q = \infty$). *Then*

$$\| I(\bar{x}) \|_{L_q(E_m)} \leqslant c \, \| f \|_p,$$

where the constant c does not depend on f.

In the one-dimensional case with $1 < p < \infty$ and $1 < q < \infty$ this lemma was proved by Hardy and Littlewood [7], and in the multidimensional case with $\alpha_1 = \cdots = \alpha_n = 1$, $1 \leqslant p \leqslant \infty$, $1 \leqslant q \leqslant \infty$ by Il'in [8].

The proof of the lemma is carried out just as in [8], with the replacement of $|x| = \sqrt{x_1^2 + \cdots + x_n^2}$ by $\rho(x)$.

LEMMA 4. *Let* $1 \leqslant p \leqslant q \leqslant \infty$, $\beta < (|\alpha| + \alpha_i)/p'$, $-\beta + \mu < |\alpha|/q$, $\mu > 0$, *and* $\mu \geqslant 0$ *for* $1 < p < q < \infty$ ($\beta = \mu = 0$ *for* $p = 1, q = \infty$), *let* $\kappa = 1 - |\alpha|(1/p - 1/q) - \mu - \sigma = 0$, *and let* $F \in L_{p,\beta,\alpha_i}(E^+_{n+1})$. Then for $0 \leqslant \epsilon < H \leqslant \infty$

$$\| I_{\varepsilon, H}(x) \|_{q, \beta-\mu} \leqslant c \, \| F \|_{p, \beta, \alpha_i},$$
$$\| I_{\varepsilon, H}(x) \|_{q, \beta-\mu} \to 0, \quad \varepsilon, H \to \infty, \qquad \| I_{\varepsilon, H}(x) \|_{q, \beta-\mu} \to 0, \quad \varepsilon, H \to 0. \tag{10}$$

PROOF. Using properties of the kernel $T(y, t)$, switching the order of integration, noting that

$$c_1 [\rho(y) + t^{1/\alpha_i}] \leqslant \max_{1 \leqslant j \leqslant n} (y_j^{1/\alpha_j}, t^{1/\alpha_i}) \leqslant c_2 [\rho(y) + t^{1/\alpha_i}],$$

where the positive constants c_1 and c_2 do not depend on (y, t) and $|\alpha| + \alpha_i/p' + \sigma > 1$, and putting $\lambda = (|\alpha| + \alpha_i)/p' + |\alpha|/q$, we obtain

$$| I_{\varepsilon, H}(x) | \leqslant \int\limits_{0 < y_j < H^{\alpha_j}} dy \int\limits_0^{H^{\alpha_i}} | F(x+y, t) | \, dt \int\limits_{\max\limits_{1 \leqslant j \leqslant n} (y_j^{1/\alpha_j}, t^{1/\alpha_i})}^{\infty} \frac{dh}{h^{|\alpha| + \frac{\alpha_i}{p'} + \sigma}}$$
$$\leqslant \int\limits_0^{\infty} \int \frac{| F(y, t) | \, dy dt}{[\rho(x-y) + t^{1/\alpha_i}]^{\lambda-\mu}}.$$

Applying Lemma 3, we obtain the required estimate (10) from this.

The second assertion of the lemma follows from the first, which makes it possible for one to apply the theorem on the passage to the limit under a Lebesgue integral sign for functions with an integrable majorant.

LEMMA 5. *Let* $f \in b^{l(r)}_{p,\beta}$, $\sigma_0 = \min_{1 \leqslant i \leqslant n} (|\alpha| + \alpha_i)/p'$, *and* $\kappa = 1 - |\alpha|(1/p - 1/q) - \mu - \mathbf{k}\alpha$. *Then, as* $H \to \infty$, $D^{\mathbf{k}} f_H(x)$ *and each of the integrals in* (4) *converges in the following sense*:

1) $L_{q,\beta-\mu}$ *for* $\kappa < 0$, $1 \leqslant p \leqslant q \leqslant \infty$, $\beta < \sigma_0$, $0 \leqslant \mu < |\alpha|/q + \beta$ *(for* $p = 1$ *or* $q = \infty$ *one can respectively put* $\beta = 0$ *or* $\mu = \beta$*), and* $\epsilon > 0$; *here, if* $k_j \leqslant r_j = s_j + \sigma_j$ *for some* j, *then*

$$\| D^{\mathbf{k}} f_H \|_{q,\, \beta-\mu} \to 0, \qquad H \to \infty;$$

2) $L_{q,\beta-\mu}$ *for* $\kappa = 0$, $1 \leqslant p \leqslant q \leqslant \infty$, $\beta < \sigma_0$, $-\beta + \mu - |\alpha|/q$, $\mu > 0$, *and* $\mu \geqslant 0$ *for* $1 < p < q < \infty$ ($\beta = \mu = 0$ *for* $p = 1$, $q = \infty$).

PROOF. The first part of assertion 1) and assertion 2) follow from (4) and Lemmas 2 and 4.

Using an integral equality from [5], which expresses one integral, under whose sign appears a derivative $D_j^{s_j} f$, in terms of another integral, under whose sign appears

$$\Delta_j^{\sigma_j}\left(\frac{t}{\sigma_j}\right) D_j^{s_j} f,$$

one can show that $D^{\mathbf{k}} f_H(x)$, $k_j < r_j$ for some j, is a linear combination of functions of the form

$$H^{-s_j+1-\mathbf{k}\alpha} \iint T\left(\frac{y}{H^{\alpha}}, \frac{t}{H^{\alpha_j}}\right) \Delta_j^{\sigma_j}\left(\frac{t}{\sigma_j}\right) D_j^{s_j} f(x+y)\, dt\, dy,$$

where $T(y, t) \in C_0^\infty(E_{n+1})$ with support in the region $0 < y_i < 1$, $i = 1, \dots, n$, $0 < t < 1$, $0 < t + y_j < 1$. From this, applying Lemma 1, we obtain

$$\| D^{\mathbf{k}} f_H \|_{q,\, \beta-\mu} \leqslant c H^{\varkappa} \| t^{-1/p - l_j + s_j} \Delta_j^{\sigma_j}(t) D_j^{s_j} f(x) \|_{p,\, \beta,\, \alpha_j} \to 0, \quad \varkappa < 0,$$

as $H \to \infty$, which also proves the second part of assertion 1).

LEMMA 6. *For each function* $f \in b_{p,\beta}^{1(\mathbf{r})}$ *there exists a polynomial of degree* $r_i - 1$ *with respect to* x_i, $i = 1, \dots, n$,

$$P_{\mathbf{r}-1}(x; f) = \sum_{\gamma \leqslant \mathbf{r}-1} c_\gamma x^\gamma, \qquad c_\gamma = 0 \qquad \text{for} \qquad \gamma\alpha \leqslant 1 - \frac{|\alpha|}{p} - \beta \tag{11}$$

such that as $H \to \infty$

$$D^{\mathbf{k}} f_H(x) \to D^{\mathbf{k}} P_{\mathbf{r}-1}(x; f), \quad \mathbf{k}\alpha > 1 - \frac{|\alpha|}{p} - \beta, \tag{12}$$

in the sense of the conditions 1) *and* 2) *of Lemma* 5.

PROOF. Put

$$c_\gamma = \lim_{H \to \infty} \frac{1}{\gamma!} D^\gamma f_H(0), \qquad \gamma\alpha > 1 - \frac{|\alpha|}{p} - \beta. \tag{13}$$

These limits exist and $c_\gamma = 0$ for $\gamma \nleqslant \mathbf{r} - 1$ by property 1) of Lemma 5, in which one must put $q = \infty$ and $\mu = \beta$. Applying the Taylor formula and the second part of property 1) of Lemma 5, it is easy to show that for $\mathbf{k}\alpha > 1 - |\alpha|/p - \beta$ on each compact set $K \subset E_n$, relation (12) is satisfied in the sense of uniform convergence. Consequently the right side of (12) also will be the limit of the left side in the sense of the conditions 1) and 2) of Lemma 5.

THEOREM 1. *For each function* $f \in b_{p,\beta}^{1(\mathbf{r})}$, $-|\alpha|/p < \beta < \sigma_0$, $1 \leqslant p \leqslant \infty$, *there exists a polynomial* $P_{\mathbf{r}-1}(x; f)$ *of the form* (11), (13), *which linearly depends on* f, *such that for any vector* k *for which* $k\alpha > 1 - |\alpha|/p - \beta$, *one has the equality*

$$D^{k}[f_{\varepsilon}(x) - P_{r-1}(x; f)] = (-1)^{|k|} \sum_{i=1}^{n} \int_{\varepsilon}^{\infty} h^{-\delta_i - k\alpha} dh \iint T_i^{(k)}\left(\frac{y}{h^{\alpha}}, \frac{t}{h^{\alpha_i}}\right) \times \Delta_i^{\sigma_i}\left(\frac{t}{\sigma_i}\right) D_i^{s_i} f(x+y)\, dt\, dy, \tag{14}$$

where $\epsilon \geqslant 0$, *if* $1 - |\alpha|/p - \beta < k\alpha < 1$; $\epsilon > 0$, *if* $1 \leqslant k\alpha$,

$$D^{k} f_0(x) = D^{k} f(x), \quad 1 - |\alpha|/p - \beta < k\alpha < 1.$$

In this connection the outer integrals of the right side of (14) *in the case* $\epsilon = 0$ $(1 - |\alpha|/p - \beta - k\alpha < 1)$ *converge to zero in* $L_{q(k), \beta - \mu(k)}$, *where* $q = q(k)$, *and* $\mu = \mu(k)$ *are determined from the conditions* $k\alpha = 1 - |\alpha|(1/p - 1/q) - \mu$, $1 \leqslant p \leqslant q \leqslant \infty$, $-\beta + \mu < |\alpha|/q$, $\mu > 0$, *and* $\mu \geqslant 0$ *for* $1 < p < q < \infty$. *The outer integrals of the right side of* (14) *converge at infinity in the sense of the conditions* 1) *and* 2) *of Lemma* 5.

The proof of this theorem consists of the application of Lemmas 4, 5, and 6 to the representation (4).

THEOREM 2. *For any function* $f \in b_{p,\beta}^{l(r)}$ *the following assertions are valid*:

a) *If* $\kappa = 0$, $1 \leqslant p \leqslant q \leqslant \infty$, $\beta < \sigma_0$, $-\beta + \mu < |\alpha|/q$, $\mu \geqslant 0$ *and* $\mu > 0$, *for* $1 < p < q < \infty$ ($\beta = \mu = 0$ *for* $p = 1$, $q = \infty$), *then for the polynomial* $P_{r-1}(x; f)$ *from* (11), (13) *and for any* $\epsilon \geqslant 0$

$$\| D^{k}[f_{\varepsilon}(x) - P_{r-1}(x; f)] \|_{q, \beta - \mu} \leqslant c \| f \|_{b_{p,\beta}^{l(r)}}, \tag{15}$$

where the constant c does not depend on f or ϵ *and* $D^{k} f_0 = D^{k} f$.

b) *If* $\kappa < 0$, $1 \leqslant p \leqslant q \leqslant \infty$, $\beta < \sigma_0$, *and* $0 \leqslant \mu < |\alpha|/q + \beta$ (*for* $p = 1$ *or* $q = \infty$ *one can respectively put* $\beta = 0$ *or* $\mu = \beta$), *then for any* $\epsilon > 0$

$$\| D^{k}[f_{\varepsilon}(x) - P_{r-1}(x; f)] \|_{q, \beta - \mu} \leqslant c \varepsilon^{\kappa} \| f \|_{b_{p,\beta}^{l(r)}}, \tag{16}$$

where the constant c does not depend on f or ϵ.

PROOF. By virtue of representation (14) it suffices to estimate each of the integrals of its right side. Inequality (15) follows from Lemma 4, and inequality (16) from Lemma 2, in which one must put $\sigma = k\alpha$.

This theorem shows that the well-defined derivatives of $f(x)$ or $f_{\epsilon}(x)$ are near "at infinity" to the corresponding derivatives of some polynomial $P_{r-1}(x; f)$ in the sense of the finiteness of the left sides of (15) and (16). In this case, for brevity we shall say that the function $f \in b_{p,\beta}^{l(r)}$ converges at infinity to the polynomial $P_{r-1}(x; f)$. If for a function $\varphi \in b_{p,\beta}^{l(r)}$ the polynomial $P_{r-1}(x; \varphi) = 0$, then we shall say that the function φ converges to zero at infinity.

As is evident from (11), if $1 - |\alpha|/p - \beta \geqslant 0$, then $P_{r-1}(x; f)$ does not contain "lowest terms" (does not contain powers for which $\gamma\alpha \leqslant 1 - |\alpha|/p - \beta$). However, the addition of any "lowest terms" does not affect estimates (15) and (16), since they vanish with the differentiation in the left side of (15) and (16).

For $f \in b_{p,\beta}^{l(r)}$ there exists a unique polynomial $P_{r-1}(x; f)$ without "lowest terms" to which it converges at infinity. Indeed, for any other polynomial $P_{r-1}^{*}(x; f) = \Sigma_{\gamma \leqslant r-1} c_{\gamma}^{*} x^{\gamma}$,

$c_\gamma = 0$ for $\gamma\alpha \leqslant 1 - |\alpha|/p - \beta$, the inclusion

$$D^{\mathbf{k}}[P_{\mathbf{r}-1}(x; f) - P^*_{\mathbf{r}-1}(x, f)] \in L_{q,\beta-\mu}, \quad \varkappa \leqslant 0,$$

must be satisfied, and this is possible only in the case of the equality $c^*_\gamma = c_\gamma$ for all $\gamma \leqslant \mathbf{r} - 1$.

REMARK 1. An inequality of the type (15), (16) is also possible for the hyperplanes $E_m = \{x: x_{m+1} = \cdots = x_n = 0\}$, $1 \leqslant m < n$. Namely, the estimates (15) and (16) are valid, in which one must replace $|\alpha|/q$ by $|\bar{\alpha}|/q$ ($|\bar{\alpha}| = \alpha_1 + \cdots + \alpha_m$), x by $\bar{x} = (x_1, \ldots, x_m, 0, \ldots, 0)$ in the left sides of the inequalities, and the metric of $L_{q,\beta-\mu}(E_n)$ by the metric of $L_{q,\beta-\mu}(E_m)$ (if $\epsilon = 0$, then by $f(\bar{x})$ is understood the trace of the function $f \in b^{\mathbf{l}(\mathbf{r})}_{p,\beta}$ on the hyperplane E_m, which exists, for example in the sense of L^{loc}_p convergence). The proof of these estimates is carried out exactly as for estimates (15) and (16).

We shall study the question of approximation, in the norm (3), of functions from $b^{\mathbf{l}(\mathbf{r})}_{p,\beta}$ by infinitely differentiable functions with compact support. Since a function $\varphi \in C^\infty_0(E_n)$ converges to zero at infinity, it is easy to show that each function $f \in b^{\mathbf{l}(\mathbf{r})}_{p,\beta}$ which is approximable with arbitrary precision in the norm (3) by functions from $C^\infty_0(E_n)$ also converges to zero at infinity. It will also be shown below that, conversely, each function $f \in b^{\mathbf{l}(\mathbf{r})}_{p,\beta}$ which converges to zero at infinity can be approximated within an arbitrary precision in the norm (3) by functions from $C^\infty_0(E_n)$. By the same token it will be established that the set of functions from $b^{\mathbf{l}(\mathbf{r})}_{p,\beta}$ which converge to zero at infinity coincides with the closure in $B^{\mathbf{l}(\mathbf{r})}_{p,\beta}$ of the set of functions $C^\infty_0(E_n)$.

LEMMA 7. *Let* $F \in L_{p,\beta,\alpha_i}(E^+_{n+1})$, $\alpha = (\alpha_1, \ldots, \alpha_n)$, $\alpha_j > 0$, $j = 1, \ldots, n$, $-|\alpha|/p < \beta < |\alpha|/p'$ *and* $1 \leqslant p \leqslant \infty$ *(for* $p = \infty$ *one can put* $\beta = 0$); *for* $p = \infty$ *one additionally requires*

$$\operatorname*{ess\,sup}_{(x,t) \in \Omega_R} |[\rho(x) + t^{1/\alpha_i}]^\beta F(x, t)| \to \infty, \quad R \to \infty, \tag{17}$$

where $\Omega_R = E^+_{n+1} \setminus E_R$, $E_R = \{(x, t): |x_j| < R, j = 1, \ldots, n, 1/R < t < R\}$,

$$F_h(x, t) = \frac{1}{h^{|\alpha|}} \int \Omega\left(\frac{y}{h^\alpha}\right) F(x + y, t)\, dy, \tag{18}$$

where $\Omega(y)$ *is a piecewise continuous finite function with support in the region* $0 < y_j < 1$, $j = 1, \ldots, n$. *Then as* $h \to \infty$

$$\| F_h(x, t) \|_{p,\beta,\alpha_i} \to 0. \tag{19}$$

PROOF. One can show that the averaging operator (18) is uniformly bounded in $L_{p,\beta,\alpha_i}(E^+_{n+1})$, $-|\alpha|/p < \beta < |\alpha|/p'$, $1 \leqslant p \leqslant \infty$ (for $p = 1$ or $p = \infty$ one can put $\beta = 0$), i.e.

$$\| F_h(x, t) \|_{p,\beta,\alpha_i} \leqslant c \| F(x, t) \|_{p,\beta,\alpha_i}, \tag{20}$$

where the constant c does not depend on F or h.

Since the set of finite bounded functions with compact support in E^+_{n+1} is dense in $L_{p,\beta,\alpha_i}(E^+_{n+1})$, $1 \leqslant p \leqslant \infty$ (for $p = \infty$ one additionally requires that functions $F \in L_{p,\beta,\alpha_i}(E^+_{n+1})$ satisfy the condition (17)), therefore, using (20), it suffices to prove (19) for finite functions bounded in E^+_{n+1}.

Thus, let $|\varphi(x, t)| \leqslant M$, $\varphi(x, t) = 0$ outside the parallelepiped $|x_j| < a$, $j = 1, \ldots, n$, $0 < t_0 < t < T < \infty$. Then

$$|\varphi_h(x, t)| \leqslant ch^{-|\alpha|} \int_{|y_j|<a} |\varphi(y, t)|\, dy \leqslant c_1 h^{-|\alpha|}. \tag{21}$$

Since $\varphi_h(x, t) = 0$ outside the parallelepiped $-a - h^{\alpha_j} < x_j < a$, $j = 1, \ldots, n$, $t_0 < t < T$, one can determine an h_0 such that, for any $h > h_0$, $\varphi_h(x, t) = 0$ outside the parallelepiped $|x_j| < 2h^{\alpha_j}$, $t_0 < t < T$. Note that the set

$$\Omega = \left\{(x, t) : \rho(x) < dh,\ t_0 < t < T,\ d = \sum_{i=1}^{n} 2^{1/\alpha_i}\right\}$$

entirely contains the parallelepiped $|x_j| < 2h^{\alpha_j}$, $t_0 < t < T$. Therefore, taking into account (21) and the fact that $h > h_0$, and passing to generalized spherical coordinates according to the formulas $x_j = p^{\alpha_j}\xi_j$, $j = 1, \ldots, n$, we obtain

$$\| \varphi_h(x, t) \|_{p, \beta, \alpha_i} \leqslant ch^{-|\alpha|} \| [\rho(x) + t^{1/\alpha_i}]^{\beta} \|_{L_p(\Omega)} \leqslant c_1 h^{\beta + |\alpha|/p - |\alpha|} \longrightarrow 0,$$

as $h \longrightarrow \infty$, which proves (19).

LEMMA 8. *Suppose that* $f \in b^{l(r)}_{p, \beta}$, $-|\alpha|/p < \beta < |\alpha|/p'$, $1 \leqslant p \leqslant \infty$ *(for* $p = \infty$ *one can put* $\beta = 0$*), and let the functions* $\varphi^{(i)}(x, t)$ *(cf. notation) satisfy* (17). *Then, as* $h \longrightarrow \infty$,

$$\| f_h(x) \|_{b^{l(r)}_{p, \beta}} = \sum_{i=1}^{n} \| \varphi_h^{(i)}(x, t) \|_{p, \beta, \alpha_i} \longrightarrow 0.$$

This lemma immediately follows from Lemma 7.

In the sequel we use the following lemma, which is due to O. V. Besov [3]; we state it in a formulation which is convenient for us.

LEMMA 9. *With* $\delta = 1 - |\alpha|/p - \beta \geqslant 0$, $1 \leqslant p \leqslant \infty$, *for a sufficiently smooth function* f *there exists a projection operator*

$$Q(x; f) = \sum_{\gamma\alpha \leqslant \delta} c_\gamma x^\gamma$$

such that

$$\| f - Q(x; f) \|_{L_p(1<|x|<2)} \leqslant c \sum_{\delta < k\alpha < 2} \| D^k f \|_{L_p\left(\frac{1}{2}<|x|<3\right)},$$

where the constant c *does not depend on* f.

LEMMA 10. *For any sufficiently smooth function* f *such that* $[\rho(x) + t^{1/\alpha_i}]^{\beta} \Delta_i^s(t) f(x) \in L_p(E_n)$, $0 < t < \infty$, $1 \leqslant p \leqslant \infty$, *one has*

$$\| [\rho(x) + t^{1/\alpha_i}]^{\beta} \Delta_i^s(t) f(x) \|_{L_p(E_n)} \leqslant ct^s \| [\rho(x) + t^{1/\alpha_i}]^{\beta} D_i^s f(x) \|_{L_p(E_n)}, \tag{22}$$

where the consant c *does not depend on* f.

PROOF. Using the inequality

$$|\Delta_i^s(t) f(x)| \leqslant t^{s-1} \int_0^{st} | D_i^s f(x_1, \ldots, x_i + \xi, \ldots, x_n) |\, d\xi$$

applying the generalized Minkowski inequality, making a change of variables according to the formula $x_i + \xi = x_i'$, and noting that

$$c_1(\rho(x) + t^{1/\alpha_i}) \leqslant \sum_{\substack{j=1 \\ j\neq i}}^{n} |x_j|^{1/\alpha_j} + |x_i - \xi| + t^{1/\alpha_i} \leqslant c_2(\rho(x) + t^{1/\alpha_i}), \quad 0 < \xi < st,$$

we obtain

$$\|[\rho(x) + t^{1/\alpha_i}]^\beta \Delta_i^s(t) f(x)\|_{L_p(E_n)} \leqslant t^{s-1} \int_0^{st} \|[\rho(x) + t^{1/\alpha_i}]^\beta D_i^s f(x_1, \ldots, x_i + \xi, \ldots$$

$$\ldots, x_n)\|_p d\xi \leqslant t^{s-1} \int_0^{st} \|[\rho(x) + t^{1/\alpha_i}]^\beta D_i^s f(x)\|_p d\xi \leqslant t^s \|[\rho(x) + t^{1/\alpha_i}]^\beta D_i^s f(x)\|_{L_p(E_n)}.$$

Note that in order that the space $b_{p,\beta}^{\mathbf{l}(\mathbf{r})}$, $1 \leqslant p \leqslant \infty$, contain the functions $\varphi \in C_0^\infty(E_n)$ ($\varphi \not\equiv 0$), it is necessary that $\beta < \omega = \min_{1\leqslant i\leqslant n}(1 - s_i/l_i)$ (for $p = \infty$ one can put $\beta = \omega$). Indeed, let $\beta \geqslant \omega$ ($\beta > \omega$ for $p = \infty$). From this it will follow that $\beta \geqslant 1 - s_j/l_j$ ($\beta > 1 - s_j/l_j$ for $p = \infty$) for some j, $1 \leqslant j \leqslant n$. For functions $\varphi \in C_0^\infty(E_n)$, supp $\varphi \subset \{x\colon |x_i| < a,$ $i = 1, \ldots, n\}$, $\varphi \not\equiv 0$, we have

$$\|\varphi\|_{b_{p,\beta}^{\mathbf{l}(\mathbf{r})}} \geqslant \|[\rho(x) + t^{1/\alpha_j}]^\beta t^{-1/p - l_j + s_j} \Delta_j^{\sigma_j}(t) D_j^{s_j}\varphi(x)\|_{L_p(t>a,\,0<x_i<a)}$$

$$\geqslant c\|t^{\beta/\alpha_j - 1/p - l_j + s_j}\|_{L_p(t>a)} = +\infty,$$

i.e. $\varphi \notin b_{p,\beta}^{\mathbf{l}(\mathbf{r})}$ if $\beta \geqslant \omega$ ($\beta > \omega$ for $p = \infty$).

LEMMA 11. *Each monomial x^γ for $\gamma\alpha \leqslant 1 - |\alpha|/p - \beta$ is appoximable by functions from $C_0^\infty(E_n)$ with arbitrary precision in $B_{p,\beta}^{\mathbf{l}(\mathbf{r})}$, $-|\alpha|/p \leqslant \beta < \omega$, $1 \leqslant p \leqslant \infty$.*

PROOF. Suppose that $r = (\Sigma_1^n (x_i^2)^{1/\alpha_i})^{1/2}$, and that $\Psi(t)$ is an infinitely differentiable function of the variable $t \in E_1$, where $\Psi(t) = 1$, $t < \delta$, $0 < \delta < 1$; $\Psi(t) = 0$, $t > 1$.

We shall show that the norm in $B_{p,\beta}^{\mathbf{l}(\mathbf{r})}$ of the difference

$$g_{\eta,\delta}(x) = x^\gamma - x^\gamma \Psi\left(\frac{\ln r}{\ln \eta}\right)$$

can be made as small as desired with a proper choice of $\eta > 0$ and δ, $0 < \delta < 1$. By this the lemma will be proved.

Since $\|g_{\eta,\delta}(x)\|_{L_p(1<|x|<2)} = 0$ for sufficiently large η, one need only estimate the terms

$$\|\varphi_{\eta,\delta}^{(i)}(x, t)\|_{p,\beta,\alpha_i} = \|t^{-1/p - l_i + s_i} \Delta_i^{\sigma_i}(t) D_i^{s_i} g_{\eta,\delta}(x)\|_{p,\beta,\alpha_i}$$

of the norm (3). Note that $\Delta_i^{\sigma_i}(t) D_i^{s_i} x^\gamma = 0$.

It is easy to show that

$$\left| D_i^s \Psi\left(\frac{\ln r}{\ln \eta}\right)\right| \leqslant \frac{c}{\rho^{\alpha_i s} \ln \eta}, \qquad s = 1, 2, \ldots, \qquad \eta^\delta < r < \eta.$$

Using the latter inequality, we obtain

$$\left| D_i^{s_i+\sigma_i}\left[x^\gamma \Psi\left(\frac{\ln r}{\ln \eta}\right)\right]\right| = \left|\frac{x^\gamma}{x^{\gamma_i}}\sum_{m=0}^{\gamma_i} c_m x^{\gamma_i - m} D_i^{s_i+\sigma_i-m}\Psi\left(\frac{\ln r}{\ln \eta}\right)\right|$$
$$\leqslant c\frac{|x^\gamma|}{\ln \eta}\sum_{m=0}^{\gamma_i}\frac{1}{|x_i|^m r^{(s_i+\sigma_i-m)\alpha_i}}, \quad h^\delta < r < \eta. \tag{23}$$

Applying Lemma 10, noting that

$$D_i^{s_i+\sigma_i} g_{\eta,\delta}(x) = D_i^{s_i+\sigma_i}\left[x^\gamma \psi\left(\frac{\ln r}{\ln \eta}\right)\right] = 0$$

outside the region $\eta^\delta < r(x) < \eta$, using (23) and the fact that $c_1 r(x) \leqslant \rho(x) \leqslant c_2 r(x)$, passing to generalized spherical coordinates (r, ξ) according to the formula $x_j = r^{\alpha_j}\xi_j$, $j = 1, \ldots, n$, and putting $\Theta = 1 - |\alpha|/p - \beta - \gamma\alpha$, we obtain

$$\begin{aligned}
&\| \varphi_{\eta,\delta}^{(i)}(x,t)\|_{L_{p,\beta,\alpha_i}(0<t<\eta^{\delta\alpha_i})}\\
&\leqslant \frac{1}{\ln \eta}\sum_{m=0}^{\gamma_i}\left\|\frac{t^{\sigma_i - 1/p - l_i + s_i}[r(x) + t^{1/\alpha_i}]^\beta x^\gamma}{x_i^m r^{(s_i+\sigma_i-m)\alpha_i}}\right\|_{L_p(\eta^\delta < r(x) < \eta,\ 0<t<\eta^{\delta\alpha_i})}\\
&\leqslant \frac{1}{\ln \eta}\| t^{\sigma_i - 1/p - l_i + s_i} r^{-\Theta - 1/p + (l_i - s_i - \sigma_i)\alpha_i}\|_{L_p(\eta^\delta < r < \eta,\ 0<t<\eta^{\delta\alpha_i})}\\
&\leqslant \begin{cases} 1/\eta^{\Theta\delta}\ln \eta, & \Theta > 0;\\ (1-\delta)^{1/p}/(\ln \eta)^{1-1/p}, & \Theta = 0.\end{cases}
\end{aligned}\tag{24}$$

Noting that

$$\begin{aligned}c_1(r(x) + t^{1/\alpha_i}) &\leqslant |x_1|^{1/\alpha_1} + \ldots + |x_i - jt|^{1/\alpha_i} + \ldots + |x_n|^{1/\alpha_n} + t^{1/\alpha_i}\\ &\leqslant c_2(r(x) + t^{1/\alpha_i}), \qquad j = 0, 1, \ldots, \sigma_i,\end{aligned}\tag{25}$$

we have

$$\varphi_{\eta,\delta}^{(i)}(x,t)_{L_{p,\beta,\alpha_i}(t>\eta^{\delta\alpha_i})} \leqslant \left\| t^{-1/p - l_i + s_i}[r(x) + t^{1/\alpha_i}]^\beta D_i^{s_i}\left[x^\gamma \Psi\left(\frac{\ln r}{\ln \eta}\right)\right]\right\|_{L_p(t>\eta^{\delta\alpha_i})}. \tag{26}$$

Let $s_i \geqslant 1$. Then, using (26), acting just as with estimate (24), noting that $\eta^\delta < r(x) < \eta$, $t > \eta^{\delta\alpha_i}$, and $r(x) \leqslant r(x) + t^{1/\alpha_i} \leqslant ct^{1/\delta\alpha_i}$, assuming that δ is sufficiently close to unity, and putting $a_\beta = \beta$ if $\beta \geqslant 0$ and $a_\beta = 0$ if $\beta < 0$, we obtain

$$\begin{aligned}
&\| \varphi_{\eta,\delta}^{(i)}(x,t)\|_{L_{p,\beta,\alpha_i}(t>\eta^{\delta\alpha_i})}\\
&\leqslant \frac{1}{\ln \eta}\| t^{-1/p - l_i + s_i + a_\beta/\delta\alpha_i} r^{-\Theta - 1/p + (l_i - s_i)\alpha_i - a_\beta}\|_{L_p(\eta^\delta < r < \eta,\ t>\eta^{\delta\alpha_i})}\\
&\leqslant \begin{cases} \dfrac{\eta^{(1-\delta)[(l_i - s_i)\alpha_i + a_\beta]}}{\eta^{\Theta\delta}\ln \eta}, & \Theta > 0;\\[2ex] \dfrac{(1-\delta)^{1/p}\eta^{(1-\delta)[(l_i-s_i)\alpha_i + a_\beta]}}{(\ln \eta)^{1-1/p}}, & \Theta = 0.\end{cases}
\end{aligned}\tag{27}$$

Let $s_i = 0$. Then, putting $d_\beta = 0$ if $\beta \geqslant 0$ and $d_\beta = \beta$ if $\beta < 0$, for δ sufficiently close to unity we shall have

$$\| \varphi_{\eta,\delta}^{(i)}(x,t)\|_{L_{p,\beta,\alpha_i}(t>\eta^{\delta\alpha_i})} \leqslant (\ln \eta)^{-1}\eta^{(1-\delta)(1-d_\beta)}. \tag{28}$$

Using estimates (24) and (26)–(28), we obtain that the norm $\|\varphi_{\eta,\delta}^{(i)}(x,t)\|_{p,\beta,\alpha_i}$

can be made as small as desired by a proper choice of η and δ. Indeed, let, for example, $\Theta > 0$, $s_i \geqslant 1$ and $1 \leqslant p \leqslant \infty$. Then from (24) we obtain that the norm

$$\| \varphi_{\eta,\delta}^{(i)} (x, t) \|_{L_{p,\beta,\alpha_i}(0<t<\eta^{\delta\alpha_i})}$$

is as small as desired for sufficiently large η and any δ. Further, let us choose η so large that $1/\ln \eta < \epsilon/2$, where $\epsilon > 0$ is an arbitrarily given number. For the η chosen we select δ, $0 < \delta < 1$, such that $c\eta^{(1-\delta)[(l_i - s_i)\alpha_i + \alpha\beta]} < 2$ (c is the constant from the right side of (27)). Then, for the η and δ chosen, we obtain from (27) that

$$\| \varphi_{\eta,\delta}^{(i)} (x, t) \|_{L_{p,\beta,\alpha_i}(t > \eta^{\delta\alpha_i})} < \varepsilon.$$

From this it follows that the norm $\|\varphi_{\eta,\delta}^{(i)}(x, t)\|_{p,\beta,\alpha_i}$, and hence also the norm of $g_{\eta,\delta}(x)$ in $B_{p,\beta}^{\mathbf{l}(\mathbf{r})}$, can be made as small as desired by a proper choice of η and δ. The remaining cases are analyzed in an analogous manner.

The lemma is proved.

LEMMA 12. *Let* $\varphi_i(x, t) = [\rho(x) + t^{1/\alpha_i}]^\beta \Psi_i(x, t) \in L_p(E_{n+1}^+)$, $i = 1, \ldots, n$, $-|\alpha|/p < \beta < \min\{\omega, \sigma_0\}$ *and* $1 \leqslant p \leqslant \infty$ *(for* $p = 1$ *or* $p = \infty$ *one can put* $\beta = 0$*), and set*

$$g_{\varepsilon, H}(x) = \sum_{i=1}^{n} \int_{\varepsilon}^{H} h^{-\delta_i} dh \iint \tau^{1/p + l_i - s_i} T_i\left(\frac{y}{h^\alpha}, \frac{\tau}{h^{\alpha_i}}\right) \Psi_i(x + y, \tau)\, d\tau\, dy,$$

where $\delta_i = 1 + |\alpha| - s_i\alpha_i + \alpha_i$ *and* $T_i(y, t)$ *are piecewise continuous functions with support respectively concentrated in the regions* $0 < y_j < 1$, $j = 1, \ldots, n$, $0 < t < 1$, $0 < t + y_i < 1$. *Then, for* $0 \leqslant \epsilon < H < \infty$,

$$\begin{aligned} \| g_{\varepsilon, H} \|_{b_{p,\beta}^{\mathbf{l}(\mathbf{r})}} &\leqslant \sum_{i=1}^{n} \int_0^1 u^{l_i - s_i} \left\{ \int_{u\varepsilon^{\alpha_i}}^{uH^{\alpha_i}} \| \varphi_i(x, t) \|_p^p \, dt \right\}^{1/p} du + \sum_{i=1}^{n} \left\{ \int_0^1 u^{p(l_i - s_i)} du \right. \\ &\times \left. \int_{u\varepsilon^{\alpha_i}}^{uH^{\alpha_i}} \| \varphi_i(x, t) \|_p^p \, dt \right\}^{1/p} \leqslant \sum_{i=1}^{n} \left\{ \int_0^{H^{\alpha_i}} \| \varphi_i(x, t) \|_p^p \, dt \right\}^{1/p} \leqslant \sum_{i=1}^{n} \| \Psi_i \|_{p,\beta,\alpha_i}. \end{aligned} \tag{29}$$

PROOF. Let $\zeta_{\epsilon,H}(t)$ be the characteristic function of the interval (ϵ, H). Transferring differentiation to the kernel, using (25), in which one can replace $r(x)$ by $\rho(x)$, and Lemma 10, and applying the generalized Minkowski inequality, we obtain

$$\begin{aligned} \| [\rho(x) + t^{1/\alpha_j}]^\beta \Delta_j^{\sigma_j}(t) D_j^{s_j} g_{\varepsilon, H}(x) \|_p &\leqslant \sum_{i=1}^{n} \left(\int_0^{t^{l_j}} + t^{\sigma_j} \int_{t^{l_j}}^{\infty} h^{-\sigma_j\alpha_j} \right) \zeta_{\varepsilon, H}(h) h^{-s_j\alpha_j - \delta_i} \\ &\times \left\| [\rho(x) + t^{1/\alpha_j}]^\beta \int_{0<y_m<h^{\alpha_m}} dy \int_0^{h^{\alpha_i}} \tau^{1/p + l_i - s_i} | \Psi(x + y, \tau) | \, d\tau \right\|_p dh. \end{aligned} \tag{30}$$

Let us first estimate the first integral with respect to h in the right side of (30). Let $\beta \geqslant 0$. Noting that for $\rho(x) > 2c_0 nt^{l_j}$ and $\rho(y) < nt^{l_j}$ (c_0 is the constant from (6)) we have $\rho(x) + t^{1/\alpha_j}/\rho(x + y) + \tau^{1/\alpha_i} \leqslant c$; and applying the generalized Minkowski inequality, we have

$$\int_0^{t^{l_j}} \zeta_{\varepsilon,H}(h)\, h^{-s_j\alpha_j-\delta_i} \Big\| [\rho(x)+t^{1/\alpha_j}]^\beta \int\limits_{0<y_m<h^{\alpha_m}} dy \int_0^{h^{\alpha_i}} \tau^{1/p+l_i-s_i}$$

$$\times |\Psi_i(x+y,\tau)|\, d\tau \Big\|_{L_p(\rho(x)>2c_0nt^{l_j})} dh \leqslant \int_0^{t^{l_j}} \zeta_{\varepsilon,H}(h)\, h^{-s_j\alpha_j-1+s_i\alpha_i-\alpha_i} dh \tag{31}$$

$$\times \int_0^{h^{\alpha_i}} \tau^{1/p+l_i-s_i} \|\varphi_i(x,\tau)\|_p\, d\tau.$$

Using the fact that for $\rho(x)+t^{1/\alpha_j} \leqslant ct^{1/\alpha_j}$, $\rho(x) < 2c_0nh$, $h < t^{l_j}$, and reasoning exactly as for the estimate (9), we obtain

$$\int_0^{t^{l_j}} \zeta_{\varepsilon,H}(h)\, h^{-s_j\alpha_j-\delta_i} \Big\| [\rho(x)+t^{1/\alpha_j}]^\beta \int\limits_{0<y_m<h^{\alpha_m}} dy \int_0^{h^{\alpha_i}} \tau^{1/p+l_i-s_i}$$

$$\times |\Psi_i(x+y,\tau)|\, d\tau \Big\|_{L_p(\rho(x)<2c_0nh)} dh \leqslant t^{\beta/\alpha_j} \int_0^{t^{l_j}} \zeta_{\varepsilon,H}(h)\, h^{-s_j\alpha_j-\delta_i}$$

$$\times \left(\int_0^{h^{\alpha_i}} \int \tau^{1+p(l_i-s_i)} |\varphi_i(x,\tau)|^p\, dx\, d\tau \right)^{1/p'} \left(\int_0^{dh} \rho^{-\beta p'+|\alpha|+\alpha_i-1} d\rho \right)^{1/p} \left(\int_0^{2c_0nh} \rho^{|\alpha|-1} d\rho \right)^{1/p} \tag{32}$$

$$\leqslant t^{\beta/\alpha_j} \int_0^{t^{l_j}} \zeta_{\varepsilon,H}(h)\, h^{-s_j\alpha_j-1+s_i\alpha_i-\alpha_i/p-\beta} \left(\int_0^{h^{\alpha_i}} \int \tau^{1+p(l_i-s_i)} |\varphi_i(x,\tau)|^p\, dx\, d\tau \right)^{1/p} dh.$$

Applying Hölder's inequality, noting that for $2c_0nh < \rho(x) < 2c_0nt^{l_j}$ and $\rho(y) < nh$ the inequalities $\rho(x)+t^{1/\alpha_j} \leqslant ct^{1/\alpha_j}$ and $\rho(x+y)+\tau^{1/\alpha_j} \geqslant nh$ are satisfied, we obtain, after easy calculations,

$$\int_0^{t^{l_j}} \zeta_{\varepsilon,H}(h)\, h^{-s_j\alpha_j-\delta_i} \Big\| [\rho(x)+t^{1/\alpha_j}]^\beta \int\limits_{0<y_m<h^{\alpha_m}} dy \int_0^{h^{\alpha_i}} \tau^{1/p+l_i-s_i} |\Psi_i(x+y,\tau)|$$

$$\times d\tau \Big\|_{L_p(2c_0nh<\rho(x)<2c_0nt^{l_j})} dh \leqslant t^{\beta/\alpha_j} \int_0^{t^{l_j}} \zeta_{\varepsilon,H}(h)\, h^{-s_j\alpha_j-1+s_i\alpha_i-\alpha_i/p-\beta} \tag{33}$$

$$\times \left(\int_0^{h^{\alpha_i}} \int \tau^{1+p(l_i-s_i)} |\varphi_i(x,\tau)|^p\, dx\, d\tau \right)^{1/p} dh.$$

Let $\beta < 0$. Considering the two cases $\rho(x) > 2c_0nh$ and $\rho(x) < 2c_0nh$ and carrying out calculations analogous to those carried out for (31) and (32), we have

$$\int_0^{t^{l_j}} \zeta_{\varepsilon,H}(h)\, h^{-s_j\alpha_j-\delta_i} \Big\| [\rho(x)+t^{1/\alpha_j}]^{\beta} \int\limits_{0<y_m<h^{\alpha_m}} dy \int_0^{h^{\alpha_i}} \tau^{1/p+l_i-s_i} |\Psi_i(x+y,\tau)|\, d\tau \Big\|_p$$

$$\times\, dh \ll \int_0^{t^{l_j}} \zeta_{\varepsilon,H}(h)\, h^{-s_j\alpha_j-1+s_i\alpha_i} \Big[h^{-\alpha_i} \int_0^{h^{\alpha_i}} \tau^{1/p+l_i-s_i} \|\varphi_i\|_p\, d\tau + h^{-\alpha_i/p} \tag{34}$$

$$\times \Big(\int_0^{h^{\alpha_i}} \int \tau^{1+p(l_i-s_i)} |\varphi_i|^p\, dx\, d\tau \Big)^{1/p} \Big]\, dh.$$

Now let us pass to an estimate of the second integral with respect to h in the right side of (30). Since for $h > t^{l_j}$, $\rho(x) > 2c_0 nh$, $\rho(y) < nh$ and $\tau < h^{\alpha_i}$ one has the inequality

$$c_1[\rho(x) + t^{1/\alpha_i}] \leqslant \rho(x+y) + \tau^{1/\alpha_i} \leqslant c_2[\rho(x) + t^{1/\alpha_j}],$$

we have, on applying the generalized Minkowski inequality,

$$t^{\sigma_j} \int_{t^{l_j}}^{\infty} \zeta_{\varepsilon,H}(h)\, h^{-s_j\alpha_j-\sigma_i-\sigma_j\alpha_j} \Big\| [\rho(x)+t^{1/\alpha_j}]^{\beta} \int\limits_{0<y_m<h^{\alpha_m}} dy \int_0^{h^{\alpha_i}} \tau^{1/p+l_i-s_i}$$

$$\times |\Psi_i(x+y,\tau)|\, d\tau \Big\|_{L_p(\rho(x)>2c_0nh)} dh \ll t^{\sigma_j} \int_{t^{l_j}}^{\infty} \zeta_{\varepsilon,H}(h)\, h^{-s_j\alpha_j-1+s_i\alpha_i-\alpha_i-\sigma_j\alpha_j}\, dh \tag{35}$$

$$\times \int_0^{h^{\alpha_i}} \tau^{1/p+l_i-s_i} \|\varphi_i(x,\tau)\|_p\, d\tau.$$

Reasoning exactly as for the estimate (32), we obtain

$$t^{\sigma_j} \int_{t^{l_j}}^{\infty} \zeta_{\varepsilon,H}(h)\, h^{-\sigma_j\alpha_j-s_j\alpha_j-\delta_i} \Big\| [\rho(x)+t^{1/\alpha_j}]^{\beta} \int\limits_{0<y_m<h^{\alpha_m}} dy \int_0^{h^{\alpha_i}} \tau^{1/p+l_i-s_i}$$

$$\times |\Psi_i(x+y,\tau)|\, d\tau \Big\|_{L_p(\rho(x)<2c_0nh)} dh \ll t^{\sigma_j} \int_{t^{l_j}}^{\infty} \zeta_{\varepsilon,H}(h)\, h^{-\sigma_j\alpha_j-s_j\alpha_j-1+s_i\alpha_i-\alpha_i/p} \tag{36}$$

$$\times \Big(\int_0^{h^{\alpha_i}} \int \tau^{1+p(l_i-s_i)} |\varphi_i(x,\tau)|^p\, dx\, d\tau \Big)^{1/p} dh.$$

Using the inequalities (30)–(36), putting $\mu_\beta = \beta$, $\beta \geqslant 0$, $\mu_\beta = 0$ if $\beta < 0$, replacing t^{l_j} by t, applying the Hardy inequality

$$\|x^{-\sigma} F(x)\|_{L_p(x>0)} \leqslant c\|x^{-\sigma+1} f(x)\|_{L_p(x>0)},$$

where

$$1 \leqslant p \leqslant \infty, \quad F(x) = \int_0^x f(t)\, dt \quad \Big(\sigma > \frac{1}{p}\Big), \quad F(x) = \int_x^{\infty} f(t)\, dt \quad \Big(\sigma < \frac{1}{p}\Big);$$

successively making the change of variables $\tau = t^{\alpha_i}u$, $t^{\alpha_i} = h$, applying the generalized Minkowski inequality, and again making the change of variables $hu = t$, we obtain

$$\Big\{ \int_0^{\infty} \| [\rho(x) + t^{1/\alpha_j}]^{\beta} \Delta_j^{\sigma_j}(t)\, D_j^{s_j} g_{\varepsilon,H}(x) \|_p^p \frac{dt}{t^{1+(l_j-s_j)p}} \Big\}^{1/p} \leqslant$$

$$\leqslant \sum_{i=1}^{n} \left\{ \int_0^\infty \left[\left(\int_0^t + t^{\sigma_j\alpha_j} \int_t^\infty h^{-\sigma_j\alpha_j} \right) \zeta_{\varepsilon, H}(h) h^{-s_j\alpha_j - 1 + s_i\alpha_i - \alpha_i} \int_0^{h^{\alpha_i}} \tau^{1/p + l_i - s_i} \|\varphi_i(x,\tau)\|_p \, d\tau dh \right]^p \right.$$

$$\times \left. \frac{dt}{t^{1+p-s_j\alpha_j p}} \right\}^{1/p} + \sum_{i=1}^{n} \left\{ \int_0^\infty \left[\left(t^{\mu_\beta} \int_0^t h^{-\mu_\beta} + t^{\sigma_j\alpha_j} \int_t^\infty h^{-\sigma_j\alpha_j} \right) \zeta_{\varepsilon, H}(h) h^{-s_j\alpha_j - 1 + s_i\alpha_i - \alpha_i/p} \right. \right.$$

$$\times \left. \left. \left(\int_0^{h^{\alpha_i}} \int \tau^{1+p(l_i - s_i)} |\varphi_i(x,\tau)|^p \, dx \, d\tau \right)^{1/p} dh \right]^p \frac{dt}{t^{1+p-s_j\alpha_j p}} \right\}^{1/p}$$

$$\leqslant \sum_{i=1}^{n} \left\{ \int_0^\infty \zeta_{\varepsilon, H}(t) t^{-p+s_i\alpha_i p - \alpha_i p - 1} \left(\int_0^{t^{\alpha_i}} \tau^{1/p + l_i - s_i} \|\varphi_i(x,\tau)\|_p \, d\tau \right)^p dt \right\}^{1/p}$$

$$+ \sum_{i=1}^{n} \left\{ \int_0^\infty \zeta_{\varepsilon, H}(t) t^{-p+s_i\alpha_i p - \alpha_i - 1} \int_0^{t^{\alpha_i}} \int \tau^{1+p(l_i - s_i)} |\varphi_i(x,\tau)|^p \, dx \, d\tau \, dt \right\}^{1/p}$$

$$\leqslant \sum_{i=1}^{n} \int_0^1 u^{1/p + l_i - s_i} \left\{ \int_0^\infty \zeta_{\varepsilon, H}(h^{l_i}) \|\varphi_i(x, hu)\|_p^p \, dh \right\}^{1/p} du$$

$$+ \sum_{i=1}^{n} \left\{ \int_0^1 u^{1+p(l_i - s_i)} du \int_0^\infty \zeta_{\varepsilon, H}(h^{l_i}) \, dh \int |\varphi_i(x, hu)|^p \, dx \right\}^{1/p}$$

$$= \sum_{i=1}^{n} \int_0^1 u^{l_i - s_i} \left\{ \int_{u\varepsilon^{\alpha_i}}^{uH^{\alpha_i}} \|\varphi_i(x,t)\|_p^p \, dt \right\}^{1/p} du + \sum_{i=1}^{n} \left\{ \int_0^1 u^{p(l_i - s_i)} du \int_{u\varepsilon^{\alpha_i}}^{uH^{\alpha_i}} \|\varphi_i(x,t)\|_p^p \, dt \right\}^{1/p}$$

$$\leqslant \sum_{i=1}^{n} \left(\int_0^{H^{\alpha_i}} \|\varphi_i(x,t)\|_p^p \, dt \right)^{1/p} \leqslant \sum_{i=1}^{n} \|\Psi_i\|_{p, \beta, \alpha_i}.$$

Q. E. D.

COROLLARY. *For $f \in b_{p,\beta}^{\mathbf{l}(\mathbf{r})}$, $-|\alpha|/p < \beta < \min\{\omega, \sigma_0\}$, $1 \leqslant p \leqslant \infty$ (for $p = 1$ or $p = \infty$ one can put $\beta = 0$), as $H \longrightarrow 0$,*

$$\|f - f_H\|_{b_{p,\beta}^{\mathbf{l}(\mathbf{r})}} \to 0. \tag{37}$$

The proof follows from the representation (4) and the estimate (29).

THEOREM 3. *Any function $f \in b_{p,\beta}^{\mathbf{l}(\mathbf{r})}$, $-|\alpha|/p < \beta < \min\{\omega, |\alpha|/p'\}$, $1 \leqslant p \leqslant \infty$ (for $p = \infty$ the functions $\varphi^{(i)}(x, t)$ satisfy (17) and one can put $\beta = 0$) which vanishes at infinity is approximable with arbitrary precision in the norm $B_{p,\beta}^{\mathbf{l}(\mathbf{r})}$ by functions from $C_0^\infty(E_n)$.*

PROOF. Suppose $f \in b_{p,\beta}^{\mathbf{l}(\mathbf{r})}$ and it vanishes at infinity, i.e. $P_{\mathbf{r}-1}(x; f) = 0$. On the basis of (37), we have, as $\epsilon \longrightarrow 0$,

$$\|f - f_\varepsilon\|_{B_{p,\beta}^{\mathbf{l}(\mathbf{r})}} \to 0. \tag{38}$$

Using the representation (4) and applying inequality (16) of Theorem 2, we obtain, for fixed $\epsilon > 0$, $\mathbf{k}\alpha > 1 - |\alpha|/p - \beta$ and $H \longrightarrow \infty$,

$$\|D^{\mathbf{k}}(f_\varepsilon - f_{\varepsilon, H})\|_{q(\mathbf{k}), \beta - \mu(\mathbf{k})} = \|D^{\mathbf{k}} f_H\|_{q(\mathbf{k}), \beta - \mu(\mathbf{k})} \leqslant cH^{\varkappa} \|f\|_{b_{p,\beta}^{\mathbf{l}(\mathbf{r})}} \to 0,$$

where $q = q(\mathbf{k}) \geqslant p$ and $\mu = \mu(\mathbf{k}) \geqslant 0$ are determined from the conditions $\kappa = 1 - |\alpha| \times$

$(1/p - 1/q) - \mu - \mathbf{k}\alpha < 0$, $\mu < |\alpha|/q + \beta$ (for $p = 1$ or $p = \infty$ one can put $\beta = 0$ and $\mu = \beta$, respectively). Thus, for any $\epsilon > 0$, the function f_ϵ is approximated with an arbitrary precision in the seminorm

$$\sum_{1-|\alpha|/p-\beta<\mathbf{k}\alpha<2} \| D^{\mathbf{k}} u \|_{q(\mathbf{k}),\, \beta-\mu(\mathbf{k})} \tag{39}$$

by the functions

$$f_{\varepsilon, H} = \sum_{i=1}^{n} \int_{\varepsilon}^{H} h^{-\delta_i} dh \iint T_i \left(\frac{y}{h^{\alpha}}, \frac{t}{h^{\alpha_i}} \right) \Delta_i^{\sigma_i} \left(\frac{t}{\sigma_i} \right) D_i^{s_i} f(x+y)\, dt\, dy. \tag{40}$$

Applying Lemma 8, we obtain that as $H \to \infty$

$$\| f_\varepsilon - f_{\varepsilon, H} \|_{b_{p,\beta}^{\mathbf{l}(\mathbf{r})}} = \| f_H \|_{b_{p,\beta}^{\mathbf{l}(\mathbf{r})}} \to 0. \tag{41}$$

Consequently the function f_ϵ is approximated with an arbitrary precision in $b_{p,\beta}^{\mathbf{l}(\mathbf{r})}$ by the functions (40).

Let us consider the functions $f_{\epsilon,H}^{[R]}$, which differ from $f_{\epsilon,H}$ only by the replacement of $\Delta_i^{\sigma_i}(t/\sigma_i) D_i^{s_i} f(y)$ by $\chi(y, t/\sigma_i, R) \Delta_i^{\sigma_i}(t/\sigma_i) D_i^{s_i} f(y)$, where $\chi(y, t, R)$ is the characteristic function of the set $E_R = \{(y, t)\colon |y_j| < R,\ j = 1, \ldots, n,\ 1/R < t < R\}$.

For fixed $\epsilon > 0$ and $H < \infty$, by virtue of Lemma 2, in which we put $\sigma = \mathbf{k}\alpha$, we obtain, for $\mathbf{k}\alpha > 1 - |\alpha|/p - \beta$ and $R \to \infty$,

$$\| D^{\mathbf{k}} (f_{\varepsilon, H} - f_{\varepsilon, H}^{[R]}) \|_{q(\mathbf{k}),\, \beta-\mu(\mathbf{k})} \leqslant c \varepsilon^{\varkappa} \sum_{i=1}^{n} \| \varphi^{(i)}(x, t) \|_{L_{p,\beta,\alpha_i}(E_{n+1}^{+} \setminus E_R)} \to 0. \tag{42}$$

It is evident from the preceding calculations that the seminorm (39) of the difference $f_\epsilon - f_{\epsilon,H}^{[R]}$, for any $\epsilon > 0$, becomes as small as desired for the proper choice of $H < \infty$ and $R < \infty$. Applying Lemma 12, we obtain, for $R \to \infty$,

$$\| f_{\varepsilon, H} - f_{\varepsilon, H}^{[R]} \|_{b_{p,\beta}^{\mathbf{l}(\mathbf{r})}} \leqslant c \sum_{i=1}^{n} \| \varphi^{(i)}(x, t) \|_{L_{p,\beta,\alpha_i}(E_{n+1}^{+} \setminus E_R)} \to 0. \tag{43}$$

Using Lemma 9 and the fact that $c_1 \leqslant \rho(x) \leqslant c_2$ for $1/2 < |x| < 3$, we have

$$\begin{aligned} &\| f_\varepsilon - f_{\varepsilon, H}^{[R]} - Q(x; f_\varepsilon - f_{\varepsilon, H}^{[R]}) \|_{L_p(1<|x|<2)} \\ &\leqslant \sum_{1-|\alpha|/p-\beta<\mathbf{k}\alpha<2} \| D^{\mathbf{k}} (f_\varepsilon - f_{\varepsilon, H}^{[R]}) \|_{L_p(1/2<|x|<2)} \\ &\leqslant \sum_{1-|\alpha|/p-\beta<\mathbf{k}\alpha<2} \| D^{\mathbf{k}} (f_\varepsilon - f_{\varepsilon, H}^{[R]}) \|_{q(\mathbf{k}),\, \beta-\mu(\mathbf{k})}. \end{aligned} \tag{44}$$

By virtue of (38)–(44), the norm

$$\| f - f_{\varepsilon, H}^{[R]} - Q(x; f_\varepsilon - f_{\varepsilon, H}^{[R]}) \|_{B_{p,\beta}^{\mathbf{l}(\mathbf{r})}}$$

is as small as desired for suitable choice of $\epsilon > 0$, $H < \infty$ and $R < \infty$. By construction $f_{\epsilon,H}^{[R]} \in C_0^\infty(E_n)$, and, on the basis of Lemma 11, the polynomial $Q(x; f_\epsilon - f_{\epsilon,H}^{[R]})$ is approximated with arbitrary precision by functions from $C_0^\infty(E_n)$ in the norm of $B_{p,\beta}^{\mathbf{l}(\mathbf{r})}$. From this we conclude that any function $f \in b_{p,\beta}^{\mathbf{l}(\mathbf{r})}$ which converges to zero at infinity is approximated with arbitrary precision in the norm of $B^{\mathbf{l}(\mathbf{r})}_{p,\beta}$ by functions from $C_0^\infty(E_n)$.

REMARK 2. The result obtained about the approximation of the functions $f \in b^{\mathbf{l}(\mathbf{r})}_{p,\beta}$ by functions from C_0^∞ can also be formulated in less precise terms of the space $b^{\mathbf{l}(\mathbf{r})}_{p,\beta}$, which is the factor space of $B^{\mathbf{l}(\mathbf{r})}_{p,\beta}$ by polynomials of degree not greater than $r_i - 1$ with respect to x_i ($i = 1, \ldots, n$). Namely, any function

$$f \in b^{\mathbf{l}(\mathbf{r})}_{p,\beta}, \quad -|\alpha|/p < \beta < \min\{\omega,\ |\alpha|/p'\},\ 1 \leqslant p \leqslant \infty$$

(for $p = \infty$ the functions $\varphi^{(i)}(x, t)$ satisfy (17) and one can put $\beta = 0$) is approximated with arbitrary precision in the seminorm of $b^{\mathbf{l}(\mathbf{r})}_{p,\beta}$ by functions from $C_0^\infty(E_n)$. The proof of this assertion is easily discovered in the proof of Theorem 3.

The analysis was carried out for functions f defined on the entire space E_n. The results obtained are also true for more general regions, due to the possibility, proved below, of the extension of functions from these regions to the entire space E_n with the preservation of differential properties and the validity of the corresponding inequalities for the norms.

Let Ω be a region of n-dimensional Euclidean space E_n such that in it we permit a translation along any vector with nonnegative coordinates, i.e. for any $x \in \Omega$ and $y = (y_1, \ldots, y_n)$, $y_i \geqslant 0$, the point $x + y \in \Omega$. Let us define the space $b^{\mathbf{l}(\mathbf{r})}_{p,\beta}(\Omega)$, $1 \leqslant p \leqslant \infty$, of functions f which have generalized derivatives with respect to x_i of order s_i in Ω with finite seminorm (1), (2), in which one must replace E_n by Ω.

THEOREM 4. *The space* $b^{\mathbf{l}(\mathbf{r})}_{p,\beta}(\Omega)$, $-|\alpha|/p < \beta < |\alpha|/p'$, $1 \leqslant p \leqslant \infty$ *(for* $p = \infty$ *one can put* $\beta = 0$ *and the functions* $\varphi^{(i)}(x, t)$ *satisfy condition* (17), *in which one must replace* E^+_{n+1} *by* $\Omega^+ = \{(x, t)\colon x \in \Omega,\ t > 0\}$), *admits a bounded linear extension from* Ω *to* E_n.

This theorem was proved for $\beta = 0$ by Besov [9].

PROOF. Put

$$f_{\varepsilon,H}(x) = \sum_{i=1}^{n} \int_{\varepsilon}^{H} h^{-\delta_i} dh \iint T_i\left(\frac{y}{h^{\alpha}}, \frac{t}{h^{\alpha_i}}\right) \varphi_i(x + y, t)\, dt\, dy,$$

where $\varphi_i(x, t) = \Delta_i^{\sigma_i}(t/\sigma_i) D_i^{s_i} f(x)$ for $x \in \Omega$, $\varphi_i(x, t) = 0$ for $x \notin \Omega$.

One can show that the space $b^{\mathbf{l}(\mathbf{r})}_{p,\beta}$ is complete. It follows from (29) that as $\epsilon \to 0$ and $H \to \infty$, $f_{\epsilon,H}(x)$ converges in the seminorm of $b^{\mathbf{l}(\mathbf{r})}_{p,\beta}$, and hence to some function $f^* \in b^{\mathbf{l}(\mathbf{r})}_{p,\beta}$. It is not difficult to see that representation (4), Lemma 8, and relation (37) also remain in force for the functions $f \in b^{\mathbf{l}(\mathbf{r})}_{p,\beta}(\Omega)$. Therefore it is evident that in the region Ω we have $f_{\epsilon,H}(x) \to f(x)$ in the seminorm of $b^{\mathbf{l}(\mathbf{r})}_{p,\beta}$ as $\epsilon \to 0$ and $H \to \infty$. On the basis of Lemma 12 we have

$$\| f_{\varepsilon,H} \|_{b^{\mathbf{l}(\mathbf{r})}_{p,\beta}} \leqslant c \| f \|_{b^{\mathbf{l}(\mathbf{r})}_{p,\beta}} (\Omega).$$

Passing to the limit as $\epsilon \to 0$ and $H \to \infty$, we obtain

$$\| f^* \|_{b^{\mathbf{l}(\mathbf{r})}_{p,\beta}} \leqslant c \| f \|_{b^{\mathbf{l}(\mathbf{r})}_{p,\beta}} (\Omega).$$

From this we also conclude that the function $f^*(x) = \lim_{\epsilon \to 0, H \to \infty} f_{\epsilon,H}(x)$, where the limit is understood in the sense of $b^{\mathbf{l}(\mathbf{r})}_{p,\beta}$, is a bounded linear extension of the function $f \in b^{\mathbf{l}(\mathbf{r})}_{p,\beta}(\Omega)$, with a replacement of Ω by E_n.

REMARK 3. Let $r(x)$ be a continuous, positive (for $x \neq 0$), α-homogeneous function of degree 1, i.e. which possesses the property $r(t^{\alpha_1}x_1, \ldots, t^{\alpha_n}x_n) = tr(x)$ for any $t > 0$. Let us define the space $\widetilde{b}^{1(\mathbf{r})}_{p,\beta}$, $1 \leqslant p \leqslant \infty$, of functions with finite seminorm (1), (2), in which one must replace $\rho(x)$ by $r(x)$. The results obtained for the space $b^{1(\mathbf{r})}_{p,\beta}$ also remain valid for the space $\widetilde{b}^{1(\mathbf{r})}_{p,\alpha}$.

BIBLIOGRAPHY

1. S. V. Uspenskiĭ, *Imbedding theorems for weighted classes*, Trudy Mat. Inst. Steklov. **60** (1961), 282–303; English transl., Amer. Math. Soc. Transl. (2) **87** (1970), 121–145. MR **25** #440; **41** #3202.

2. S. L. Sobolev, *The density of functions with compact support in the space $L_p^{(m)}(E_n)$*, Dokl. Akad. Nauk SSSR **149** (1963), 40–43 = Soviet Math. Dokl. **4** (1963), 313–316. MR **26** #6758.

3. O. V. Besov, *The behavior of differentiable functions at infinity, and the density of functions with compact support*, Trudy Mat. Inst. Steklov. **105** (1969), 3–14 = Proc. Steklov Inst. Math. **105** (1969), 1–15. MR **42** #6603.

4. Ju. S. Nikol'skiĭ, *Asymptotic behavior at infinity of differentiable functions of weight classes*, Trudy Mat. Inst. Steklov. **117** (1972), 244–255 = Proc. Steklov Inst. Math. **117** (1972), 287–300.

5. V. P. Il'in, *Integral representations of differentiable functions and their application to questions of extension of functions of the classes $W_p^{(l)}(G)$*, Sibirsk. Mat. Ž. **8** (1967), 573–586 = Siberian Math. J. **8** (1967), 421–432. MR **36** #5683.

6. O. V. Besov and P. I. Lizorkin, *Singular integral operators and sequences of convolutions in L_p-spaces*, Mat. Sb. **73** (115) (1967), 65–88 = Math. USSR Sb. **2** (1967), 57–76. MR **36** #1936.

7. G. H. Hardy, J. E. Littlewood and G. Pólya, *Inequalities*, Cambridge Univ. Press, New York, 1934.

8. V. P. Il'in, *Some integral inequalities and their applications in the theory of differentiable functions of several variables*, Mat. Sb. **54** (96) (1961), 331–380. (Russian) MR **25** #157.

9. O. V. Besov, *On the density of functions with compact support in $L_{p,\theta}$ and the extension of functions*, Trudy Mat. Inst. Steklov. **89** (1967), 18–30 = Proc. Steklov Inst. Math. **89** (1967), 16–31. MR **35** #7118.

Translated by J. D. FABREY

Trudy Mat. Inst. Steklov.
131 (1974)

Proc. Steklov Inst. Math.
131 (1974)

IMBEDDING CLASSES OF FUNCTIONS WITH DOMINANT MIXED MODULUS OF SMOOTHNESS

UDC 517.518.22

M. K. POTAPOV

ABSTRACT. In this paper it is shown that

$$\omega_{k_1\dots k_n}\left(f(x_1,\dots,x_n),\frac{1}{l_1+1},\dots,\frac{1}{l_n+1}\right)_q$$
$$\leqslant C\left\{\sum_{\nu_1=l_1+1}^{\infty}\dots\sum_{\nu_n=l_n+1}^{\infty}\prod_{i=1}^{n}\nu_i^{\theta(1/p-1/q)-1}\omega^{\theta}_{k_1\dots k_n}\left(f,\frac{1}{\nu_1},\dots,\frac{1}{\nu_n}\right)_p\right\}^{1/\theta},$$

where $1\leqslant p<q\leqslant\infty$, the constant C does not depend upon ν_i or f, and the number θ is defined by the rule

$$\theta=\begin{cases}1 & \text{for } q=\infty,\\ q & \text{for } q<\infty,\end{cases}$$

With the help of this inequality, imbedding theorems are proved for classes of functions which have a mixed modulus of smoothness and satisfy some other conditions.

Moreover, this paper shows the unimprovability of the imbedding theorems obtained.

Bibliography: 8 items.

I. It is convenient to conduct the study of periodic functions of several variables whose modulus of smoothness in the coordinate directions has some order or another by means of direct and inverse theorems of the approximation of these functions by trigonometric polynomials (see, for example, [1]).

In [2], S. M. Nikol'skiĭ introduced a class of functions whose mixed modulus of smoothness satisfies multiple Hölder conditions, and pointed out that the attempt to obtain direct and inverse theorems concerning the approximation of these functions by trigonometric polynomials was not successful. Therefore he proposed a new method for investigating this class of functions, which consists in the replacement of approximation theorems by theorems on the possibility of representing these functions by series whose terms are trigonometric polynomials.

The imbedding theorems obtained by this method in [2] are sharp. However, the attempt to use this method to prove imbedding theorems for certain classes of functions whose mixed modulus of smoothness satisfies different conditions either does not result in the desired goal, or results in theorems which are not sharp (see, for example, Remark 2 at the end of this paper). Therefore the question arose concerning the investigation of such classes by some other method.

In [3] the present author proposed to study such classes of functions by means of "angular" approximation. In this paper we make more precise some results of [3], and prove imbedding theorems on the basis of these results. In this connection it is shown that

---

AMS (MOS) subject classifications (1970). Primary 46E35; Secondary 26A15, 42A08, 41A50.

the imbedding theorems are sharp not only for the classes of functions considered in [2], but also for other classes of functions, for which the methods of [2] give less precise results. We note that in making the results of [3] more precise, a theorem of P. L. Ul'janov [4] will play a large role.

II. We shall say that $f \in L_p([0, 2\pi]^n)$ if $f = f(x_1, \dots, x_n)$ is a measurable function of n variables, 2π-periodic in each variable, and if $\|f\|_p < \infty$ for some given number p in the interval $1 \leqslant p \leqslant \infty$.

We shall further say that $f \in L_p^0([0, 2\pi]^n)$ if $f \in L_p([0, 2\pi]^n)$ and $\int_0^{2\pi} f dx_i = 0$ for $i = 1, \dots, n$.

In what follows, we will not distinguish equivalent functions, i.e. functions which differ from each other only on a set of n-dimensional measure zero.

By $C, C_1, C_2, \dots$ we shall denote positive constants, generally different in different formulas.

We shall further let

$$\Delta_{h_i}^{k_i} F = \sum_{\nu_i=0}^{k_i} (-1)^{k_i-\nu_i} C_{k_i}^{\nu_i} F(x_1, \dots, x_{i-1}, x_i + \nu_i h_i, x_{i+1}, \dots, x_n),$$

$$\Delta_{h_{i_1} h_{i_2} \dots h_{i_j}}^{k_{i_1} k_{i_2} \dots k_{i_j}} F = \Delta_{h_{i_1}}^{k_{i_1}} \varphi,$$

where $\varphi = \Delta_{h_{i_2} \dots h_{i_j}}^{k_{i_2} \dots k_{i_j}} F$ for $j \geqslant 2$.

By $\omega_{k_{i_1} \dots k_{i_m}}(f, \delta_{i_1}, \dots, \delta_{i_m})$ we shall denote the mixed m-dimensional ($m \leqslant n$) modulus of smoothness of order k_{i_j} in the variable X_{i_j} ($1 \leqslant j \leqslant m$) of functions $f \in L_p([0, 2\pi]^n)$, i.e.

$$\omega_{k_{i_1} \dots k_{i_m}}(f, \delta_{i_1}, \dots, \delta_{i_m})_p = \sup_{\substack{|h_{i_j}| \leqslant \delta_{i_j} \\ j=1, \dots, m}} \| \Delta_{h_{i_1} \dots h_{i_m}}^{k_{i_1} \dots k_{i_m}} f \|_p.$$

Let the function $T_{l_i} \in L_p([0, 2\pi]^n)$ be a trigonometric polynomial of order l_i in the variable x_i. We shall call

$$Y_{l_{i_1} \dots l_{i_m}}(f)_p = \inf_{\substack{T_{l_{i_j}} \\ j=1, \dots, m}} \left\| f - \sum_{j=1}^m T_{l_{i_j}} \right\|_p, \qquad l_{i_j} = 0, 1, 2, \dots,$$

the *best m-dimensional "angular" approximation* of the function f in the variables $x_{i_1}, \dots, x_{i_m}$.

The simplest properties of "angular" approximation were investigated by the author in [3]. In particular, the following inequalities were proved there:

$$Y_{\nu_{i_1} \dots \nu_{i_m}}(f)_p \leqslant C_1 \omega_{k_{i_1} \dots k_{i_m}} \left(f, \frac{1}{\nu_{i_1} + 1}, \dots, \frac{1}{\nu_{i_m} + 1} \right)_p, \tag{1}$$

$$\omega_{k_{i_1} \dots k_{i_m}} \left(f, \frac{1}{\nu_{i_1} + 1}, \dots, \frac{1}{\nu_{i_m} + 1} \right)_p$$
$$\leqslant C_2 \prod_{j=1}^m (\nu_{i_j} + 1)^{-k_{i_j}} \sum_{\mu_{i_1}=0}^{\nu_{i_1}} \dots \sum_{\mu_{i_m}=0}^{\nu_{i_m}} \prod_{i=1}^m (\mu_{i_j} + 1)^{k_{i_j}-1} Y_{\mu_{i_1} \dots \mu_{i_m}}(f)_p, \tag{2}$$

$$Y_{2^{N_1-1}\dots 2^{N_n-1}}(f)_q \leqslant C_3 \sum_{\mu_1=N_1}^{\infty} \dots \sum_{\mu_n=N_n}^{\infty} \prod_{i=1}^{n} 2^{\mu_i(1/p-1/q)} Y_{2^{\mu_1-1}\dots 2^{\mu_n-1}}(f)_p, \tag{3}$$

where $1 \leqslant p \leqslant q \leqslant \infty$ and the constants C_1, C_2, and C_3 do not depend upon f, ν_i, or N_i.

In what follows we shall make inequality (3) exact for $q < \infty$. For this we shall depend upon the following theorem of P. L. Ul'janov [4]:

If $f \in L_p([0, 2\pi]^1)$, $1 \leqslant p < q < \infty$, then

$$\int_0^{2\pi} |f(x)|^q \, dx \leqslant C \left\{ \left(\int_0^{2\pi} |f(x)| \, dx \right)^q + \int_0^{2\pi} t^{-q/p} \| \Delta_t^1 f(x) \|_p^q \, dt \right\}, \tag{4}$$

where the constant C does not depend upon f.

We note that Ul'janov actually obtained, not inequality (4), but the inequality

$$\|f\|_q^q \leqslant C_1 \left\{ \|f\|_1^q + \sum_{m=1}^{\infty} m^{q/p-2} \omega_1^q \left(f, \frac{1}{m} \right)_p \right\}, \tag{5}$$

where C_1 is also independent of f. However, in [5] it is shown that the correctness of the inequality (5) follows from that of (4).

III. LEMMA 1. *If $f(x, y) \in L_p([0, 2\pi]^2)$, $1 \leqslant p < q < \infty$, then*

$$\int_0^{2\pi}\int_0^{2\pi} |f(x, y)|^q \, dx \, dy \leqslant C \left\{ \left(\int_0^{2\pi}\int_0^{2\pi} |f(x, y)| \, dx \, dy \right)^q \right.$$
$$+ \int_0^{2\pi} t_1^{-q/p} \| \Delta_{t_1}^1 f(x, y) \|_p^q \, dt_1 + \int_0^{2\pi} t_2^{-q/p} \| \Delta_{t_2}^1 f(x, y) \|_p^q \, dt_2$$
$$\left. + \int_0^{2\pi}\int_0^{2\pi} (t_1 t_2)^{-q/p} \| \Delta_{t_1 t_2}^{11} f(x, y) \|_p^q \, dt_1 dt_2 \right\},$$

where the constant C is independent of f.

PROOF. Applying Ul'janov's theorem to the inner integral, we obtain that

$$I = \int_0^{2\pi} \left[\int_0^{2\pi} |f(x, y)|^q \, dx \right] dy \leqslant C_1 \left\{ \int_0^{2\pi} \left[\int_0^{2\pi} |f(x, y)| \, dx \right]^q dy \right.$$
$$\left. + \int_0^{2\pi} t^{-q/p} \int_0^{2\pi} \| \Delta_t^1 f(x, y) \|_{p_x}^q \, dy \, dt \right\} = C_1 \left\{ I_1 + \int_0^{2\pi} t^{-q/p} I_2(t) \, dt \right\},$$

where

$$\| \varphi(x, y) \|_{p_x} = \left(\int_0^{2\pi} |\varphi(x, y)|^p \, dx \right)^{1/p},$$

$$I_1 = \int_0^{2\pi} \left(\int_0^{2\pi} |f(x, y)| \, dx \right)^q dy, \qquad I_2(t) = \int_0^{2\pi} \| \Delta_t^1 f(x, y) \|_{p_x}^q \, dy.$$

Letting $F(y) = \int_0^{2\pi} |f(x, y)| \, dx$, we consider $I_1 = \int_0^{2\pi} |F(y)|^q \, dy$. Again applying Ul'janov's theorem, we have

$$I_1 \leqslant C_2 \left\{ \left(\int_0^{2\pi} |F|\, dy \right)^q + \int_0^{2\pi} u^{-q/p} \| \Delta_u^1 F \|_{p_y}^q \, du \right\}$$

$$= C_2 \left\{ \left(\int_0^{2\pi} \int_0^{2\pi} |f(x, y)|\, dx\, dy \right)^q + \int_0^{2\pi} u^{-q/p} I_3^q(u)\, du \right\}.$$

We now estimate

$$I_3(u) = \| \Delta_u^1 F \|_{p_y} = \left(\int_0^{2\pi} |F(y+u) - F(y)|^p \, dy \right)^{1/p} = \left(\int_0^{2\pi} I_4^p(y, u)\, dy \right)^{1/p}.$$

It is clear that

$$I_4(y, u) = |F(y+u) - F(y)| = \left| \int_0^{2\pi} [\, |f(x, y+u)| - |f(x, y)|\,] dx \right|$$

$$\leqslant \int_0^{2\pi} |f(x, y+u) - f(x, y)|\, dx = \int_0^{2\pi} |\Delta_u^1 f(x, y)|\, dx.$$

Applying Hölder's inequality, we obtain

$$I_4(y, u) \leqslant C_3 \; \| \Delta_u^1 f(x, y) \|_{p_x},$$

and consequently

$$I_3(u) \leqslant C_4 \left(\int_0^{2\pi} \| \Delta_u^1 f(x, y) \|_{p_x}^p \, dy \right)^{1/p} = C_4 \| \Delta_u^1 f(x, y) \|_p.$$

This means

$$I_1 \leqslant C_5 \left\{ \left(\int_0^{2\pi} \int_0^{2\pi} |f(x, y)|\, dx\, dy \right)^q + \int_0^{2\pi} u^{-q/p} \| \Delta_u^1 f(x, y) \|_p^q \, du \right\}.$$

Letting $\Phi(y, t) = \| \Delta_t^1 f(x, y) \|_{p_x}$, we consider $I_2(t) = \int_0^{2\pi} |\Phi(y, t)|^q \, dy$. Once more applying Ul'janov's theorem, we have

$$I_2(t) \leqslant C_6 \left\{ \left(\int_0^{2\pi} |\, \Phi(y, t)\, dy \right)^q + \int_0^{2\pi} u^{-q/p} \| \Delta_u^1 \Phi \|_{p_y}^q \, du \right\}$$

$$= C_6 \left\{ I_5^q(t) + \int_0^{2\pi} u^{-q/p} I_6^q(u, t)\, du \right\},$$

where

$$\| \psi(y, t) \|_{p_y} = \left(\int_0^{2\pi} |\psi(y, t)|^p \, dy \right)^{1/p},$$

$$I_5(t) = \int_0^{2\pi} |\Phi(y, t)|\, dy, \qquad I_6(u, t) = \| \Delta_u^1 \Phi(y, t) \|_{p_y}.$$

We now consider

$$I_5(t) = \int_0^{2\pi} |\Phi(y, t)|\, dy = \int_0^{2\pi} \left(\int_0^{2\pi} |\Delta_t^1 f(x, y)|^p \, dx \right)^{1/p} dy.$$

Applying Hölder's inequality to the inner integral, we obtain

$$I_5(t) \leqslant C_7 \left(\int_0^{2\pi} \int_0^{2\pi} |\Delta_t^1 f(x, y)|^p \, dx\, dy \right)^{1/p} = C_7 \| \Delta_t^1 f(x, y) \|_p.$$

We now estimate

$$I_6(u,t) = \| \Delta_u^1 \Phi \|_{p_y} = \left(\int_0^{2\pi} | \Phi(y+u,t) - \Phi(y,t) |^p \, dy \right)^{1/p} = \left(\int_0^{2\pi} I_7^p(y,u,t) \, dy \right)^{1/p}.$$

We make a preliminary estimate of

$$I_7(y,u,t) = | \Phi(y+u,t) - \Phi(y,t) | = \left(\int_0^{2\pi} | \Delta_t^1 f(x, y+u) |^p dx \right)^{1/p} - \left(\int_0^{2\pi} | \Delta_t^1 f(x,y) |^p \, dx \right)^{1/p}.$$

It is clear that

$$I_7(y,u,t) \leqslant \left(\int_0^{2\pi} | \Delta_t^1 f(x,y+u) - \Delta_t^1 f(x,y) |^p \, dx \right)^{1/p} = \left(\int_0^{2\pi} | \Delta_{tu}^{11} f(x,y) |^p dx \right)^{1/p}.$$

Therefore

$$I_6(u,t) \leqslant \left(\int_0^{2\pi} \int_0^{2\pi} | \Delta_{tu}^{11} f(x,y) |^p \, dx \, dy \right)^{1/p} = \| \Delta_{tu}^{11} f(x,y) \|_p.$$

This means

$$I_2(t) \leqslant C_8 \left\{ \| \Delta_t^1 f(x,y) \|_p + \int_0^{2\pi} u^{-q/p} \| \Delta_{tu}^{11} f(x,y) \|_p^q \, du \right\}.$$

Now substituting the estimates for I_1 and I_2 into the estimate for I, we obtain Lemma 1.

IV. *If $f \in L_p^0([0, 2\pi]^2)$, $1 \leqslant p \leqslant q < \infty$, then*

$$\| f \|_q \leqslant C_1 \left\{ \int_0^{2\pi} \int_0^{2\pi} (t_1 t_2)^{-q/p} \| \Delta_{t_1 t_2}^{11} f(x,y) \|_p^q \, dt_1 \, dt_2 \right\}^{1/q},$$

$$\| f \|_q \leqslant C_2 \left\{ \sum_{\nu_1=0}^{\infty} \sum_{\nu_2=0}^{\infty} 2^{(\nu_1+\nu_2)(1/p-1/q)q} Y_{2^{\nu_1}-1\, 2^{\nu_2}-1}^q (f)_p \right\}^{1/q},$$

where the constants C_1 and C_2 do not depend upon f.

PROOF. Since $\int_0^{2\pi} f(x,y)\,dx = 0$, we have

$$\Delta_{t_2}^1 f(x,y) = -\frac{1}{2\pi} \int_0^{2\pi} \Delta_{t_1 t_2}^{11} f(x,y) \, dt_1.$$

Therefore, applying Hölder's inequality, we obtain

$$I_1 = \| \Delta_{t_2}^1 f(x,y) \|_p \leqslant C_3 \int_0^{2\pi} \| \Delta_{t_1 t_2}^{11} f(x,y) \|_p \, dt_1.$$

Once more applying Hölder's inequality, and taking into account that $t \leqslant 2\pi$, we have

$$I_1 \leqslant C_4 \left(\int_0^{2\pi} \| \Delta_{t_1 t_2}^{11} f(x,y) \|_p^q \, dt_1 \right)^{1/q} \leqslant C_5 \left(\int_0^{2\pi} t_1^{-q/p} \| \Delta_{t_1 t_2}^{11} f(x,y) \|_p^q \, dt_1 \right)^{1/q}.$$

Analogously, it is shown that

$$\| \Delta^1_{t_1} f(x, y) \|_p \leqslant C_6 \left(\int_0^{2\pi} t_2^{-q/p} \| \Delta^{11}_{t_1 t_2} f(x, y) \|_p^q \, dt_2 \right)^{1/q},$$

$$\| f \|_1 \leqslant C_7 \left\{ \int_0^{2\pi} \int_0^{2\pi} (t_1 t_2)^{-q/p} \| \Delta^{11}_{t_1 t_2} f(x, y) \|_p^q \, dt_1 \, dt_2 \right\}^{1/q}.$$

Now applying Lemma 1, we obtain the first inequality of Lemma 2. Applying Lemma 5 of [3], we obtain the second inequality of Lemma 2.

LEMMA 3. *If* $f(x, y) \in L_p^0([0, 2\pi]^2)$, $1 \leqslant p < q < \infty$, *then*

$$Y_{2^{N_1}-1\, 2^{N_2}-1}(f)_q \leqslant C \left\{ \sum_{\mu_1=N_1}^{\infty} \sum_{\mu_2=N_2}^{\infty} 2^{(\mu_1+\mu_2)(1/p-1/q)q} Y^q_{2^{\mu_1}-1\, 2^{\mu_2}-1}(f)_p \right\}^{1/q}, \tag{6}$$

where the constant C *does not depend upon* f *or* N_i.

PROOF. a) First let $1 < p < q < \infty$. Evidently

$$I = Y^q_{2^{N_1}-1\, 2^{N_2}-1}(f)_q \leqslant \| f - u_{2^{N_1}-1\, 2^{N_2}-1}(f) \|_q^q,$$

where $u_{l_1 l_2}(f) = S_{l_1 \infty}(f) + S_{\infty l_2}(f) - S_{l_1 l_2}(f)$, $S_{l_1 \infty}(f)$ is the l_1th partial sum in x of the Fourier series of $f(x, y)$, $S_{\infty l_2}(f)$ is the l_2th partial sum in y of the Fourier series of $f(x, y)$, and $S_{l_1 l_2}(f)$ is the partial sum of order l_1 in x and l_2 in y of the Fourier series of $f(x, y)$. Applying Lemma 2, we obtain

$$I \leqslant C_1 \left(\sum_{l_1=0}^{\infty} \sum_{l_2=0}^{\infty} 2^{(l_1+l_2)(1/p-1/q)q} Y^q_{2^{l_1}-1\, 2^{l_2}-1}(f - u_{2^{N_1}-1\, 2^{N_2}-1}(f)) \right)$$
$$= C_1 \left(\sum_{l_1=0}^{N_1} \sum_{l_2=0}^{N_2} + \sum_{l_1=0}^{N_1} \sum_{l_2=N_2+1}^{\infty} + \sum_{l_1=N_1+1}^{\infty} \sum_{l_2=0}^{N_2} + \sum_{l_1=N_1+1}^{\infty} \sum_{l_2=N_2+1}^{\infty} \right)$$
$$= C_1 (I_1 + I_2 + I_3 + I_4).$$

We now estimate

$$I_5 = Y_{2^{l_1}-1\, 2^{l_2}-1}(f - u_{2^{N_1}-1\, 2^{N_2}-1})_p.$$

It is clear that

$$I_5 \leqslant \| f - u_{2^{N_1}-1\, 2^{N_2}-1} - u_{2^{l_1}-1\, 2^{l_2}-1}(f - u_{2^{N_1}-1\, 2^{N_2}-1}) \|_p.$$

We consider

$$I_6 = f - u_{2^{N_1}-1\, 2^{N_2}-1}(f) - u_{2^{l_1}-1\, 2^{l_2}-1}(f - u_{2^{N_1}-1\, 2^{N_2}-1}).$$

It is easy to verify that

$$I_6 = \begin{cases} f - u_{2^{N_1}-1\, 2^{N_2}-1}(f) & \text{for } l_1 \leqslant N_1 \text{ and } l_2 \leqslant N_2; \\ f - u_{2^{N_1}-1\, 2^{l_2}-1}(f) & \text{for } l_1 \leqslant N_1 \text{ and } l_2 > N_2; \\ f - u_{2^{l_1}-1\, 2^{N_2}-1}(f) & \text{for } l_1 > N_1 \text{ and } l_2 \leqslant N_2; \\ f - u_{2^{l_1}-1\, 2^{l_2}-1}(f) & \text{for } l_1 > N_1 \text{ and } l_2 > N_2. \end{cases}$$

In [3] it is show that for $1 < p < \infty$

$$\| f - u_{l_1 l_2}(f) \|_p \leqslant C_2 Y_{l_1 l_2}(f)_p,$$

where the constant C_2 depends only upon p. Applying this estimate to the estimate for I_5 and the formula for I_6, we have

$$I_1 \leqslant C_3 Y^q_{2^{N_1}-1 2^{N_2}-1}(f)_p \sum_{l_1=0}^{N_1} \sum_{l_2=0}^{N_2} 2^{(l_1+l_2)(1/p-1/q)q} \leqslant C_4 2^{(N_1+N_2)(1/p-1/q)q} Y^q_{2^{N_1}-1 2^{N_2}-1}(f)_p,$$

$$I_2 \leqslant C_5 \sum_{l_1=0}^{N} \sum_{l_2=N_2+1}^{\infty} 2^{(l_1+l_2)(1/p-1/q)q} Y^q_{2^{N_1}-1 2^{l_2}-1}(f)_p \leqslant C_6 \sum_{l_2=N_2+1}^{\infty} 2^{(N_1+l_2)(1/p-1/q)q} Y^q_{2^{N_1}-1 2^{l_2}-1}(f)_p.$$

Analogously,

$$I_3 \leqslant C_7 \sum_{l_1=N_1+1}^{\infty} 2^{(l_1+N_2)(1/p-1/q)q} Y^q_{2^{l_1}-1 2^{N_2}-1}(f)_p,$$

$$I_4 \leqslant C_8 \sum_{l_1=N_1+1}^{\infty} \sum_{l_2=N_2+1}^{\infty} 2^{(l_1+l_2)(1/p-1/q)q} Y^q_{2^{l_1}-1 2^{l_2}-1}(f)_p.$$

Substituting the estimates for I_1, I_2, I_3 and I_4 into the estimate for I, we obtain (6) for $p > 1$.

b) Now let $p = 1$. We select a number p_1 such that $1 < p_1 < q < \infty$. Applying only that part of Lemma 3 which has been proved for p, we obtain

$$I = Y^q_{2^{N_1}-1 2^{N_2}-1}(f)_q \leqslant C_9 \sum_{l_1=N_1}^{\infty} \sum_{l_2=N_2}^{\infty} 2^{q(l_1+l_2)(1/p_1-1/q)} Y^q_{2^{l_1}-1 2^{l_2}-1}(f)_{p_1}.$$

Making use, then, of inequality (3), we have

$$I \leqslant C_{10} \sum_{l_1=N_1}^{\infty} \sum_{l_2=N_2}^{\infty} 2^{(l_1+l_2)(1/p_1-1/q)q} \left(\sum_{\nu_1=l_1}^{\infty} \sum_{\nu_2=l_2}^{\infty} 2^{(\nu_1+\nu_2)(1-1/p_1)} Y_{2^{\nu_1}-1 2^{\nu_2}-1}(f_1) \right)^q.$$

Finally, applying Hölder's inequality, we obtain

$$I \leqslant C_{11} \sum_{l_1=N_1}^{\infty} \sum_{l_2=N_2}^{\infty} \prod_{i=1}^{2} 2^{ql_i(1/p_1-1/q-\varepsilon)} \sum_{\nu_1=l_1}^{\infty} \sum_{\nu_2=l_2}^{\infty} \prod_{i=1}^{2} 2^{\nu_i q(1-1/p_1+\varepsilon)} Y^q_{2^{\nu_1}-1 2^{\nu_2}-1}(f)_1,$$

where $0 < \epsilon < 1/p_1 - 1/q$. Changing the order of summation in the right-hand side of this inequality, we arrive at

$$I \leqslant C_{12} \sum_{\nu_1=N_1}^{\infty} \sum_{\nu_2=N_2}^{\infty} \prod_{i=1}^{2} 2^{\nu_i q(1-1/q)} Y^q_{2^{\nu_1}-1 2^{\nu_2}-1}(f)_1,$$

which proves Lemma 3 for $p = 1$.

LEMMA 4. *If $f \in L_p^0([0, 2\pi]^2)$, then*

$$\omega_{k_1}(f, \delta_1)_p \leqslant C_{13}\, \omega_{k_1 k_2}(f, \delta_1, 1)_p,$$
$$\omega_{k_2}(f, \delta_2)_p \leqslant C_{14}\, \omega_{k_1 k_2}(f, 1, \delta_2)_p,$$

where the constants C_{13} and C_{14} do not depend upon f.

The proof of this lemma is contained in [3].

V. THEOREM 1. *If* $f \in L_p^0([0, 2\pi]^n)$, $1 \leqslant p < q \leqslant \infty$, *if* $k_i' \geqslant k_i$, *and if the number* θ *is chosen according to the rule*

$$\theta = \begin{cases} 1 & \text{for} \quad q = \infty, \\ q & \text{for} \quad q < \infty, \end{cases}$$

then

$$\omega_{k_1'\ldots k_n'}\left(f, \frac{1}{N_1+1}, \ldots, \frac{1}{N_n+1}\right)_q \leqslant C\left\{\sum_{l_1=N_1+1}^{\infty} \cdots \sum_{l_n=N_n+1}^{\infty} \prod_{i=1}^{n} l_i^{\theta(1/p-1/q)-1} \omega_{k_1\ldots k_n}^{\theta}\left(f, \frac{1}{l_1}, \ldots, \frac{1}{l_n}\right)_p\right\}^{1/\theta},$$

where C does not depend upon f or N_i.

PROOF. For simplicity, assume $n = 2$. It is possible to write inequality (2) (see [3], p. 380) in the following form:

$$I = \omega_{k_1'k_2'}\left(f, \frac{1}{2^{\mu_1}}, \frac{1}{2^{\mu_2}}\right)_q \leqslant C_1 \prod_{i=1}^{2} 2^{-k_i'\mu_i} \sum_{l_1=0}^{\mu_1} \sum_{l_2=0}^{\mu_2} \prod_{i=1}^{2} 2^{k_i' l_i} Y_{2^{l_1}-1 2^{l_2}-1}(f)_q.$$

Applying Lemma 3 if $q < \infty$ or inequality (3) if $q = \infty$, we obtain

$$I \leqslant C_2 \prod_{i=1}^{2} 2^{-k_i'\mu_i} \sum_{l_1=0}^{\mu_1} \sum_{l_2=0}^{\mu_2} \prod_{i=1}^{2} 2^{k_i' l_i} \left\{\sum_{\nu_1=l_1}^{\infty} \sum_{\nu_2=l_2}^{\infty} \prod_{i=1}^{2} 2^{\theta\nu_i(1/p-1/q)} Y^{\theta}_{2^{\nu_1}-1 2^{\nu_2}-1}(f)_p\right\}^{1/\theta}.$$

Then, applying Hölder's inequality, we have

$$I \leqslant C_3 \prod_{i=1}^{2} 2^{-(k_i'-\varepsilon_i)\mu_i} \left\{\sum_{l_1=0}^{\mu_1} \sum_{l_2=0}^{\mu_2} \prod_{i=1}^{2} 2^{l_i\theta(k_i'-\varepsilon_i)} \times \sum_{\nu_1=l_1}^{\infty} \sum_{\nu_2=l_2}^{\infty} \prod_{i=1}^{2} 2^{\nu_i\theta(1/p-1/q)} Y^{\theta}_{2^{\nu_1}-1 2^{\nu_2}-1}(f)_p\right\}^{1/\theta},$$

where $0 < \epsilon_i < k_i' - k_i + 1/p - 1/q$.

We now estimate

$$I_1 = \sum_{\nu_1=l_1}^{\infty} \sum_{\nu_2=l_2}^{\infty} \prod_{i=1}^{2} 2^{\nu_i\theta(1/p-1/q)} Y^{\theta}_{2^{\nu_1}-1 2^{\nu_2}-1}(f)_p.$$

Taking (1) into account, we obtain

$$\begin{aligned} I_1 &\leqslant C_4 \sum_{\nu_1=l_1}^{\infty} \sum_{\nu_2=l_2}^{\infty} \prod_{i=1}^{2} 2^{\nu_i\theta(1/p-1/q)} \omega^{\theta}_{k_1k_2}\left(f, \frac{1}{2^{\nu_1}}, \frac{1}{2^{\nu_2}}\right)_p \\ &= C_4\left(\sum_{\nu_1=l_1}^{\mu_1} \sum_{\nu_2=l_2}^{\mu_2} + \sum_{\nu_1=l_1}^{\mu_1} \sum_{\nu_2=\mu_2+1}^{\infty} + \sum_{\nu_1=\mu_1+1}^{\infty} \sum_{\nu_2=l_2}^{\mu_2} + \sum_{\nu_1=\mu_1+1}^{\infty} \sum_{\nu_2=\mu_2+1}^{\infty}\right) \\ &= C_4(I_2 + I_3 + I_4 + I_5). \end{aligned}$$

Using properties of the modulus of smoothness, we have

$$I_2\leqslant C_5\sum_{\nu_1=l_1}^{\mu_1}\sum_{\nu_2=l_2}^{\mu_2}\prod_{i=1}^{2}2^{\nu_i\theta(1/p-1/q)}\omega^{\theta}_{k_1k_2}\left(f,\frac{2^{\mu_1-\nu_1}}{2^{\mu_1+1}},\frac{2^{\mu_2-\nu_2}}{2^{\mu_2+1}}\right)_p$$

$$\leqslant C_6\omega^{\theta}_{k_1k_2}\left(f,\frac{1}{2^{\mu_1+1}},\frac{1}{2^{\mu_2+1}}\right)_p\sum_{\nu_1=l_1}^{\mu_1}\sum_{\nu_2=l_2}^{\mu_2}\prod_{i=1}^{2}2^{\nu_i\theta(1/p-1/q)+(\mu_i+\nu_i)k_i\theta}$$

$$\leqslant C_7\prod_{i=1}^{2}2^{\mu_i\theta k_i-l_i\theta(k_i-1/p+1/q)}\omega^{\theta}_{k_2k_2}\left(f,\frac{1}{2^{\mu_1+1}},\frac{1}{2^{\mu_2+1}}\right)_p$$

$$\leqslant C_8\prod_{i=1}^{2}2^{(\mu_i-l_i)(k_i-1/p+1/q)|\theta}\prod_{i=1}^{2}2^{(\mu_i+1)(1/p-1/q)\theta}\omega^{\theta}_{k_1k_2}\left(f,\frac{1}{2^{\mu_1+1}},\frac{1}{2^{\mu_2+1}}\right)_p$$

$$\leqslant C_8\prod_{i=1}^{2}2^{(\mu_i-l_i)(k_i-1/p+1/q)\theta}\sum_{\nu_1=\mu_1+1}^{\infty}\sum_{\nu_2=\mu_2+1}^{\infty}\prod_{i=1}^{2}2^{\nu_i\theta(1/p-1/q)}\omega^{\theta}_{k_1k_2}\left(f,\frac{1}{2^{\nu_1}},\frac{1}{2^{\nu_2}}\right)_p$$

$$=C_8\prod_{i=1}^{2}2^{(\mu_i-l_i)(k_i-1/p+1/q)\theta}I_5.$$

Analogously, it is shown that

$$I_3\leqslant C_9 2^{(\mu_1-l_1)(k_1-1/p+1/q)\theta}I_5,\qquad I_4\leqslant C_{10}2^{(\mu_2-l_2)(k_2-1/p+1/q)\theta}I_5.$$

Taking the estimates for I_1, I_2 and I_3 into account, we obtain that

$$I_1\leqslant C_{11}\prod_{i=1}^{2}2^{(\mu_i-l_i)(k_i-1/p+1/q)\theta}I_5.$$

Therefore

$$I\leqslant C_{12}\prod_{i=1}^{2}2^{-(k_i-\varepsilon_i)\mu_i}I_5^{1/\theta}\left\{\sum_{l_1=0}^{\mu_1}\sum_{l_2=0}^{\mu_2}\prod_{i=1}^{2}2^{l_i\theta(k_i-\varepsilon)+(\mu_i-l_i)(k_i-1/p+1/q)\theta}\right\}^{1/\theta}.$$

Since $k_i'-k_i+1/p-1/q-\epsilon>0$, we obtain from this that

$$I_i\leqslant C_{14}I_5^{1/\theta}.$$

For the given N_i we now select numbers μ_i such that $2^{\mu_i-1}<N_i+1\leqslant 2^{\mu_i}$; then, applying only that part of the inequality which has been proved, we have

$$I_7=\omega_{k_1'k_2'}\left(f,\frac{1}{N_1+1},\frac{1}{N_2+1}\right)_q\leqslant\omega_{k_1'k_2'}\left(f,\frac{1}{2^{\mu_1}},\frac{1}{2^{\mu_2}}\right)_q$$

$$\leqslant C_{15}\left\{\sum_{\nu_1=\mu_1+1}^{\infty}\sum_{\nu_2=\mu_2+1}^{\infty}\prod_{i=1}^{2}2^{\nu_i\theta(1/p-1/q)}\omega^{\theta}_{k_1k_2}\left(f,\frac{1}{2^{\nu_1}},\frac{1}{2^{\nu_2}}\right)_p\right\}^{1/\theta}.$$

Now it is easy to show by standard techniques that

$$I_7\leqslant C_{16}\left\{\sum_{\nu_1=2^{\mu_1}}^{\infty}\sum_{\nu_2=2^{\mu_2}}^{\infty}\prod_{i=1}^{2}(\nu_1\nu_2)^{\theta(1/p-1/q)-1}\omega^{\theta}_{k_1k_2}\left(f,\frac{1}{\nu_1},\frac{1}{\nu_2}\right)_p\right\}^{1/\theta},$$

from which Theorem 1 follows.

VI. We shall say that a function $\beta=\beta(t_1,\dots,t_n)$ is a function of type MH if it satisfies the following conditions:

1. β is a function of n variables, nonnegative and continuous on the cube $0 \leqslant t_i \leqslant 1$, $i = 1, \ldots, n$.

2. $\beta(t_1, \ldots, t_{i-1}, W_i, t_{i+1}, \ldots, t_n) \leqslant C_1 \beta(t_1, \ldots, t_{i-1}, V_i, t_{i+1}, \ldots, t_n)$ for all t_j ($j = 1, 2, \ldots, i-1, i+1, \ldots, n$) and for $0 \leqslant W_i \leqslant V_i \leqslant 1$.

3. $\beta(t_1, \ldots, t_{i-1}, 2V_i, t_{i+1}, \ldots, t_n) \leqslant C_2 \beta(t_1, \ldots, t_{i-1}, V_i, t_{i+1}, \ldots, t_n)$ for all t_j ($j = 1, 2, \ldots, i-1, i+1, \ldots, n$) and for $0 \leqslant V_i \leqslant 1/2$, where the constants C_1 and C_2 do not depend upon V_i, W_i, or t_i.

We shall say that $f \in S_p H(\beta, k)$ if it satisfies the following conditions:

1) $f \in L_p([0, 2\pi]^n)$.

2) β is a function of type MH.

3) $k = (k_1, \ldots, k_n)$, where the k_i are given natural numbers.

4) For all collections of indices $i_1, \ldots, i_n$ such that $1 \leqslant i_j \leqslant n$, $1 \leqslant j \leqslant m \leqslant n$, we have

$$\omega_{k_{i_1} \ldots k_{i_m}}(f, \delta_{i_1}, \ldots, \delta_{i_m})_p \leqslant C_3 \beta_{i_1 \ldots i_m}(\delta_{i_1}, \ldots, \delta_{i_m}),$$

where $\beta_{i_1, \ldots, i_m}(\delta_{i_1}, \ldots, \delta_{i_m}) = \beta(\delta_1^*, \ldots, \delta_n^*)$, with $\delta_{i_j}^* = \delta_{i_j}$ for the given collection of indices $i_1, \ldots, i_m$, $\delta_i^* = 1$ for $i \neq i_j$, and the constant C_3 does not depend upon f or δ_i.

THEOREM 2. *In order for the imbedding $S_p H(\beta_1, k) \subset S_q H(\beta_2, k')$ to hold, where $1 \leqslant p < q \leqslant \infty$ and $k_i' \geqslant k_i$, it suffices that*

$$\left\{ \sum_{\nu_1=\mu_1+1}^{\infty} \cdots \sum_{\nu_n=\mu_n+1}^{\infty} \prod_{i=1}^{n} \nu^{\theta(1/p-1/q)-1} \beta_1^{\theta}\left(\frac{1}{\nu_1}, \ldots, \frac{1}{\nu_n}\right) \right\}^{1/\theta} \leqslant C_4 \beta_2 \left(\frac{1}{\mu_1+1}, \ldots, \frac{1}{\mu_n+1}\right), \tag{7}$$

where the constant C_4 does not depend upon μ_i, and where the number θ is defined by the rule

$$\theta = \begin{cases} 1 & \text{for } q = \infty, \\ q & \text{for } q < \infty. \end{cases}$$

PROOF. Again for simplicity we assume $n = 2$. From the condition $f \in L_p^0([0, 2\pi]^2)$ it follows that

$$f(x, y) = F(x, y) + \varphi(x) + \psi(y) = A,$$

where $F \in L_p^0([0, 2\pi]^2)$, $\varphi \in L_p^0([0, 2\pi]^1)$, $\psi \in L_p^0([0, 2\pi]^1)$, and A is a constant.

In what follows we shall write

$$\|\varphi(x)\|_p = \left(2\pi \int_0^{2\pi} |\varphi(x)|^p dx\right)^{1/p} = \left(\int_0^{2\pi}\int_0^{2\pi} |\varphi(x)|^p dx\, dy\right)^{1/p}$$

$$\|\psi(y)\|_p = \left(2\pi \int_0^{2\pi} |\psi(y)|^p dy\right)^{1/p} = \left(\int_0^{2\pi}\int_0^{2\pi} |\psi(y)|^p dx\, dy\right)^{1/p}.$$

Evidently

$$\omega_{k_1 k_2}(F, \delta_1, \delta_2)_p = \omega_{k_1 k_2}(f, \delta_1, \delta_2)_p \leqslant C_5 \beta_1(\delta_1, \delta_2).$$

Applying Lemma 4 to the function $F(x, y)$, we obtain

$$\omega_{k_1}(F,\ \delta_1)_p \leqslant C_6\beta_1(\delta_1,\ 1),\quad \omega_{k_2}(F,\ \delta_2)_p \leqslant C_7\beta_1(1,\ \delta_2).$$

It is easily seen that

$$\omega_{k_1}(\varphi,\ \delta_1)_p \leqslant \omega_{k_1}(f,\ \delta_1)_p + \omega_{k_1}(F,\ \delta_1)_p \leqslant C_8\beta_1(\delta_1,\ 1).$$
$$\omega_{k_2}(\psi,\ \delta_2)_p \leqslant \omega_{k_1}(f,\ \delta_2)_p + \omega_{k_2}(F,\ \delta_2)_p \leqslant C_9\beta_1(1,\ \delta_2).$$

Now applying Theorem 1 to $F(x, y)$, and taking condition (7) of Theorem 2 into account, we obtain that for $k_i' \geqslant k_i$

$$\omega_{k_1'k_2'}\left(F, \frac{1}{\mu_1}, \frac{1}{\mu_2}\right)_q \leqslant C_{10}\beta_2\left(\frac{1}{\mu_1}, \frac{1}{\mu_2}\right). \tag{8}$$

Again applying Lemma 4, we obtain

$$\omega_{k_1'}\left(\varphi, \frac{1}{\mu_1}\right)_q \leqslant C_{11}\beta_2\left(\frac{1}{\mu_1}, 1\right),\quad \omega_{k_2'}\left(F, \frac{1}{\mu_2}\right)_q \leqslant C_{12}\beta_2\left(1, \frac{1}{\mu_2}\right). \tag{9}$$

Applying Theorem 1 to the function $\varphi(x)$, we have

$$\omega_{k_1'}\left(\varphi, \frac{1}{\mu_1}\right)_q \leqslant C_{12}\left\{\sum_{\nu_1=\mu_1+1}^{\infty} \nu_1^{\theta(1/p-1/q)-1}\beta_1^{\theta}\left(\frac{1}{\nu_1}, 1\right)\right\}^{1/\theta}.$$

Since it is evident that

$$\sum_{\nu_1=\mu_1+1}^{\infty} \nu_1^{\theta(1/p-1/q)-1}\beta_1^{\theta}\left(\frac{1}{\nu_1}, 1\right) \leqslant \sum_{\nu_1=\mu_1+1}^{\infty}\sum_{\nu_2=1}^{\infty}(\nu_1\nu_2)^{\theta(1/p-1/q)-1}\beta_1^{\theta}\left(\frac{1}{\nu_1}, \frac{1}{\nu_2}\right),$$

on using condition (7) of Theorem 2 we obtain

$$\omega_{k_1'}\left(\varphi, \frac{1}{\mu_1}\right)_q \leqslant C_{13}\beta_2\left(\frac{1}{\mu_1}, 1\right). \tag{10}$$

Analogously,

$$\omega_{k_2'}\left(\psi, \frac{1}{\mu_2}\right)_q \leqslant C_{14}\beta_2\left(1, \frac{1}{\mu_2}\right). \tag{11}$$

It is easy to see that (8)–(11) imply

$$\begin{aligned}
&\omega_{k_1'k_2'}\left(f, \frac{1}{\mu_1}, \frac{1}{\mu_2}\right)_q = \omega_{k_1'k_2'}\left(F, \frac{1}{\mu_1}, \frac{1}{\mu_2}\right)_q \leqslant C_{15}\beta_2\left(\frac{1}{\mu_1}, \frac{1}{\mu_2}\right),\\
&\omega_{k_1'}\left(f, \frac{1}{\mu_1}\right)_q \leqslant \omega_{k_1'}\left(F, \frac{1}{\mu_1}\right)_q + \omega_{k_1'}\left(\varphi, \frac{1}{\mu_1}\right)_q \leqslant C_{16}\beta_2\left(\frac{1}{\mu_1}, 1\right),\\
&\omega_{k_2'}\left(f, \frac{1}{\mu_2}\right)_q \leqslant \omega_{k_2'}\left(F, \frac{1}{\mu_2}\right)_q + \omega_{k_2'}\left(\psi, \frac{1}{\mu_2}\right)_q \leqslant C_{17}\beta_2\left(1, \frac{1}{\mu_2}\right).
\end{aligned} \tag{12}$$

Taking (12) into account for all $\mu_1 \geqslant 1$ and $\mu_2 \geqslant 1$, since the constants $C_5 - C_{17}$ do not depend upon f or μ_i, Theorem 2 is proved.

VII. THEOREM 3. *In order for the imbedding $S_pH(\beta_1, 1) \subset S_qH(\beta_2, 1)$ to obtain, where $1 \leqslant p < q \leqslant \infty$, $\beta_1 = \Pi_1^n h_1(\delta_i)$, $\beta_2 = \Pi_1^n \alpha_i(\delta_i)$, and h_i and α_i are the moduli of continuity, it is necessary and sufficient that*

$$\left\{\sum_{\nu_1=\mu_i+1}^{\infty} \nu_i^{\theta(1/p-1/q)-1}h_i^{\theta}\left(\frac{1}{\nu_i}\right)\right\}^{1/\theta} = O\left\{\alpha_i\left(\frac{1}{\mu_i+1}\right)\right\} \tag{13}$$

for all $i = 1, 2, \ldots, n$, *where* θ *is a number defined by the rule*

$$\theta = \begin{cases} 1 & \text{for} \quad q = \infty, \\ q & \text{for} \quad q < \infty. \end{cases}$$

PROOF. The sufficiency follows from Theorem 2. We show the necessity of condition (13). Assume that (13) is not satisfied for one i, and let, for simplicity, this be $i = 1$, so that for $i = 2, \ldots, n$ condition (13) is satisfied. If (13) is satisfied for a given i, then it is easy to construct a function $\psi_i(x_i)$ of one variable x_i, which satisfies the condition

$$\psi_i \in L_p^0([0, 2\pi]^1), \quad \omega_1(\psi_i, \delta_i)_p \leqslant C_1 h_i(\delta_i), \quad \omega_1(\psi_i, \delta_i)_q \leqslant C_2 \alpha_i(\delta_i).$$

At the same time, it has been shown by Ul'janov [4] and V. A. Andrienko [6] that if for a given i condition (13) is not satisfied, then it is possible to construct a function $\varphi_i(x_i)$ of one variable x_i such that

$$\varphi_i \in L_p^0([0, 2\pi]^1), \quad \omega_1(\varphi_i, \delta_i)_p \leqslant C_3 h_i(\delta_i), \quad \omega_1(\varphi_i, \delta_i)_q \neq O\{\alpha_i(\delta_i)\}.$$

It is easy to see that for the function $f^* = \varphi_1(x_1)\Pi_2^n \psi_i(x_i)$ we have

$$f^* \in S_p H\left(\prod_{i=1}^n h_i, 1\right),$$

but $f^* \notin S_q H(\Pi_1^n \alpha_i, 1)$, and this completes the proof of Theorem 3.

VIII. REMARKS. 1. If in Theorem 2 we assume $\beta_1 = \Pi_1^n \delta_i^{\alpha_i}$ and $\beta_2 = \Pi_1^n \delta_i^{\alpha_i - 1/p + 1/q}$, where $\alpha_i < 1/p - 1/q$, then we obtain the imbedding theorem of S. M. Nikol'skiĭ [2], and from Theorem 3, its inexactness.

2. Special cases of Theorem 2 are contained in the following imbedding theorem: if $\gamma_i > 1/q$, $q < \infty$, then

$$S_p H\left(\prod_{i=1}^n \delta_i^{1/p - 1/q}\left[\ln \frac{2}{\delta_i}\right]^{-\gamma_i}\right) \subset S_q H\left(\prod_{i=1}^n \left[\ln \frac{2}{\delta_i}\right]^{-\gamma_i + 1/q}\right),$$

and the unimprovability of this theorem follows from Theorem 3.

We note that by the method presented in [2] it is possible to obtain the imbedding only for $\gamma_i > 1$, and it will turn out to be of the form

$$S_p H\left(\prod_{i=1}^n \delta_i^{1/p - 1/q}\left[\ln \frac{2}{\delta_i}\right]^{-\gamma_i}\right) \subset S_q H\left(\prod_{i=1}^n \left[\ln \frac{2}{\delta_i}\right]^{-\gamma_i + 1}\right).$$

3. Theorems 2 and 3 for the cases $1 < p < q \leqslant 2$, $1 = p < q < \infty$, and $1 \leqslant p < q = \infty$ were proved earlier by other methods in [7] and [8].

BIBLIOGRAPHY

1. S. M. Nikol'skiĭ, *Inequalities for entire functions of exponential type and their application to the theory of differentiable functions of several variables*, Trudy Mat. Inst. Steklov. **38** (1951), 244–278; English transl., Amer. Math. Soc. Transl. (2) **80** (1969), 1–38. MR **14**, 32.

2. ———, *Functions with dominant mixed derivative satisfying a multiple Hölder condition*, Sibirsk. Mat. Ž. **4** (1963), 1342–1364; English transl., Amer. Math. Soc. Transl. (2) **102** (1973), 27–51. MR **28** #3322.

3. M. K. Potapov, *On "angular" approximation*, Proc. Conf. Constructive Function Theory (Budapest, 1969), Akad. Kiadó, Budapest, 1972, pp. 371–399. (Russian)

4. P. L. Ul'janov, *The imbedding of certain function classes* H_p^ω, Izv. Akad. Nauk SSSR Ser. Mat. **32** (1968), 649–686 = Math. USSR Izv. **2** (1968), 601–638. MR **37** #6749.

5. M. K. Potapov, *The imbedding and equality of certain classes of functions*, Izv. Akad. Nauk SSSR Ser. Mat. **33** (1969), 840–860 = Math. USSR Izv. **3** (1969), 795–814. MR **40** #6254.

6. V. A. Andrienko, *Necessary conditions for imbedding the function classes* H_p^ω, Mat. Sb. **78** (120) (1969), 280–300 = Math. USSR Sb. **7** (1969), 273–292. MR **38** #6347.

7. M. K. Potapov, *Angular approximation and embedding theorems*, Math. Balkanica **2** (1972), 183–198. (Russian) RŽMat. **1973** #8Б83.

8. ———, *The Hardy-Littlewood and Marcinkiewicz-Littlewood-Paley theorems, angular approximation, and embedding of some classes of functions*, Mathematica (Cluj) **14** (1972), 339–362. (Russian) RŽMat. **1974** #10Б58.

Translated by C. V. and G. A. COMISKY

Trudy Mat. Inst. Steklov.
131 (1974)

Proc. Steklov Inst. Math.
131 (1974)

STRUCTURAL AND CONSTRUCTIVE CHARACTERIZATIONS OF SOME CLASSES OF FUNCTIONS

UDC 517.5

M. K. POTAPOV

ABSTRACT. This paper solves the problem of the equivalence of structural and constructive characterizations of some classes of functions that are defined in terms of the $L_{p,\mu}$ metric, where $1 \leqslant p < \infty$, $\mu \geqslant -1/p$, and

$$\| f \|_{p,\mu} = \left(\int_{-1}^{1} | f(x) \sqrt{1-x^2}^{\mu} |^p dx \right)^{1/p} < \infty.$$

It is shown, in particular, that the existence of a sequence of algebraic polynomials $P_n(x)$ such that

$$\left\| \frac{f(x) - P_n(x)}{(\sqrt{1-x^2} + 1/n)^{\beta}} \right\|_{p,\mu} \leqslant \frac{C_1}{n^{\alpha}},$$

where $0 < \alpha \leqslant 1$, $0 \leqslant \beta \leqslant 1$, is equivalent to having the function $f(x)$ satisfy the condition

$$\left\| \frac{f(x) - f(x, t, \mu)}{(\sqrt{1-x^2} + |\sin t|)^{\beta}} \right\|_{p,\mu} \leqslant C_2 |\sin t|^{\alpha}$$

when $\mu \geqslant -1/p$, where

$$f(x, t, \mu) = \frac{1}{\gamma_m} \int_0^{\pi} f(x \cos t + \cos \lambda \sin t \sqrt{1-x^2}) (\sin \lambda)^{\mu} d\lambda, \quad \gamma_m = \int_0^{\pi} (\sin \lambda)^{\mu} d\lambda;$$

when $\mu = -1/p$ the condition is

$$\left\| \frac{f(x \cos t + \sin t \sqrt{1-x^2}) - 2f(x) + f(x \cos t - \sin t \sqrt{1-x^2})}{(\sqrt{1-x^2} + |\sin t)^{\beta}} \right\|_{p,\mu} \leqslant C_3 |\sin t|^{\alpha}.$$

When $\beta = 0$ we obtain a structural characterization of the class of functions for which the best approximation by algebraic polynomials satisfies $E_n(f)_{p,\mu} \leqslant C_4/n^{\alpha}$.

Bibliography: 16 items.

Introduction

It is well known [1] that if $f(x)$ is a periodic function and $f(x) \in H_p^{r+\alpha}$, where $1 \leqslant p \leqslant \infty$, $r = 0, 1, 2, \ldots$, $0 < \alpha \leqslant 1$, this is equivalent to having $E_n(f)_p^* \leqslant C/n^{r+\alpha}$, where $E_n(f)_p^*$ is the best approximation to $f(x)$ by trigonometric polynomials of degree at most n. However, if $f(x)$ is not periodic and $f(x) \in H_r^{r+\alpha}$, we can only state that $E_n(f)_p \leqslant C/n^{r+\alpha}$, where $E_n(f)_p$ is the best approximation to $f(x)$ by algebraic polynomials of degree at most n, and the converse is not completely valid.

As was first shown by Nikol'skiĭ [2], in the C metric ($C = L_{\infty}$) the direct theorems on best approximation of nonperiodic functions $f(x) \in H_p^{r+\alpha}$ can be sharpened. He also conjectured that in the sharpened form their converses are correct. Subsequently A. F. Timan [3], [4] sharpened and generalized Nikol'skiĭ's result, and Dzjadyk [5], [6] showed that it has a valid converse.

AMS (MOS) subject classifications (1970). Primary 41A10, 41A25, 41A50.

Consequently for the C metric we have necessary and sufficient conditions for a nonperiodic $f(x)$ to belong to $H_p^{r+\alpha}$. These conditions can be stated as follows: A necessary and sufficient condition that for $p = \infty$ $(C = L_\infty)$ a nonperiodic function $f(x) \in H_p^{r+\alpha}$ is that for each $n = 1, 2, \dots$ there is an algebraic polynomial $P_n(x)$ such that

$$\left\| \frac{f(x) - P_n(x)}{(\sqrt{1 - x^2} + 1/n)^{r+\alpha}} \right\|_p \leqslant \frac{C}{n^{r+\alpha}}. \qquad (*)$$

When $p = \infty$, inequality $(*)$ provides a constructive characterization of the class $H_p^{r+\alpha}$. It is natural to ask the following questions:

1. What is a constructive characterization of $H_p^{r+\alpha}$ for $1 \leqslant p < \infty$?

2. What is a structural characterization of the class of functions satisfying $(*)$ for $1 \leqslant p < \infty$?

3. What is a structural characterization of the class of functions satisfying $E_n(f)_p \leqslant C/n^{r+\alpha}$ for $1 \leqslant p < \infty$?

The first question is still open. The second was answered by Lebed′ and Potapov in [7]–[10]. The third question was answered for $r = 0$, $1 \leqslant p < \infty$, in [10], and for $r = 0, 1, 2, \dots, p = 2$, by Židkov [11]. It should be emphasized that when $r = 0$, $p = 2$, the structural characterizations obtained in these papers are different. We also note that this question was discussed for $p = \infty$ in [12].

Questions similar to the second and third can be formulated not only for L_p but for weighted L_p spaces.

For convenience we shall suppose that $f(x)$ is defined on $[-1, 1]$, and we say that $f(x) \in L_{p,\mu}$ if $f(x)$ is a measurable function and

$$\|f\|_{p,\mu} = \left(\int_{-1}^{1} |f(x) \sqrt{1 - x^2}^{\mu}|^p \, dx\right)^{1/p} < \infty.$$

The fololwing questions arise.

4. What is a structural characterization of the class of functions that satisfy the following condition for $1 \leqslant p < \infty$: For each $n = 1, 2, \dots$ there is an algebraic polynomial such that

$$\left\| \frac{f(x) - P_n(x)}{(\sqrt{1 + x^2} + 1/n)^{r+\alpha}} \right\|_{p,\mu} \leqslant \frac{C}{n^{r+\alpha}}?$$

5. What is a structural characterization of the class of functions that satisfy the condition

$$E_n(f)_{p,\mu} \leqslant \frac{C}{n^{r+\alpha}},$$

where $1 \leqslant p < \infty$ and $E_n(f)_{p,\mu}$ is the best approximation to $f(x)$ by algebraic polynomials of degree at most n in the $L_{p,\mu}$ metric?

The fourth question is answered for $\mu = -1/p$ in [10]. The fifth question is also answered in [10] for $\mu = -1/p$, $r = 0$; and for $p = 2$, $\mu > -1/2$ in Rafal'son's paper [13].

In the present paper we answer the second through the fifth questions in all the remaining open cases. These answers are obtained as special cases of more general theorems that also are of independent interest.

§1. Statement of the general theorems

We say that $f(x)$ has an rth derivative $f^{(r)}(x)$ in $(-1, 1)$ if it has an absolutely continuous $(r-1)$th derivative on each interval $[a, b] \subset (-1, 1)$. This implies that almost everywhere on $[a, b]$ there is a finite rth derivative, which is summable on this interval.

We identify equivalent functions, i.e. functions which differ on sets of measure zero.

We use $C, C_1, C_2, \ldots$ to denote constants which, in general, will be different in different formulas.

We use $R_n(x)$, $P_n(x)$ and $Q_n(x)$ to denote algebraic polynomials of degree at most n.

We put

$$f(x, t, \mu) = \frac{1}{\gamma_\mu}\int_0^\pi f(x\cos t + \cos\lambda \sin t\sqrt{1-x^2})(\sin\lambda)^\mu d\lambda,$$

where $\mu > -1$, $\gamma_\mu = \int_0^\pi (\sin\lambda)^\mu d\lambda$.

The change of variable $\lambda = \pi - \lambda_1$ shows that $f(x, t, \mu) = f(x, -t, \mu)$.

If $f(x)$ has an rth derivative $f^{(r)}(x) = \varphi(x)$ in $(-1, 1)$, the function $\varphi(x, t, \mu)$ will be denoted by $f^{(r)}(x, t, \mu)$.

We put

$$f(x, t) = f(x\cos t - \sin t\sqrt{1-x^2}).$$

If $f(x)$ has an rth derivative $f^{(r)}(x) = \psi(x)$ in $(-1, 1)$, the function $\psi(x, t)$ will be denoted by $f^{(r)}(x, t)$.

The object of this paper is to prove Theorems 1 and 2 and corollaries that answer the second through the fifth questions. We note that Corollary 1 answers the fourth question, and hence also the second. However, we have not succeeded in answering the fifth question by a corollary similar to Corollary 1, and hence we have to answer this question differently depending on the hypotheses on α, μ and r.

THEOREM 1. *Let the numbers* α, ρ, r, μ, p *and* δ *satisfy* $1 \leqslant p < \infty$, $\mu \geqslant -1/p$, $0 < \alpha \leqslant 1$, $\rho < r + \alpha$, $\delta < \mu + r + 1/p$, $\delta < \alpha - \rho < 2 + \mu + r + 2/p$ *and* $r \geqslant 0$, r *an integer, and let* $f(x) \in L_{p,\mu}$. *A necessary and sufficient condition for* $f(x)$ *to have an* r*th derivative* $f^{(r)}(x)$ *in* $(-1, 1)$ *satisfying*

$$\left\| [f^{(r)}(x) - f^{(r)}(x, t, \mu + r - \delta)]\sqrt{1-x^2}^{\,2\mu+r-\delta}\left(\sqrt{1-x^2} + |\sin t|\right)^{\rho+\delta} \right\|_p \leqslant C_1 |\sin t|^\alpha$$

is that there is a sequence of polynomials $P_n(x)$ *such that*

$$\left\| [f(x) - P_n(x)]\sqrt{1-x^2}^{\,2\mu}\left(\sqrt{1-x^2} + \frac{1}{n}\right)^\rho \right\|_p \leqslant \frac{C_2}{n^{r+\alpha}},$$

where C_1 and C_2 are independent of f, t and n.

Theorem 1 is a corollary of Theorems 3 and 4.

THEOREM 2. *Let the numbers α, ρ, r, μ, p and δ satisfy $1 \leqslant p < \infty$, $\mu \geqslant -1/p$, $0 \leqslant \mu + r - \delta + 1/p$, $0 < \alpha \leqslant 1$, $r + \mu + 1/p < \alpha - \rho < r + \mu + 2/p + 2$, and $r \geqslant 0$, r an integer; and let $f \in L_{p,\mu}$. A necessary and sufficient condition for $f(x)$ to have an rth derivative $f^{(r)}(x)$ in $(-1, 1)$ satisfying*

$$\| [f^{(r)}(x, t) - f^{(r)}(x)] \sqrt{1-x^2}^{\mu+r-\delta} (\sqrt{1-x^2} + |\sin t|)^{\rho+\delta} \|_p \leqslant C_1 |\sin t|^\alpha$$

for $\alpha < 1$, or

$$\| [f^{(r)}(x, t) - 2f^{(r)}(x) + f^{(r)}(x, -t)] \sqrt{1-x^2}^{\mu+r-\delta} (\sqrt{1-x^2} + |\sin t|)^{\rho+\delta} \|_p \leqslant C_2 |\sin t|^\alpha$$

for $\alpha < 2$ is that there is a sequence of polynomials $P_n(x)$ such that

$$\left\| [f(x) - P_n(x)] \sqrt{1-x^2}^{\mu} \left(\sqrt{1-x^2} + \frac{1}{n} \right)^\rho \right\|_p \leqslant \frac{C_3}{n^{r+\alpha}},$$

where C_1, C_2 and C_3 are independent of f, t and n.

Theorem 2 is a corollary of Theorems 3 and 5.

COROLLARY 1. *Let $1 \leqslant p < \infty$, $\mu \geqslant -1/p$, $0 < \alpha \leqslant 1$ and $r \geqslant 0$, r an integer; then a structural characterization of the class of functions satisfying*

$$\left\| \frac{f(x) - P_n(x)}{(\sqrt{1-x^2} + 1/n)^{r+\alpha}} \right\|_{p,\mu} \leqslant \frac{C_1}{n^{r+\alpha}}$$

is as follows: $f(x)$ has a rth derivative $f^{(r)}(x)$ in $(-1, 1)$ and this derivative satisfies

$$\left\| \frac{f^{(r)}(x) - f^{(r)}(x, t, \mu)}{(\sqrt{1-x^2} + |\sin t|)^\alpha} \right\|_{p,\mu} \leqslant C_2 |\sin t|^\alpha$$

for $0 < \alpha \leqslant 1$ and $\mu > -1/p$;

$$\left\| \frac{f^{(r)}(x) - f^{(r)}(x, t)}{(\sqrt{1-x^2} + |\sin t|)^\alpha} \right\|_{p,\mu} \leqslant C_3 |\sin t|^\alpha$$

for $0 < \alpha < 1$ and $\mu = -1/p$; and

$$\left\| \frac{f^{(r)}(x, t) - 2f^{(r)}(x) + f^{(r)}(x, -t)}{\sqrt{1-x^2} + |\sin t|} \right\|_{p,\mu} \leqslant C_4 |\sin t|$$

for $\alpha = 1$ and $\mu = -1/p$.

COROLLARY 2. *Let $1 \leqslant p < \infty$, $\mu \geqslant -1/p$, $0 < \alpha \leqslant 1$, and r a positive integer. Then a structural characterization of the class of functions satisfying*

$$E_n(f)_{p,\mu} \leqslant \frac{C_1}{n^{r+\alpha}}$$

is as follows: $f(x)$ has an rth derivative $f^{(r)}(x)$ in $(-1, 1)$, and this derivative satisfies

$$\| f^{(r)}(x) - f^{(r)}(x, t, \mu + r) \|_{p,\mu+r} \leqslant C_2 |\sin t|^\alpha.$$

COROLLARY 3. *Let* $1 \leqslant p < \infty$, $\mu > -1/p$ *and* $0 < \alpha \leqslant 1$. *Then a structural characterization of the class of functions satisfying*

$$E_n(f)_{p,\mu} \leqslant \frac{C_1}{n^\alpha}$$

is as follows:

$$\|f(x) - f(x, t, \mu)\|_{p,\mu} \leqslant C_2 |\sin t|^\alpha.$$

COROLLARY 4. *Let* $1 \leqslant p < \infty$ *and* $0 \leqslant \mu + 1/p < \alpha \leqslant 1$; *then a structural characterization of the class of functions satisfying*

$$E_n(f)_{p,\mu} \leqslant \frac{C_1}{n^\alpha}$$

is as follows: *for* $\alpha < 1$,

$$\|f(x) - f(x, t)\|_{p,\mu} \leqslant C_2 |\sin t|^\alpha,$$

and for $\alpha = 1$,

$$\| f(x, t) - 2f(x) + f(x, -t)\|_{p,\mu} \leqslant C_3 |\sin t|.$$

§2. Lemmas

Let positive integers m and q be given. It is known that the function

$$\gamma(t) = \left(\frac{\sin mt/2}{\sin t/2}\right)^{2q+4}$$

is an even trigonometric polynomial of order $(q + 2)(m - 1)$. It is easily seen that the polynomial

$$K_{m,\mu} = \frac{\gamma(t)}{\gamma_{m\mu}},$$

where $\mu \geqslant -1$, $\gamma_{m\mu} = \int_{-\pi}^{\pi} \gamma(t) |\sin t|^\mu dt$, satisfies the conditions

$$\int_{-\pi}^{\pi} K_{m,\mu}(t) |\sin t|^\mu dt = 1, \qquad \int_{-\pi}^{\pi} |t|^\gamma K_{m,\mu}(t) |\sin t|^\mu dt \leqslant C/m^\gamma,$$

for $-1 < \gamma + \mu < 2q + 3$, where C is independent of m.

LEMMA 1. *If* $f(x)$ *is summable on* $[-1, 1]$, *then*

$$Q(x) = \int_{-\pi}^{\pi} f(x, t) K_{m,0}(t)\, dt$$

is an algebraic polynomial of degree $(q + 2)(m - 1)$.

This is known (cf., for example, [14]).

LEMMA 2. *If* $f(x)$ *is summable on* $[-1, 1]$ *with weight* $\sqrt{1 - x^2}^{\mu}$, *the function*

$$Q(x) = \int_{-\pi}^{\pi} K_{m,\mu+1}(t) f(x, t, \mu) |\sin t|^{\mu+1} dt,$$

where $\mu > -1$, *is an algebraic polynomial of degree* $(q+2)(m-1)$.

PROOF. Since $f(x, t, \mu) = f(x, -t, \mu)$, we have

$$Q(x) = 2\int_0^\pi K_{m,\mu+1}(t) f(x, t, \mu) (\sin t)^{\mu+1} dt.$$

Since

$$K_{m,\mu+1}(t) = \sum_{k=0}^{(q+2)(m-1)} a_k (\cos t)^k,$$

we have

$$K_{m,\mu+1} = \sum_{k=0}^{(q+2)(m-1)} b_k P_{k,\mu}(\cos t),$$

where $P_{k,\mu}(x)$ are algebraic polynomials of degree k, orthogonal with weight $\sqrt{1-x^2}^{\mu}$. Hence

$$Q(x) = \frac{2}{\gamma_{m\mu}} \sum_{k=0}^{(q+2)(m-1)} b_k \varphi(x),$$

where

$$\varphi(x) = \int_0^\pi\int_0^\pi P_{k,\mu}(\cos t) f(x\cos t + \cos\lambda \sin t \sqrt{1-x^2}) (\sin\lambda)^\mu (\sin t)^{\mu+1} d\lambda dt.$$

Making the change of variables $y = \cos\lambda$, $z = \cos t$, we have

$$\varphi(x) = \int_{-1}^{1}\int_{-1}^{1} P_{k,\mu}(z) f(xz + y\sqrt{1-x^2}\sqrt{1-z^2}) \sqrt{1-z^2}^{\mu} \sqrt{1-y^2}^{\mu-1} dz\, dy.$$

Making the further changes of variable

$$z = xv + u\sqrt{1-x^2}\sqrt{1-v^2},$$

$$y = \frac{v\sqrt{1-x^2} - ux\sqrt{1-x^2}}{\sqrt{(1-v^2)(1-u^2) + (\sqrt{1-x^2}\, v - ux\sqrt{1-v^2})^2}},$$

we obtain

$$\varphi(x) = \int_{-1}^{1}\int_{-1}^{1} P_{k,\mu}(xv + u\sqrt{1-x^2}\sqrt{1-v^2}) f(v) \sqrt{1-v^2}^{\mu} \sqrt{1-u^2}^{\mu-1} du\, dv$$

$$= \int_1^1 f(v)\sqrt{1-v^2}^{\mu}\left[\int_{-1}^{1} P_{k,\mu}(xv + u\sqrt{1-x^2}\sqrt{1-v^2})\sqrt{1-u^2}^{\mu-1} du\right] dv.$$

If we apply the composition theorem for ultraspherical polynomials to $P_{k,\mu}$ and use the orthogonality of these polynomials with weight $\sqrt{1-x^2}^{\mu}$, we obtain

$$\int_{-1}^{1} P_{k,\mu}(xv + u\sqrt{1-x^2}\sqrt{1-v^2})\sqrt{1-u^2}^{\mu-1} du = C_{k,\mu} P_{k,\mu}(x) P_{k,\mu}(v),$$

where $C_{k,\mu}$ depend only on k and μ.

Using this equation and

$$\int_{-1}^{1} f(v)\, P_{k,\mu}(v) \sqrt{1-v^{2\mu}}\, dv = a_{k,\mu},$$

where $a_{k,\mu}$ are the Fourier coefficients of $f(x)$ with respect to the polynomials orthogonal with the weight $\sqrt{1-x^{2\mu}}$, we obtain

$$\varphi_k(x) = c_{k,\mu}\, a_{k,\mu}\, P_{k,\mu}(x),$$

i.e. $\varphi(x)$ is an algebraic polynomial of degree k, and hence $Q(x)$ is an algebraic polynomial of degree $(q+2)(m-1)$.

This establishes Lemma 2.

LEMMA 3. *If $1 \leqslant p < \infty$, $\mu > -2/p$, $\alpha + \beta \geqslant 0$, ρ is an arbitrary real number, and $Q_n(x)$ is an algebraic polynomial of degree n, then*

$$\left\| Q_n(x) \sqrt{1-x^{2\mu}} \left(\sqrt{1-x^2} + \frac{1}{n}\right)^{\rho} \right\|_p \leqslant C_1 n^{\alpha+\beta} \left\| Q_n(x) \sqrt{1-x^{2\alpha+\mu}} \left(\sqrt{1-x^2} + \frac{1}{n}\right)^{\beta+\rho} \right\|_p,$$

$$\left\| Q_n'(x) \left(\sqrt{1-x^2} + \frac{1}{n}\right)^{\rho+1} \right\|_p \leqslant C_2 n \left\| Q_n(x) \left(\sqrt{1-x^2} + \frac{1}{n}\right)^{\rho} \right\|_p,$$

where C_1 and C_2 are independent of n.

Lemma 3 is proved in [15].

LEMMA 4. *If $R = x\cos t + \cos\lambda \sqrt{1-x^2} \sin t$, and ρ is an arbitrary real number, then*

$$\left(\sqrt{1-x^2} + \frac{1}{n}\right)^{\rho} \leqslant C_1 [1 + (n|\sin t|)^{|\rho|}] \left(\sqrt{1-R^2} + \frac{1}{n}\right)^{\rho},$$

$$\left(\sqrt{1-x^2} + \frac{1}{n}\right)^{\rho} \leqslant C_2 [1 + (n|\sin t|)^{-\rho}] (\sqrt{1-x^2} + |\sin t|)^{\rho},$$

where C_1 and C_2 are independent of x, λ, t and n.

PROOF. Consider

$$1 - R^2 = 1 - x^2 + \sin^2 t\, [x^2 - \cos^2\lambda\, (1-x^2)] - 2\sin t \sqrt{1-x^2}\, (x\cos\lambda\cos t).$$

It is easily seen that

$$-2|\sin t| \sqrt{1-x^2} \leqslant -2\sin t \sqrt{1-x^2}\, (x\cos\lambda\cos t) \leqslant 2|\sin t| \sqrt{1-x^2},$$

$$-1 \leqslant x^2 - \cos^2\lambda\, (1-x^2) \leqslant 1.$$

Applying the right-hand inequality, we obtain

$$1 - R^2 \leqslant 1 - x^2 + \sin^2 t + 2|\sin t| \sqrt{1-x^2} = (\sqrt{1-x^2} + |\sin t|)^2,$$

and hence

$$\sqrt{1-R^2} \leqslant \sqrt{1-x^2} + |\sin t|.$$

Applying the left-hand inequality, we have

$$1-R^2 \geqslant 1-x^2-\sin^2 t-2|\sin t|\sqrt{1-x^2}=(\sqrt{1-x^2}-|\sin t|)^2-2\sin^2 t.$$

Therefore

$$1-R^2+2\sin^2 t \geqslant (\sqrt{1-x^2}-|\sin t|)^2.$$

Using this inequality, we obtain

$$(\sqrt{1-R^2}+2|\sin t|)^2=1-R^2+4|\sin t|\sqrt{1-R^2}+4\sin^2 t$$
$$\geqslant 1-R^2+2\sin^2 t \geqslant (\sqrt{1-x^2}-|\sin t|)^2.$$

Hence

$$|\sqrt{1-x^2}-|\sin t|| \leqslant \sqrt{1-R^2}+2|\sin t|,$$

but then

$$\sqrt{1-x^2} \leqslant |\sqrt{1-x^2}-|\sin t||+|\sin t| \leqslant \sqrt{1-R^2}+3|\sin t|.$$

Thus we have shown that

$$\sqrt{1-x^2} \leqslant 3(\sqrt{1-R^2}+|\sin t|), \quad \sqrt{1-R^2} \leqslant \sqrt{1-x^2}+|\sin t|.$$

Applying these inequalities, we have

$$\sqrt{1-x^2}+\frac{1}{n} \leqslant 3\left(\sqrt{1-R^2}+|\sin t|+\frac{1}{n}\right)$$
$$\leqslant 3(1+n|\sin t|)\left(\sqrt{1-R^2}+\frac{1}{n}\right),$$
$$\sqrt{1-R^2}+\frac{1}{n} \leqslant \sqrt{1-x^2}+|\sin t|+\frac{1}{n}$$
$$\leqslant (1+n|\sin t|)\left(\sqrt{1-x^2}+\frac{1}{n}\right).$$

It is now easy to see that for $\rho \geqslant 0$

$$\left(\sqrt{1-x^2}+\frac{1}{n}\right)^\rho \leqslant [3(1+n|\sin t|)]^{|\rho|}\left(\sqrt{1-R^2}+\frac{1}{n}\right)^\rho,$$

and for $\rho < 0$

$$\left(\sqrt{1-x^2}+\frac{1}{n}\right)^\rho = \left(\sqrt{1-R^2}+\frac{1}{n}\right)^\rho \left(\frac{\sqrt{1-R^2}+1/n}{\sqrt{1-x^2}+1/n}\right)^{|\rho|}$$
$$\leqslant [1+n|\sin t|]^{|\rho|}\left(\sqrt{1-R^2}+\frac{1}{n}\right)^\rho.$$

From these inequalities we obtain the first inequality of Lemma 4.

We now establish the second inequality. Since it is clear that

$$\sqrt{1-x^2}+\frac{1}{n} \leqslant (\sqrt{1-x^2}+|\sin t|)\left(1+\frac{1}{n|\sin t|}\right),$$
$$\sqrt{1-x^2}+|\sin t| \leqslant \left(\sqrt{1-x^2}+\frac{1}{n}\right)(1+n|\sin t|),$$

we have for $\rho \geqslant 0$

$$\left(\sqrt{1-x^2}+\frac{1}{n}\right)^\rho \leqslant \left(1+\frac{1}{n\,|\sin t|}\right)^\rho (\sqrt{1-x^2}+|\sin t|)^\rho$$
$$\leqslant C\,[1+(n\,|\sin t|)^{-\rho}]\,(\sqrt{1-x^2}+|\sin t|)^\rho,$$

and for $\rho < 0$

$$\left(\sqrt{1-x^2}+\frac{1}{n}\right)^\rho = \frac{1}{(\sqrt{1-x^2}+1/n)^{|\rho|}} \leqslant \left(\frac{1+n\,|\sin t|}{\sqrt{1-x^2}+|\sin t|}\right)^{|\rho|}$$
$$\leqslant C\,[1+(n\,|\sin t|)^{|\rho|}]\,(\sqrt{1-x^2}+|\sin t|)^{-|\rho|}\,.$$

From this we obtain the second inequality of Lemma 4.

LEMMA 5. *Let $\mu > -2/p$, $\alpha > \delta_1 + \delta_2 \geqslant 0$, $\alpha > \rho + \delta_1 + \delta_2$ and $1 \leqslant p < \infty$, and let $f \in L_{p,\mu}$. If there is a sequence of polynomials $P_n(x)$ such that*

$$\left\| [f(x)-P_n(x)]\,\sqrt{1-x^2}^{\,\mu+\delta_1}\left(\sqrt{1-x^2}+\frac{1}{n}\right)^{\rho+\delta_2}\right\|_p \leqslant \frac{C_1}{n^\alpha}\,,$$

then there is a sequence of polynomials $R_n(x)$ such that

$$\left\| [f(x)-R_n(x)]\,\sqrt{1-x^2}^{\,\mu}\left(\sqrt{1-x^2}+\frac{1}{n}\right)^{\rho}\right\|_p \leqslant \frac{C_2}{n^{\alpha-\delta_1-\delta_2}}\,,$$

where C_1 and C_2 are independent of f and n. When $n = 2^k$ we have $R_{2^k}(x) = P_{2^k}(x)$.

PROOF. Consider the sequence of polynomials $Q_{2^k}(x)$, where $Q_{2^k}(x) = P_{2^k}(x) - P_{2^{k-1}}(x)$ for $k \geqslant 1$ and $Q_1(x) = P_1(x)$. It is clear that for $k \geqslant 0$

$$\left\| Q_{2^k}(x)\,\sqrt{1-x^2}^{\,\mu+\delta_1}\left(\sqrt{1-x^2}+\frac{1}{2^k}\right)^{\rho+\delta_2}\right\|_p \leqslant \frac{C_3}{2^{k\alpha}}\,.$$

Applying Lemma 3 and this inequality, we obtain

$$\left\| Q_{2^k}(x)\,\sqrt{1-x^2}^{\,\mu}\left(\sqrt{1-x^2}+\frac{1}{2^k}\right)^{\rho}\right\|_p$$
$$\leqslant C_4 2^{k(\delta_1+\delta_2)}\left\| Q_{2^k}(x)\,\sqrt{1-x^2}^{\,\mu+\delta_1}\left(\sqrt{1-x^2}+\frac{1}{2^k}\right)^{\rho+\delta_2}\right\|_p \leqslant C_5 2^{k(\delta_1+\delta_2-\alpha)}.$$

It then follows that the series $\Sigma_{k=0}^\infty Q_{2^k}(x)$ converges to $f(x)$ in $L_p([a, b])$ for every $[a, b] \subset (-1, 1)$.

Now consider

$$I = \left\| [f(x)-P_{2^N}(x)]\,\sqrt{1-x^2}^{\,\mu}\left(\sqrt{1-x^2}+\frac{1}{2^N}\right)^{\rho}\right\|_p\,.$$

It is clear that

$$I \leqslant \sum_{k=N+1}^{\infty}\left\| Q_{2^k}(x)\,\sqrt{1-x^2}^{\,\mu}\left(\sqrt{1-x^2}+\frac{1}{2^N}\right)^{\rho}\right\|_p\,.$$

If $\rho \leqslant 0$, we have

$$\left(\sqrt{1-x^2}+\frac{1}{2^N}\right)^\rho \leqslant \left(\sqrt{1-x^2}+\frac{1}{2^n}\right)^\rho\,.$$

Hence

$$I \leqslant C_9 \sum_{k=N+1}^{\infty} \left\| Q_{2^k}(x) \sqrt{1-x^2}^{\mu} \left(\sqrt{1-x^2} + \frac{1}{2^k} \right)^{\rho} \right\|_p$$

$$\leqslant C_6 \sum_{k=N+1}^{\infty} 2^{k(\delta_1+\delta_2-\alpha)} \leqslant C_7 2^{N(\delta_1+\delta_2-\alpha)}.$$

If, on the other hand, $\rho > 0$, we have

$$\left(\sqrt{1-x^2} + \frac{1}{2^k} \right)^{\rho} \leqslant C_8 \left(\sqrt{1-x^2} + \frac{1}{2^k} \right)^{\rho} + C_8 2^{-N\rho}.$$

Hence

$$I \leqslant C_9 \sum_{k=N+1}^{\infty} \left\| Q_{2^k}(x) \sqrt{1-x^2}^{2\mu} \left(\sqrt{1-x^2} + \frac{1}{2^k} \right)^{\rho} \right\|_p$$

$$+ C_9 2^{-N\rho} \sum_{k=N+1}^{\infty} \| Q_{2^k}(x) \sqrt{1-x^2}^{2\mu} \|_p.$$

Again applying Lemma 3, we have

$$I \leqslant C_{10} 2^{-N(\alpha-\delta_1-\delta_2)} + C_{10} 2^{-N\rho} \sum_{k=N+1}^{\infty} 2^{k\rho} \left\| Q_{2^k}(x) \sqrt{1-x^2}^{2\mu} \left(\sqrt{1-x^2} + \frac{1}{2^k} \right)^{\rho} \right\|_p$$

$$\leqslant C_{10} 2^{-N(\alpha-\delta_1-\delta_2)} + C_{11} 2^{-N\rho} \sum_{k=N+1}^{\infty} 2^{k(\rho+\delta_1+\delta_2-\alpha)} \leqslant C_{12} 2^{-N(\alpha-\delta_1-\delta_2)}.$$

Thus under the hypotheses of the lemma we have

$$I \leqslant C_{13} 2^{-N(\alpha-\delta_1-\delta_2)}.$$

If we now take $R_n(x) = P_{2^k}(x)$ for $2^{k-1} < n \leqslant 2^k$, we obtain Lemma 5.

LEMMA 6. *Let* $1 \leqslant p < \infty$, $\mu \geqslant -1/p$, $\alpha > 0$, $\delta \geqslant 0$, *and* ρ *any real number, and let* $f \in L_{p,\mu}$. *If* $f(x)$ *has a derivative* $f'(x)$ *in* $(-1, 1)$ *that satisfies*

$$\left\| f'(x) \sqrt{1-x^2}^{2\mu+1-\delta} \left(\sqrt{1-x^2} + \frac{1}{n} \right)^{\rho+\sigma} \right\|_p \leqslant \frac{C_1}{n^{\alpha}},$$

then there is a sequence of polynomials $P_n(x)$ *such that*

$$\left\| [f(x) - P_n(x)] \sqrt{1-x^2}^{2\mu} \left(\sqrt{1-x^2} + \frac{1}{n} \right)^{\rho} \right\|_p \leqslant \frac{C_2}{n^{1+\alpha}},$$

where C_1 *and* C_2 *are independent of* f *and* n.

PROOF. 1) First let $\mu > -1/p$. Take a positive integer q such that $q > (\mu + |\rho|)/2$. For each n take a positive integer m such that

$$\frac{n}{q+2} < m \leqslant \frac{n}{q+2} + 1. \qquad (*)$$

Consider the polynomial $Q(x)$ defined in Lemma 2. Then $Q(x)$ has degree $(q + 2)(m - 1) \leqslant n$. For all n satisfying $(*)$, take $P_n(x)$ to be $Q(x)$. Then

$$f(x) - P_n(x) = 2 \int_0^{\pi} K_{m\mu+1}(t) [f(x) - f(x, t, \mu)] (\sin t)^{\mu+1} dt$$

$$= 2 \int_0^{\pi} K_{m\mu+1}(t) \int_0^{\pi} \int_0^{t} f'(R) R_1 (\sin t)^{\mu+1} (\sin \lambda)^{\mu} du d\lambda dt,$$

where

$$R = x\cos u + \cos\lambda \sin u\sqrt{1-x^2},\ R_1 = x\sin u - \cos\lambda\cos u\sqrt{1-x^2}.$$

In addition, by applying Minkowski's inequality and then Lemma 4, we have

$$I = \left\| [f(x) - P_n(x)] \sqrt{1-x^2}^{2\mu} \left(\sqrt{1-x^2} + \frac{1}{n}\right)^{\rho} \right\|_p \leqslant C_3 \int_0^{\pi} K_{m\mu+1}(t)$$

$$\times \int_0^t [1 + (n|\sin u|)^{|\rho|}] \int_0^{\pi} \left\| f'(R) R_1 \sqrt{1-x^2}^{2\mu} \left(\sqrt{1-R^2} + \frac{1}{n}\right)^{\rho} \right\|_p$$

$$\times (\sin\lambda)^{\mu} (\sin t)^{\mu+1}\, du\, d\lambda\, dt.$$

Now applying Hölder's inequality we obtain

$$I \leqslant C_4 \int_0^{\pi} K_{m\mu+1}(t) \int_0^t [1 + (n|\sin u|)^{|\rho|}]\, I_1(u) (\sin t)^{\mu+1}\, du\, dt,$$

where

$$I_1^p(u) = \int_0^{\pi}\int_{-1}^{1} \left| f'(R) R_1 \sqrt{1-x^2}^{2\mu} \left(\sqrt{1-R^2} + \frac{1}{n}\right)^{\rho} (\sin\lambda)^{\mu} \right|^p dx\, d\lambda.$$

Setting $y = \cos\lambda$, we have

$$I_1^p(u) = \int_{-1}^{1}\int_{-1}^{1} \left| f'(x\cos u + y\sin u\sqrt{1-x^2})(x\sin u - y\cos u\sqrt{1-x^2}) \right.$$

$$\left. \times \left(\sqrt{1-(x\cos u + y\sin u\sqrt{1-x^2})^2} + \frac{1}{n}\right)^{\rho} \right|^p \sqrt{1-x^2}^{2\mu p} \sqrt{1-y^2}^{2\mu p - 1}\, dx\, dy.$$

Changing variables by the formulas

$$z = x\cos u + y\sin u\sqrt{1-x^2},$$

$$v = \frac{x\sin u - y\cos u\sqrt{1-x^2}}{\sqrt{(1-x^2)(1-y^2) + (x\sin u - y\cos u\sqrt{1-x^2})^2}},$$

we obtain

$$I_1^p(u) = \int_{-1}^{1}\int_{-1}^{1} \left| f'(z)\sqrt{1-z^2}\, v \left(\sqrt{1-z^2} + \frac{1}{n}\right)^{\rho} \right|^p$$

$$\times \sqrt{1-z^2}^{2\mu p} \sqrt{1-v^2}^{2\mu p - 1}\, dz\, dv = \int_{-1}^{1} \left| f'(z) \sqrt{1-z^2}^{2\mu+1} \left(\sqrt{1-z^2} + \frac{1}{n}\right)^{\rho} \right|^p dz$$

$$\times \int_{-1}^{1} |v|^p \sqrt{1-v^2}^{2\mu p - 1}\, dv \leqslant C_5 \int_{-1}^{1} \left| f'(z) \sqrt{1-z^2}^{2\mu+1-\delta} \left(\sqrt{1-z^2} + \frac{1}{n}\right)^{\rho+\delta} \right|^p dz \leqslant \frac{C_6}{n^{\alpha p}}.$$

Hence

$$I \leqslant C_7 n^{-\alpha} \int_0^{\pi} K_{m\mu+1}(t) \int_0^t [1 + (n|\sin u|)^{|\rho|}] (\sin t)^{\mu+1}\, du\, dt \leqslant \frac{C_8}{n^{1+\alpha}},$$

and Lemma 6 is established for $\mu > -1/p$.

2) Now let $\mu = -1/p$. Take q and m as in 1), and consider the polynomial $Q(x)$ de-

fined in Lemma 1. As in 1), we take $P_n(x)$ to be $Q(x)$ for values of n satisfying (*). Then, putting $x = \cos\theta$, we have

$$f(\cos\theta) - P_n(\cos\theta) = -\int_0^\pi K_{m0}(t)\,[f(\cos(\theta+t)) - 2f(\cos\theta)$$

$$+ f(\cos(\theta - t))]\,dt = \int_0^\pi K_{m0}(t)\int_0^t [f'(\cos(\theta+u))\sin(\theta+u)$$

$$- f'(\cos(\theta-u))\sin(\theta-u)]\,du\,dt.$$

If we then apply Minkowski's inequality, we obtain

$$I = \left\| [f(x) - P_n(x)]\sqrt{1-x^2}^{\,-1/p}\left(\sqrt{1-x^2}+\frac{1}{n}\right)^\rho\right\|_p$$

$$= \frac{1}{2}\left(\int_{-\pi}^{\pi}\left|[f(\cos\theta) - P_n(\cos\theta)]\left(|\sin\theta| + \frac{1}{n}\right)^\rho\right|^p d\theta\right)^{1/p}$$

$$\leqslant \frac{1}{2}\int_0^\pi K_{m0}(t)\int_0^t\left(\int_{-\pi}^{\pi}|f'(\cos(\theta+u))\sin(\theta+u) - f'(\cos(\theta-u))\sin(\theta-u)|^p\right.$$

$$\times\left.\left(|\sin\theta| + \frac{1}{n}\right)^{\rho p}d\theta\right)^{1/p}du\,dt \leqslant \int_0^\pi K_{m0}(t)\int_0^t\left(\int_{-\pi}^{\pi}|f'(\cos(\theta+u))\sin(\theta+u)|^p\right.$$

$$\times\left.\left(|\sin\theta| + \frac{1}{n}\right)^{\rho p}d\theta\right)^{1/p}du\,dt.$$

Making the change of variable $\theta_1 = \theta + u$ in the inner integral, and using the inequality

$$\left(|\sin(\theta_1 - u)| + \frac{1}{n}\right)^\rho \leqslant C_9\,[1 + (n|\sin u|)^{|\rho|}]\left(|\sin\theta_1| + \frac{1}{n}\right)^\rho,$$

we have

$$I \leqslant C_{10}\int_0^\pi K_{m0}(t)\int_0^t\left[1 + (n|\sin t|)^{|\rho|}\left(\int_{-\pi}^{\pi}|f'(\cos\theta_1)\sin\theta_1|^p\left(|\sin\theta_1|\right.\right.\right.$$

$$\left.\left.\left.+\frac{1}{n}\right)^{\rho p}d\theta_1\right)^{1/p}du\,dt \leqslant \frac{C_{11}}{n}\left(\int_{-1}^{1}\left|f'(x)\sqrt{1-x^2}^{\,1-1/p}\left(\sqrt{1-x^2}+\frac{1}{n}\right)^\rho\right|^p dx\right)^{1/p}$$

$$\leqslant \frac{C_{11}}{n}\left\|f'(x)\sqrt{1-x^2}^{\,1-1/p-\delta}\left(\sqrt{1-x^2}+\frac{1}{n}\right)^{\rho+\delta}\right\|_p \leqslant \frac{C_{12}}{n^{1+\alpha}}.$$

This completes the proof of Lemma 6.

§3. Proofs of the theorems

THEOREM 3. *Let α, r, μ, p, ρ and δ satisfy $1 \leqslant p < \infty$, $\mu \geqslant -1/p$, $\alpha > 0$, $\rho < r + \alpha$, $\delta < \alpha - \rho$, $\delta < \mu + r + 2/p$, and r a positive integer, and let $f \in L_{p,\mu}$. A necessary and sufficient condition that $f(x)$*

1) *has an rth derivative in $(-1, 1)$ for which there is a sequence of polynomials $R_n(x)$ such that*

$$\left\|[f^{(r)}(x) - R_n(x)]\sqrt{1-x^2}^{\,\mu+r-\delta}\left(\sqrt{1-x^2}+\frac{1}{n}\right)^{\rho+\delta}\right\|_p \leqslant \frac{C_1}{n^\alpha} \qquad (1)$$

is that

2) *there is a sequence of polynomials $P_n(x)$ such that*

$$\left\| [f(x) - P_n(x)] \sqrt{1-x^2}^{2\mu} \left(\sqrt{1-x^2} + \frac{1}{n} \right)^{\rho} \right\|_p \leqslant \frac{C_2}{n^{r+\alpha}}, \tag{2}$$

where C_1 and C_2 are independent of $f(x)$ and n.

PROOF. Let us show that (1) implies (2).

a) We first suppose that $\delta \geqslant 0$. Put

$$\mu_1 = \mu + r - 1,$$

$$\varphi_1(x) = f^{(r-1)}(x) - \int_0^x R_{n-1}(t)\,dt.$$

Then we have the hypotheses of Lemma 6, and hence for each n there is a polynomial $T_n(x)$ such that

$$\left\| [\varphi_1(x) - T_n(x)] \sqrt{1-x^2}^{2\mu_1} \left(\sqrt{1-x^2} + \frac{1}{n} \right)^{\rho} \right\|_p \leqslant \frac{C_3}{n^{\alpha+1}}.$$

Since $Q_n(x) = T_n(x) - \int_0^x R_{n-1}(t)dt$ is a polynomial of degree at most n, the preceding inequality can be rewritten in the form

$$\left\| [f^{(r-1)}(x) - Q_n(x)] \sqrt{1-x^2}^{2\mu+r-1} \left(\sqrt{1-x^2} + \frac{1}{n} \right)^{\rho} \right\|_p \leqslant \frac{C_3}{n^{\alpha+1}}.$$

Repeating this process $r - 1$ times, we obtain (2) for any $n > r$. For $n \leqslant r$ inequality (2) is obvious.

b) We now consider the case when $\delta < 0$. Put $\rho_0 = \rho + \delta$, $\mu_0 = \mu - \delta$ and $\delta_0 = 0$; then (1) can be rewritten as follows:

$$\left\| [f^{(r)}(x) - R_n(x)] \sqrt{1-x^2}^{2\mu_0+r-\delta_0} \left(\sqrt{1-x^2} + \frac{1}{n} \right)^{\rho_0+\delta_0} \right\|_p \leqslant \frac{C_4}{n^{\alpha}}.$$

We now have the hypotheses of case a), and hence this inequality implies that for each n there is a polynomial $Q_n(x)$ such that

$$\left\| [f(x) - Q_n(x)] \sqrt{1-x^2}^{2\mu-\delta} \left(\sqrt{1-x^2} + \frac{1}{n} \right)^{\rho+\delta} \right\|_p \leqslant \frac{C_5}{n^{r+\alpha}}.$$

Applying Lemma 5, we obtain (2). Hence we have shown that (1) implies (2).

We now establish the converse, i.e., that (2) implies (1). Consider the polynomials $Q_{2^k}(x)$, where $Q_1(x) = P_1(x)$ and $Q_{2^k}(x) = P_{2^k}(x) - P_{2^{k-1}}(x)$ for $R \geqslant 1$. It is clear that

$$\left\| Q_{2^k}(x) \sqrt{1-x^2}^{2\mu} \left(\sqrt{1-x^2} + \frac{1}{2^k} \right)^{\rho} \right\|_p \leqslant \frac{C_6}{2^{k(r+\alpha)}}.$$

Applying Lemma 3, we obtain

$$\begin{aligned} & Q_{2^k}^{(r)}(x) \sqrt{1-x^2}^{2\mu+r-\delta} \left(\sqrt{1-x^2} + \frac{1}{2^k} \right)^{\rho+\delta} \Big\|_p \\ & \leqslant C_7 \left\| Q_{2^k}^{(r)}(x) \left(\sqrt{1-x^2} + \frac{1}{2^k} \right)^{\rho+\mu+r} \right\|_p \leqslant C_8 2^{kr} \left\| Q_{2^k}(x) \left(\sqrt{1-x^2} + \frac{1}{2^k} \right)^{\rho+\mu} \right\|_p \\ & \qquad \leqslant C_9 2^{kr} \left\| Q_{2^k}(x) \sqrt{1-x^2}^{2\mu} \left(\sqrt{1-x^2} + \frac{1}{2^k} \right)^{\rho} \right\|_p \leqslant \frac{C_{10}}{2^{k\alpha}}. \end{aligned} \tag{3}$$

It follows from (2) that for each $[a, b] \subset (-1, 1)$ the series $\Sigma_{k=0}^{\infty} Q_{2^k}(x)$ converges to $f(x)$ in $L_p([a, b])$. Using (3) we see that $\Sigma_{k=0}^{\infty} Q_{2^k}^{(r)}(x)$ also converges in $L_p([a, b])$. But then by Nikol′skiĭ's lemma [16] this series converges to $f^{(r)}(x)$. Thus we have shown that $f(x)$ has an rth derivative almost everywhere on $(-1, 1)$ and that (4) converges to it in $L_p([a, b])$ for every $[a, b] \subset (-1, 1)$.

Now consider

$$I = \left\| [f^{(r)}(x) - P_{2^N}^{(r)}(x)] \sqrt{1-x^2}^{\mu+r-\delta} \left(\sqrt{1-x^2} + \frac{1}{2^N}\right)^{\rho+\delta} \right\|_p .$$

It is clear that

$$I = \sum_{k=N+1}^{\infty} \left\| Q_{2^k}^{(r)}(x) \sqrt{1-x^2}^{\mu+r-\delta} \left(\sqrt{1-x^2} + \frac{1}{2^N}\right)^{\rho+\delta} \right\|_p .$$

When $\rho + \delta \leqslant 0$ we have

$$\left(\sqrt{1-x^2} + \frac{1}{2^N}\right)^{\rho+\delta} \leqslant \left(\sqrt{1-x^2} + \frac{1}{2^k}\right)^{\rho+\delta} .$$

Hence when $\rho + \delta \leqslant 0$, if we apply (3) we obtain

$$I \leqslant \sum_{k=N+1}^{\infty} \left\| Q_{2^k}^{(r)}(x) \sqrt{1-x^2}^{\mu+r-\delta} \left(\sqrt{1-x^2} + \frac{1}{2^k}\right)^{\rho+\delta} \right\|_p \leqslant C_{11} \sum_{k=N+1}^{\infty} 2^{-k\alpha} \leqslant C_{12} 2^{-N\alpha} .$$

However, if $\rho + \delta > 0$ we have

$$\left(\sqrt{1-x^2} + \frac{1}{2^N}\right)^{\rho+\delta} \leqslant C_{13} \left(\sqrt{1-x^2} + \frac{1}{2^k}\right)^{\rho+\delta} [1 + 2^{(k-N)(\rho+\delta)}].$$

Hence, when $\rho + \delta > 0$, if we apply (3) we have

$$I \leqslant C_{14} \sum_{k=N+1}^{\infty} [1 + 2^{(k-N)(\rho+\delta)}] \left\| Q_{2^k}^{(r)}(x) \sqrt{1-x^2}^{\mu+r-\delta} \left(\sqrt{1-x^2} + \frac{1}{2^k}\right)^{\rho+\delta} \right\|_p \leqslant C_{15} \left\{ \sum_{k=N+1}^{\infty} 2^{-k\alpha} + 2^{-N(\rho+\delta)} \sum_{k=N+1}^{\infty} 2^{-k(\alpha-\rho-\delta)} \right\} \leqslant C_{16} 2^{-N\alpha}.$$

Thus $I \leqslant C_{17} 2^{-N\alpha}$ for $\alpha > \rho + \delta$, and this completes the proof of the theorem.

THEOREM 4. *Let* α, μ, p *and* ρ *satisfy* $1 \leqslant p < \infty$, $\mu > -1/p$, $0 < \alpha < 2$ and $\alpha - 2 - \mu - 2/p < \rho < \alpha + \mu + 1$, *and let* $f \in L_{p,\mu}$. *A necessary and sufficient condition that* $f(x)$

1) *satisfies*

$$\| [f(x) - f(x, t, \mu)] \sqrt{1-x^2}^{2\mu} (\sqrt{1-x^2} + |\sin t|)^{\rho} \|_p \leqslant C_1 |\sin t|^{\alpha}$$

is that

2) *there is a sequence of polynomials* $P_n(x)$ *such that*

$$\left\| [f(x) - P_n(x)] \sqrt{1-x^2}^{\mu} \left(\sqrt{1-x^2} + \frac{1}{n} \right)^{\rho} \right\|_p \leqslant \frac{C_2}{n^{\alpha}},$$

where C_1 and C_2 are independent of f, t and n.

PROOF. Let us show that 1) implies 2). Take a positive integer q such that $q > (\alpha + \mu - 3)/2$ and $q > (\alpha + \mu - \rho - 3)/2$. For each n take an integer m such that

$$\frac{n}{q+2} < m \leqslant \frac{n}{q+2} + 1, \qquad (*)$$

and consider the polynomial $Q(x)$ defined in Lemma 2. Then $Q(x)$ has degree $(q + 2)(m - 1) \leqslant n$. For all n satisfying $(*)$ we take $P_n(x)$ to be $Q(x)$. Then

$$f(x) - P_n(x) = \int_{-\pi}^{\pi} K_{m\mu}(t) [f(x) - f(x, t, \mu)] |\sin t|^{\mu} dt.$$

Applying Minkowski's inequality, we have

$$I = \left\| [f(x) - P_n(x)] \sqrt{1-x^2}^{\mu} \left(\sqrt{1-x^2} + \frac{1}{n} \right)^{\rho} \right\|_p$$

$$\leqslant \int_{-\pi}^{\pi} K_{m\mu}(t) |\sin t|^{\mu} \left\| [f(x) - f(x, t, \mu)] \sqrt{1-x^2}^{\mu} \left(\sqrt{1-x^2} + \frac{1}{n} \right)^{\rho} \right\|_p dt.$$

If we now apply Lemma 4 and then 1), we obtain

$$I \leqslant C_3 \int_{-\pi}^{\pi} K_{m\mu}(t) |\sin t|^{\mu} [1 + (n |\sin t|)^{-\rho}] \| [f(x) - f(x, t, \mu)] \sqrt{1-x^2}^{\mu}$$

$$\times (\sqrt{1-x^2} + |\sin t|)^{\rho} \|_p \, dt \leqslant C_4 \int_{-\pi}^{\pi} K_{m\mu}(t) |\sin t|^{\mu+\alpha} [1 + (n |\sin t|)^{-\rho}] \, dt \leqslant \frac{C_5}{n^{\alpha}}.$$

We have therefore shown that 1) implies 2). We now establish the converse, i.e. that 2) implies 1). Consider the polynomials $Q_{2^k}(x)$, where $Q_1(x) = P_1(x)$, and $Q_{2^k}(x) = P_{2^k}(x) - P_{2^{k-1}}(x)$ for $k \geqslant 1$. It is clear that

$$\left\| Q_{2^k}(x) \sqrt{1-x^2}^{\mu} \left(\sqrt{1-x^2} + \frac{1}{2^k} \right)^{\rho} \right\|_p \leqslant C_6 2^{-k\alpha}.$$

We shall need an inequality for

$$I = \| [f(x) - f(x, t, \mu)] \sqrt{1-x^2}^{\mu} (\sqrt{1-x^2} + |\sin t|)^{\rho} \|_p.$$

It is clear that

$$I \leqslant \left\| \left[f(x) - \sum_{k=0}^{N} Q_{2^k}(x) \right] \sqrt{1-x^2}^{\mu} (\sqrt{1-x^2} + |\sin t|)^{\rho} \right\|_p$$

$$+ \left\| \left[f(x, t, \mu) - \sum_{k=0}^{N} Q_{2^k}(x, t, \mu) \right] \sqrt{1-x^2}^{\mu} (\sqrt{1-x^2} + |\sin t|)^{\rho} \right\|_p$$

$$+ \left\| \sum_{k=0}^{N} [Q_{2^k}(x) - Q_{2^k}(x, t, \mu)] \sqrt{1-x^2}^{\mu} (\sqrt{1-x^2} + |\sin t|)^{\rho} \right\|_p = I_1 + I_2 + I_3.$$

Choose N so that $2^{-N} \leqslant |\sin t| < 2^{1-N}$ It is then clear that

$$(\sqrt{1-x^2}+|\sin t|)^\rho \leqslant C_7\left(\sqrt{1-x^2}+\frac{1}{2^N}\right)^\rho.$$

Hence

$$I_1 \leqslant C_8\left\|[f(x)-P_{2^N}(x)]\sqrt{1-x^2}^{2\mu}\left(\sqrt{1-x^2}+\frac{1}{2^N}\right)^\rho\right\|_p \leqslant C_9 2^{-N\alpha} \leqslant C_9|\sin t|^\alpha.$$

We now estimate I_2:

$$I_2 \leqslant C_{10}\left\|[f(x,t,\mu)-P_{2^N}(x,t,\mu)]\sqrt{1-x^2}^{2\mu}\left(\sqrt{1-x^2}+\frac{1}{2^N}\right)^\rho\right\|_p$$

$$= C_{10}\left\|\sqrt{1-x^2}^{2\mu}\left(\sqrt{1-x^2}+\frac{1}{2^N}\right)^\rho\left\{\int_0^\pi [f(R)-P_{2^N}(R)](\sin\lambda)^\mu\,d\lambda\right\}\right\|_p,$$

where $R = x\cos t + \cos\lambda \sin t\sqrt{1-x^2}$.

Applying Hölder's inequality to the inner integral, we have

$$I_2 \leqslant C_{11}\left(\int_{-1}^{1}\int_0^\pi |f(R)-P_{2^N}(R)|^p\left(\sqrt{1-x^2}+\frac{1}{2^N}\right)^{\rho p}\right.$$

$$\left.\times\sqrt{1-x^2}^{2\mu p}|\sin\lambda|^{\mu p}\,d\lambda\,dx\right)^{1/p}.$$

Applying Lemma 4 and putting $y = \cos\lambda$, we obtain

$$I_2 \leqslant C_{12}\left(\int_{-1}^{1}\int_{-1}^{1}|f(R)-P_{2^N}(R)|^p\left(\sqrt{1-R^2}+\frac{1}{2^N}\right)^{\rho p}\sqrt{1-x^2}^{2\mu p}\right.$$

$$\left.\times\sqrt{1-y^2}^{2\mu p-1}\,dx\,dy\right)^{1/p}.$$

Making the change of variables

$$R = x\cos t + y\sin t\sqrt{1-x^2},$$

$$z = \frac{x\sin t - y\cos t\sqrt{1-x^2}}{\sqrt{(1-x^2)(1-y^2)+(x\sin t - y\cos t\sqrt{1-x^2})^2}},$$

we have

$$I_2 \leqslant C_{12}\left(\int_{-1}^{1}\int_{-1}^{1}|f(R)-P_{2^N}(R)|^p\left(\sqrt{1-R^2}+\frac{1}{2^N}\right)^{\rho p}\sqrt{1-R^2}^{2\mu p}\right.$$

$$\left.\times\sqrt{1-z^2}^{2\mu p-1}\,dR\,dz\right)^{1/p} = C_{12}\left\|[f(R)-P_{2^N}(R)]\left(\sqrt{1-R^2}+\frac{1}{2^N}\right)^\rho\right.$$

$$\left.\times\sqrt{1-R^2}^{2\mu}\right\|_p\left(\int_{-1}^{1}\sqrt{1-z^2}^{2\mu p-1}\,dz\right)^{1/p} \leqslant C_{13}2^{-N\alpha} \leqslant C_{13}|\sin t|^\alpha.$$

Now consider $I_4 = Q_{2^k}(x,t,\mu) - Q_{2^k}(x)$. Since $f(x,t,\mu) = f(x,-t,\mu)$, it is easily seen that

$$I_4 = \frac{1}{2}[Q_{2^k}(x,t,\mu) - 2Q_{2^k}(x) + Q_{2^k}(x,-t,\mu)] =$$

$$= \frac{1}{2}\int_0^\pi\int_0^t\int_{-u}^u [Q''_{2^k}(R) R_1^2 - 2RQ'_{2^k}(R)] (\sin\lambda)^\mu \, dv\, du\, d\lambda,$$

where

$$R = x\cos v + \cos\lambda \sin v \sqrt{1-x^2},\quad R_1 = x\sin v - \cos\lambda\cos v\sqrt{1-x^2}.$$

We now estimate I_3. First of all, it is clear that

$$I_3 \leqslant C_{14}\sum_{k=0}^N \left\| I_4 \sqrt{1-x^2}^{\mu}\left(\sqrt{1-x^2}+\frac{1}{2^N}\right)^\rho\right\|_p .$$

Applying Minkowski's inequality, we have

$$I_3 \leqslant C_{14}\sum_{k=0}^N \int_0^t\int_{-u}^u I_5(v)\,dv\,du,$$

where

$$I_5^p(v) = \int_0^\pi \left\| [Q''_{2^k}(R)R_1^2 - 2RQ'_{2^k}(R)]\sqrt{1-x^2}^{\mu}\left(\sqrt{1-x^2}+\frac{1}{2^N}\right)^\rho\right\|_p (\sin\lambda)^\mu\, d\lambda.$$

If we now apply Hölder's inequality to $I_5(v)$, we obtain

$$I_5^p(v) \leqslant C_{15}\int_0^\pi\int_{-1}^1 |Q''_{2^k}(R)R_1^2 - 2RQ'_{2^k}(R)|^p \times \left(\sqrt{1-x^2}+\frac{1}{2^N}\right)^{\rho p}\sqrt{1-x^2}^{\mu p}|\sin\lambda|^{\mu p}\,dx\,d\lambda.$$

If we apply Lemma 4 and put $y = \cos\lambda$, we have

$$I_5^p(v) \leqslant C_{16}\int_{-1}^1\int_{-1}^1 |Q''_{2^k}(R)R_1^2 - 2RQ'_{2^k}(R)|^p \times \left(\sqrt{1-R^2}+\frac{1}{2^N}\right)^{\rho p}\sqrt{1-x^2}^{\mu p}\sqrt{1-y^2}^{\mu p-1}\,dx\,dy.$$

Making the change of variables

$$R = x\cos v + y\sin v\sqrt{1-x^2},$$
$$z = \frac{x\sin v - y\cos v\sqrt{1-x^2}}{\sqrt{(1-x^2)(1-y^2)+(x\sin v - y\cos v\sqrt{1-x^2})^2}},$$

we obtain

$$I_5(v) \leqslant C_{16}\left(\int_{-1}^1\int_{-1}^1 |Q''_{2^k}(R)(\sqrt{1-R^2}z)^2 - 2RQ'_{2^k}(R)|^p\left(\sqrt{1-R^2}+\frac{1}{2^N}\right)^{\rho p} \times \sqrt{1-R^2}^{\mu p}\sqrt{1-z^2}^{\mu p-1}\,dR\,dz\right)^{1/p}$$
$$\leqslant C_{17}\left\{\left\| Q''_{2^k}(R)\sqrt{1-R^2}^{2+\mu}\left(\sqrt{1-R^2}+\frac{1}{2^N}\right)^\rho\right\|_p + \right.$$

$$+\left\| Q'_{2^k}(R)\sqrt{1-R^2}^{\mu}\left(\sqrt{1-R^2}+\frac{1}{2^N}\right)^{\rho}\right\|_p\Big\} = C_{17}\{I_6+I_7\}.$$

We now estimate I_7. Take μ_1 so that $\mu > \mu_1 > -2/p$. Then it is clear that

$$I_7 \leqslant \left\| Q'_{2^k}(R)\sqrt{1-R^2}^{\mu_1}\left(\sqrt{1-R^2}+\frac{1}{2^N}\right)^{\rho+\mu-\mu_1}\right\|_p.$$

If $\rho + \mu - \mu_1 \geqslant 0$, we have

$$\left(\sqrt{1-R^2}+\frac{1}{2^N}\right)^{\rho+\mu-\mu_1} \leqslant \left(\sqrt{1-R^2}+\frac{1}{2^k}\right)^{\rho+\mu-\mu_1}.$$

Hence

$$I_7 = \left\| Q'_{2^k}(R)\sqrt{1-R^2}^{\mu_1}\left(\sqrt{1-R^2}+\frac{1}{2^k}\right)^{\rho+\mu-\mu_1}\right\|_p.$$

If we now apply Lemma 3, we have

$$\begin{aligned} I_7 &\leqslant C_{18}2^k\left\| Q'_{2^k}(R)\left(\sqrt{1-R^2}+\frac{1}{2^k}\right)^{\rho+\mu+1}\right\|_p \\ &\leqslant C_{19}2^{2k}\left\| Q_{2^k}(R)\left(\sqrt{1-R^2}+\frac{1}{2^k}\right)^{\rho+\mu}\right\|_p \\ &\leqslant C_{20}2^{2k}\left\| Q_{2^k}(R)\sqrt{1-R^2}^{\mu}\left(\sqrt{1-R^2}+\frac{1}{2^k}\right)^{\rho}\right\|_p \leqslant C_{21}2^{k(2-\alpha)}. \end{aligned}$$

If, however, $\rho + \mu - \mu_1 < 0$, we have

$$\left(\sqrt{1-R^2}+\frac{1}{2^N}\right)^{\rho+\mu-\mu_1} \leqslant \left[2^{k-N}\left(\sqrt{1-R^2}+\frac{1}{2^k}\right)\right]^{\rho+\mu-\mu_1}.$$

Hence, proceeding as before, we have

$$\begin{aligned} I_7 &\leqslant C_{22}2^{(k-N)(\rho+\mu-\mu_1)}\left\| Q'_{2^k}(R)\sqrt{1-R^2}^{\mu_1}\left(\sqrt{1-R^2}+\frac{1}{2^k}\right)^{\rho+\mu-\mu_1}\right\|_p \\ &\leqslant C_{23}2^{(k-N)(\rho+\mu-\mu_1)}2^{k(2-\alpha)} = C_{24}2^{k(2-\alpha+\rho+\mu-\mu_1)}2^{-N(\rho+\mu-\mu_1)}. \end{aligned}$$

We can estimate I_6 similarly.

It is now easy to estimate I_3. For $\rho + \mu - \mu_1 \geqslant 0$ and $\alpha < 2$,

$$I_3 \leqslant C_{25}t^2\sum_{k=0}^{N}2^{k(2-\alpha)} \leqslant C_{26}t^2 2^{N(2-\alpha)},$$

and for $0 > \rho + \mu - \mu_1 > \alpha - 2$,

$$I_3 \leqslant C_{27}t^2 2^{-N(\rho+\mu-\mu_1)}\sum_{k=0}^{N}2^{k(2-\alpha+\rho+\mu-\mu_1)} \leqslant C_{28}t^2 2^{N(2-\alpha)}.$$

Thus $I_3 \leqslant C_{29}t^2 2^{N(2-\alpha)}$ for ρ such that $\rho > \alpha - 2 - \mu + \mu_1$. Since we may suppose without loss of generality that $|t| \leqslant \pi/2$, we have $t^2 \leqslant C_{30}|\sin t|^2$, and hence $I_3 \leqslant C_{31}|\sin t|^{\alpha}$. Since this inequality holds for $\rho > \alpha - 2 - \mu + \mu_1$ and $\mu_1 > -2/p$, it follows that the inequality $I_3 \leqslant C_{32}|\sin t|^\alpha$ holds for $\rho > \alpha - 2 - \mu - 2/p$.

Comparing the estimates for I_1, I_2, and I_3, we find that $I \leqslant C_{33}|\sin t|^\alpha$, and the proof of the theorem is complete.

THEOREM 5. *Let α, μ, p and ρ satisfy $1 \leqslant p < \infty$, $0 \leqslant \mu + 1/p < \alpha$ and $2 - \alpha - \mu - 2/p < \rho < \alpha - \mu - 1/p$, and let $f \in L_{p,\mu}$. A necessary and sufficient condition that $f(x)$*

1) *satisfies*

$$\| [f(x, t) - f(x)] \sqrt{1 - x^2}^{\mu} (\sqrt{1 - x^2} + |\sin t|)^\rho \|_p \leqslant C_1 |\sin t|^\alpha$$

for $\alpha < 1$, or

$$\| [f(x, t) - 2f(x) + f(x, -t)] \sqrt{1 - x^2}^{\mu} (\sqrt{1 - x^2} + |\sin t|)^\rho \|_p \leqslant C_2 |\sin t|^\alpha$$

for $\alpha < 2$ is that

2) *there is a sequence of polynomials $P_n(x)$ such that*

$$\left\| [f(x) - P_n(x)] \sqrt{1 - x^2}^{\mu} \left(\sqrt{1 - x^2} + \frac{1}{n}\right)^\rho \right\|_p \leqslant \frac{C_3}{n^\alpha},$$

where C_1, C_2 and C_3 are independent of f, t and n.

PROOF. We shall prove the theorem only for $\alpha < 2$, since the proof for $\alpha < 1$ is similar. Let us show that 1) implies 2).

Take a positive integer q such that $q > (\alpha - 3)/2$ and $q > (\alpha - \rho - 3)/2$. For each n take a positive integer m such that

$$\frac{n}{q+2} < m \leqslant \frac{n}{q+2} + 1, \qquad (*)$$

and consider the polynomial $Q(x)$ defined in Lemma 1. Then $Q(x)$ has degree $(q + 2)(m - 1) \leqslant n$. For n satisfying $(*)$ take $P_n(x)$ to be $Q(x)$. Then

$$f(x) - P_n(x) = \int_{-\pi}^{\pi} K_{m,0}(t) [f(x) - f(x, t)]\, dt$$

$$= -\int_0^{\pi} K_{m,0}(t) [f(x, t) - 2f(x) + f(x, -t)]\, dt.$$

Applying Minkowski's inequality, we have

$$I = \left\| [f(x) - P_n(x)] \sqrt{1 - x^2}^{\mu} \left(\sqrt{1 - x^2} + \frac{1}{n}\right)^\rho \right\|_p$$

$$\leqslant \int_0^{\pi} K_{m,0}(t) \left\| [f(x, t) - 2f(x) + f(x, -t)] \sqrt{1 - x^2}^{\mu} \left(\sqrt{1 - x^2} + \frac{1}{n}\right)^\rho \right\|_p dt.$$

Applying Lemma 4, we obtain

$$I \leqslant C_4 \int_0^{\pi} K_{m,0}(t)\,[1+(n|\sin t|)^{-\rho}]\,\|[f(x,t)-2f(x)+f(x,-t)]$$
$$\times \sqrt{1-x^2}^{\mu}\,(\sqrt{1-x^2}+|\sin t|)^{\rho}\|_p\,dt$$
$$\leqslant C_5 \int_0^{\pi} K_{m,0}(t)\,|\sin t|^{\alpha}\,[1+(n|\sin t|)^{-\rho}]\,dt \leqslant \frac{C_6}{n^{\alpha}}.$$

Thus we have shown that 1) implies 2). We now establish the converse, i.e., that 2) implies 1). Consider the polynomials $Q_{2^k}(x)$, where $Q_1(x) = P_1(x)$, and $Q_{2^k}(x) = P_{2^k}(x) - P_{2^{k-1}}(x)$ for $k \geqslant 1$. It is clear that

$$\left\| Q_{2^k}(x)\,\sqrt{1-x^2}^{\mu}\left(\sqrt{1-x^2}+\frac{1}{2^k}\right)^{\rho}\right\| \leqslant \frac{C_7}{2^{k\alpha}}.$$

We need to estimate

$$I = \|[f(x,t)-2f(x)+f(x,-t)]\,\sqrt{1-x^2}^{\mu}\,(\sqrt{1-x^2}+|\sin t|)^{\rho}\|_p.$$

Putting $x = \cos\theta$ and choosing N so that $2^{-N} \leqslant |\sin t| < 2^{1-N}$, we have

$$I \leqslant C_8\Bigg(\int_{-\pi}^{\pi} |f(\cos(\theta+t)) - 2f(\cos\theta)$$
$$+ f(\cos(\theta-t))|^p\,|\sin\theta|^{\mu p+1}\left(|\sin\theta|+\frac{1}{2^N}\right)^{\rho p} d\theta\Bigg)^{1/p}$$
$$\leqslant C_8\Bigg\{\Bigg(\int_{-\pi}^{\pi} |f(\cos(\theta+t))$$
$$- \sum_{k=0}^{N} Q_{2^k}(\cos(\theta+t))|^p\,|\sin\theta|^{\mu p+1}\left(|\sin\theta|+\frac{1}{2^N}\right)^{\rho p} d\theta\Bigg)^{1/p}$$
$$+\Bigg(\int_{-\pi}^{\pi} |f(\cos\theta) - \sum_{k=0}^{N} Q_{2^k}(\cos\theta)|^p\,|\sin\theta|^{\mu p+1}\left(|\sin\theta|+\frac{1}{2^N}\right)^{\rho p} d\theta\Bigg)^{1/p}$$
$$+\Bigg(\int_{-\pi}^{\pi} |f(\cos(\theta-t)) - \sum_{k=0}^{N} Q_{2^k}(\cos(\theta-t))|^p\,|\sin\theta|^{\mu p+1}\left(|\sin\theta|+\frac{1}{2^N}\right)^{\rho p} d\theta\Bigg)^{1/p}$$
$$+\Bigg(\int_{-\pi}^{\pi}\Bigg|\sum_{k=0}^{N}[Q_{2^k}(\cos(\theta+t)) - 2Q_{2^k}(\cos\theta) + Q_{2^k}(\cos(\theta-t))]\Bigg|^p$$
$$\times |\sin\theta|^{\mu p+1}\left(|\sin\theta|+\frac{1}{2^N}\right)^{\rho p} d\theta\Bigg)^{1/p}\Bigg\} = C_8(I_1+I_2+I_3+I_4).$$

It is clear that

$$I_2 \leqslant C_9\left\|[f(x) - P_{2^N}(x)]\,\sqrt{1-x^2}^{\mu}\left(\sqrt{1-x^2}+\frac{1}{2^N}\right)^{\rho}\right\|_p \leqslant C_{10}2^{-N\alpha} \leqslant C_{10}|\sin t|^{\alpha}.$$

Making the change of variable $\theta_1 = \theta + t$, we obtain

$$I_1^p = \int_{-\pi}^{\pi} |f(\cos\theta) - P_{2^N}(\cos\theta)|^p |\sin(\theta+t)|^{\mu p+1}\left(|\sin(\theta+t)| + \frac{1}{2^N}\right)^{\rho p} d\theta.$$

It is easily seen that

$$I_1^p \leqslant C_{11} \int_{-\pi}^{\pi} |f(\cos\theta) - P_{2^N}(\cos\theta)|^p \left(|\sin\theta| + \frac{1}{2^N}\right)^{\rho p+\mu p+1} d\theta$$
$$= C_{12}\left\| [f(x) - P_{2^N}(x)]\sqrt{1-x^2}^{-1/p}\left(\sqrt{1-x^2} + \frac{1}{2^N}\right)^{\rho+\mu+1/p}\right\|_p^p.$$

Now applying Lemma 5, we have

$$I_1 \leqslant C_{13} 2^{-N\alpha} \leqslant C_{13} |\sin t|^\alpha.$$

Similarly we can estimate I_3: $I_3 \leqslant C_{14}|\sin t|^\alpha$.

Now we estimate I_4:

$$I_4 \leqslant \sum_{k=0}^{N}\left(\int_{-\pi}^{\pi}\left|\int_0^t\int_{-u}^{u} [Q''_{2^k}(\cos(\theta+v))\sin^2(\theta+v)\right.\right.$$
$$\left.\left. - \cos(\theta+v)\,Q'_{2^k}(\cos(\theta+v))]\,dv\,du\right|^p |\sin\theta|^{\mu p+1}\left(|\sin\theta| + \frac{1}{2^N}\right)^{\rho p} d\theta\right)^{1/p}.$$

Applying Minkowski's inequality, we have

$$I_4 \leqslant \sum_{k=0}^{N}\int_0^t\int_{-u}^{u}\left\{\left(\int_{-\pi}^{\pi} |Q''_{2^k}(\cos(\theta+v))\sin^2(\theta+v)|^p\right.\right.$$
$$\left.\times |\sin\theta|^{\mu p+1}\left(|\sin\theta| + \frac{1}{2^N}\right)^{\rho p} d\theta\right)^{1/p}$$
$$\left.+\left(\int_{-\pi}^{\pi} |Q'_{2^k}(\cos(\theta+v))|^p |\sin\theta|^{\mu p+1}\left(|\sin\theta| + \frac{1}{2^N}\right)^{\rho p} d\theta\right)^{1/p}\right\} dv\,du$$
$$= \sum_{k=0}^{N}\int_0^t\int_{-u}^{u}\{I_5(v) + I_6(v)\}\,dv\,du.$$

We now estimate $I_6(v)$. It is clear that

$$I_6^p(v) = \int_{-\pi}^{\pi} |Q'_{2^k}(\cos\theta)|^p |\sin(\theta-v)|^{\mu p+1}\left(|\sin(\theta-v)| + \frac{1}{2^N}\right)^{\rho p} d\theta$$
$$\leqslant C_{15}\int_{-\pi}^{\pi} |Q'_{2^k}(\cos\theta)|^p \left(|\sin\theta| + \frac{1}{2^N}\right)^{\rho p+\mu p+1} d\theta.$$

We choose μ_1 so that $-1/p > \mu_1 > -2/p$; then $-1 < \mu_1 p + 1 < 0$, and hence

$$I_6^p(v) \leqslant C_{15}\int_{-\pi}^{\pi} |Q'_{2^k}(\cos\theta)|^p |\sin\theta|^{\mu_1 p+1}\left(|\sin\theta| + \frac{1}{2^N}\right)^{\rho p+\mu p-\mu_1 p} d\theta$$
$$= C_{16}\left\| Q'_{2^k}(x)\sqrt{1-x^2}^{\mu_1}\left(\sqrt{1-x^2} + \frac{1}{2^N}\right)^{\rho+\mu-\mu_1}\right\|_p^p.$$

If $\rho + \mu - \mu_1 \geqslant 0$, we have

$$I_6^p \leqslant C_{16} \left\| Q'_{2^k}(x) \sqrt{1-x^2}^{2\mu_1} \left(\sqrt{1-x^2} + \frac{1}{2^k} \right)^{\rho+\mu-\mu_1} \right\|_p^p .$$

Now applying Lemma 3, we obtain

$$\begin{aligned} I_6 &\leqslant C_{17} \left\| Q'_{2^k}(x) \left(\sqrt{1-x^2} + \frac{1}{2^k} \right)^{\rho+\mu} \right\|_p \\ &\leqslant C_{18} 2^k \left\| Q'_{2^k}(x) \left(\sqrt{1-x^2} + \frac{1}{2^k} \right)^{\rho+\mu+1} \right\|_p \\ &\leqslant C_{19} 2^{2k} \left\| Q_{2^k}(x) \left(\sqrt{1-x^2} + \frac{1}{2^k} \right)^{\rho+\mu} \right\|_p \\ &\leqslant C_{20} 2^{2k} \left\| Q_{2^k}(x) \sqrt{1-x^2}^{2\mu} \left(\sqrt{1-x^2} + \frac{1}{2^k} \right)^{\rho} \right\|_p \leqslant C_{21} 2^{k(2-\alpha)}. \end{aligned}$$

If, on the other hand, $\rho + \mu - \mu_1 < 0$, we have

$$I_6^p \leqslant C_{22} 2^{(k-N)p(\rho+\mu-\mu_1)} \left\| Q'_{2^k}(x) \sqrt{1-x^2}^{2\mu_1} \left(\sqrt{1-x^2} + \frac{1}{2^k} \right)^{\rho+\mu-\mu_1} \right\|_p^p .$$

Arguing as above, we obtain

$$I_6 \leqslant C_{23} 2^{(k-N)(\rho+\mu-\mu_1)} 2^{k(2-\alpha)} = C_{23} 2^{k(2-\alpha+\rho+\mu-\mu_1)} 2^{-N(\rho+\mu-\mu_1)}.$$

A similar estimate holds for I_5.

It is now easily seen that for $\mu - \mu_1 + \rho \geqslant 0$ and $\alpha < 2$

$$I_4 \leqslant C_{24} t^2 \sum_{k=0}^{N} 2^{k(2-\alpha)} \leqslant C_{25} t^2 2^{N(2-\alpha)},$$

and for $\mu - \mu_1 + \rho < 0$ but $\rho > -2 + \alpha - \mu + \mu_1$

$$I_4 \leqslant C_{26} t^2 2^{-N(\rho+\mu-\mu_1)} \sum_{k=0}^{N} 2^{k(2-\alpha+\rho+\mu-\mu_1)} \leqslant C_{27} t^2 2^{-N(2-\alpha)}.$$

Without loss of generality we may suppose that $|t| \leqslant \pi/2$. But then for $\rho > -2 + \alpha - \mu + \mu_1$ and $\mu_1 > -2/p$ we have $I_4 \leqslant C_{28} |\sin t|^\alpha$. This shows that the inequality $I_4 \leqslant C_{29} |\sin t|^\alpha$ holds for $\rho > -2 + \alpha - \mu - 2/p$.

Combining the estimates for I_1, I_2, I_3 and I_4 we obtain $I \leqslant C_{30} |\sin t|^\alpha$, which completes the proof of the theorem.

BIBLIOGRAPHY

1. S. M. Nikol'skiĭ, *Inequalities for entire functions of exponential type and their application to the theory of differentiable functions of several variables*, Trudy Mat. Inst. Steklov. 38 (1951), 244–278; English transl., Amer. Math. Soc. Transl. (2) 80 (1969), 1–38. MR 14, 32.

2. ———, *On the best approximation by polynomials of functions satisfying a Lipschitz condition*, Izv. Akad. Nauk SSSR Ser. Mat. 10 (1946), 295–322. (Russian) MR 8, 153;

3. A. F. Timan, *The approximation by ordinary polynomials of functions satisfying a Lipschitz condition*, Dokl. Akad. Nauk SSSR 77 (1951), 969–972. (Russian) MR 12, 823.

4. ———, *A strengthening of Jackson's theorem on the best approximation of continuous functions by polynomials on a finite segment of the real axis*, Dokl. Akad. Nauk SSSR 78 (1951), 17–20. (Russian) MR 12, 823.

5. V. K. Dzjadyk, *Constructive characterization of functions satisfying the condition* Lip α $(0 < \alpha < 1)$ *on a finite segment of the real axis*, Izv. Akad. Nauk SSSR Ser. Mat. 20 (1956), 623–642. (Russian) MR 18, 392.

6. V. K. Dzjadyk, *Approximation of functions by ordinary polynomials on a finite interval of the real axis*, Izv. Akad. Nauk SSSR Ser. Mat. 22 (1958), 337–354. (Russian) MR 20 #3406.

7. M. K. Potapov, *On Jackson type theorems in the L_p metric*, Dokl. Akad. Nauk SSSR 111 (1956), 1185–1188. (Russian) MR 18, 803.

8. ———, *Some questions of best approximation in the L_p metric*, Candidate's Dissertation, Moscow State Univ., Moscow, 1956. (Russian) RŽMat. 1957 #3925.

9. G. K. Lebed', *Certain problems of the approximation of functions of one variable by algebraic polynomials*, Dokl. Akad. Nauk SSSR 118 (1958), 239–242. (Russian) MR 21 #248.

10; M. K. Potapov, *On the approximation of nonperiodic functions by algebraic polynomials*, Vestnik Moskov. Univ. Ser. I Mat. Meh. 1960, no. 4, 14–25. (Russian) MR 23 #A3406.

11. G. V. Židkov, *Constructive characterization of a class of nonperiodic functions*, Dokl. Akad. Nauk SSSR 169 (1966), 1002–1005 = Soviet Math. Dokl. 7 (1966), 1036–1040. MR 34 #1764.

12. A. L. Fuksman, *Structural characterization of functions such that $E_n(f; -1, 1) \leqslant Mn^{-(k+\alpha)}$*, Uspehi Mat. Nauk 20 (1965), no. 4 (124), 187–190. (Russian) MR 32 #1492.

13. S. Z. Rafal'son, *The approximation of functions by Fourier-Jacobi sums*, Izv. Vysš. Učebn. Zaved. Matematika 1968, no. 4 (71), 54–62. (Russian) MR 37 #3259.

14. S. B. Stečkin, *On the order of the best approximations of continuous functions*, Izv. Akad. Nauk SSSR Ser. Mat. 15 (1951), 219–242. (Russian) MR 13, 29.

14. M. K. Potapov, *Some inequalities for polynomials and their derivatives*, Vestnik Moskov. Univ. Ser. I Mat. Meh. 1960, no. 2, 10–20. (Russian) MR 23 #A3412.

16. S. M. Nikol'skiĭ, *Approximation of functions of several variables and embedding theorems*, "Nauka", Moscow, 1969; English transl., Die Grundlehren der math. Wissenschaften, Band 205, Springer-Verlag, Berlin and New York, 1974. MR 46 #9714.

Translated by R. P. BOAS

PROPERTIES OF SOLUTIONS OF A CLASS OF SECOND-ORDER QUASILINEAR ELLIPTIC EQUATIONS IN DIVERGENCE FORM

UDC 517.518.22

G. N. JAKOVLEV

ABSTRACT. In a bounded region G of n-dimensional euclidean space we examine quasilinear elliptic equations of the form

$$-\operatorname{div} a(u_x) = f(x),\quad x \in G, \tag{1}$$

where $a(u_x)$ is a vector-valued function with components $a^i(u_x)$, $i = 1, \ldots, n$; u_x is the gradient of the desired solution $u = u(x)$.

It is assumed that the vector-valued function $a(y)$ is continuous in E^n and satisfies the following ellipticity condition: for any y and z from E^n

$$(a(y) - a(z),\ y - z) \geqslant 0.$$

Conditions are imposed on the growth of the functions $a^i(y)$: there exist numbers $\mu > 0$ and $p_1 > 1, \ldots, p_n > 1$ such that for all $y \in E^n$

$$|a^i(y)| \leqslant \mu \left(1 + \sum_{j=1}^{n} |y_j|^{p_j}\right)^{1-1/p_i} \qquad i=1,\ldots,n.$$

We study the weak solutions $u = u(x)$ of equation (1), which in G have first generalized derivatives u_{x_j} in the sense of Sobolev and, in addition, $u_{x_j} \in L_{p_j}(G)$, $j = 1, \ldots, n$.

Bibliography: 6 items.

Let G be a bounded region in the n-dimensional space E^n of points $x = (x_1, \ldots, x_n)$. In G we examine a quasilinear elliptic equation of the form

$$-\operatorname{div} a(u_x) = f(x), \tag{0.1}$$

where $a(u_x)$ is a vector-valued function with components $a^i(u_x)$, $i = 1, \ldots, n$, and $u_x = u_x(x)$ is the gradient of the desired solution $u(x)$, i.e. $u_x = (u_{x_1}, \ldots, u_{x_n})$, where u_{x_j} is the derivative with respect to x_j, $j = 1, \ldots, n$. The divergence in (0.1) is the divergence of $a(u_x(x))$ as a function of $x \in G$.

The present paper is devoted to the study of the weak solutions of equation (0.1). It is a development of the preceding papers [1] and [2], wherein, in particular, we presented the history of this question.

§1. Fundamental definitions and notation. Statement of the problem

We assume that the vector-valued function $a(y)$ is continuous for all $y = (y_1, \ldots, y_n) \in E^n$ and satisfies the following condition: for any y and z from E^n

$$\Phi(y, z) \equiv (a(y) - a(z),\ y - z) \geqslant 0, \tag{1.1}$$

AMS (MOS) subject classifications (1970). Primary 35D05, 35D10, 35J60.

where $(\cdots, \cdots)$ is the scalar product of the corresponding vectors. Condition (1.1) is called the monotonicity condition.

We remark that if the function $a(y)$ is continuously differentiable and if equation (0.1) satisfies an ellipticity condition of the form: for all ξ and y from E^n

$$A(y;\xi) = \sum_{i,j=1}^{n} a^i_{y_j}(y)\,\xi_i\xi_j \geqslant 0, \tag{1.2}$$

then equation (0.1) satisfies the monotonicity condition (1.1).

In addition, throughout the paper we assume the fulfillment of the following condition: there exist numbers $\mu > 0$ and $p_1 > 1, \ldots, p_n > 1$ such that the inequalities

$$|a^i(y)| \leqslant \mu\left(1 + \sum_{j=1}^{n} |y_j|^{p_j}\right)^{1-1/p_i}, \qquad i = 1, \ldots, n, \tag{1.3}$$

are fulfilled for all $y \in E^n$.

By $L^1_{\mathbf{p}}(G)$, $\mathbf{p} = (p_1, \ldots, p_n)$, $1 < p_i < \infty$, $i = 1, \ldots, n$, we denote the space of functions $u = u(x)$ locally summable in G, having in G the generalized derivatives u_{x_j} from $L_{p_j}(G)$, $1, \ldots, n$, i.e. the space of functions for which the seminorm

$$\sum_{j=1}^{n} \| u_{x_j} \|_{p_j} \tag{1.4}$$

(where $\|\ldots\|_{p_j}$ is the norm of the function in $L_{p_j}(G)$) has a meaning and is finite.

A function $u(x) \in L^1_{\mathbf{p}}(G)$ is called a weak solution of equation (0.1) if the equality

$$(a(u_x),\ v_x)_G = (f,\ v)_G, \tag{1.5}$$

where $(\cdots, \cdots)_G$ is the integral over G of the scalar product of the corresponding vectors, is fulfilled for any function $v(x) \in \overset{\circ}{C}{}^\infty(G)$.

It is easy to see that the inequalities

$$\| v \|_{p_j} \leqslant d \,\| v_{x_j} \|_{p_j}, \qquad j = 1, \ldots, n, \tag{1.6}$$

where d is the diameter of the region G, are fulfilled for any function $v(x) \in \overset{\circ}{C}{}^\infty(G)$. Thus (1.4) is the norm on $\overset{\circ}{C}{}^\infty(G)$.

By $\overset{\circ}{L}{}^1_{\mathbf{p}}(G)$ we denote the closure in norm (1.4) of $\overset{\circ}{C}{}^\infty(G)$. Obviously inequalities (1.6) are fulfilled for any function $v(x)$ of $\overset{\circ}{L}{}^1_{\mathbf{p}}(G)$, and therefore, if $v(x) \in \overset{\circ}{L}{}^1_{\mathbf{p}}(G)$, then $v(x) \in L_{p_0}(G)$, where $p_0 = \max\{p_1, \ldots, p_n\}$.

Note that if $u(x)$ is a weak solution of equation (0.1) in which $a(y)$ satisfies conditions (1.3) and

$$f(x) \in L_{q_0}(G), \quad 1/q_0 + 1/p_0 = 1, \tag{1.7}$$

then (1.5) is fulfilled for any function $v(x) \in \overset{\circ}{L}{}^1_{\mathbf{p}}(G)$. However, if $f(x)$ is a generalized function of the form

$$f(x) = -\sum_{j=1}^{n} f^j_{x_j}(x), \tag{1.8}$$

where

$$f^j(x) \in L_{q_j}(G), \quad 1/p_j + 1/q_j = 1, \quad j = 1, \ldots, n,$$

then the right-hand side of (1.5) should be regarded as a functional. Then, in accord with the definition of derivatives of generalized functions, (1.5) takes the form

$$(a\,(u_x),\ v_x)_G = (F,\ v_x)_G, \tag{1.9}$$

where $F(x)$ is a vector-valued function with components $f^j(x), j = 1, \ldots, n$.

Let a function $g(x) \in L^1_{\mathrm{p}}(G)$ be given. A weak solution $u(x)$ of equation (0.1), satisfying the condition

$$u\,(x) - g\,(x) \in \mathring{\mathscr{L}}^1_{\mathrm{p}}\,(G) \tag{1.10}$$

is called a solution of the Dirichlet problem (0.1), (1.10).

It can be shown that under natural requirements a solution from $L^1_{\mathrm{p}}(G)$ of the Dirichlet problem (0.1), (1.10) exists. The important question is whether this solution has second-order generalized derivatives. Examples (see [1] and [2]) show that weak solutions of an equation of form (0.1) may not have second-order generalized derivatives in Sobolev's sense even for sufficiently smooth functions $a(y)$, $f(x)$ and $g(x)$.

For equations of form (0.1), satisfying only the monotonicity condition (1.1), we shall show in the present paper that under certain additional conditions the weak solutions are such that the function $a(u_x(x))$ has partial derivatives in Sobolev's sense [3], and consequently a weak solution satisfies equation (0.1) almost everywhere.

In [1] and [2] we studied weak solutions of a Dirichlet problem of type (0.1), (1.10) for the case when $p = p_1 = \cdots = p_n \geqslant 2$, while equation (0.1) satisfied the following ellipticity condition: for any $y \in E^n$ and $\xi \in E^n$

$$\sum_{i,\,j=1}^{n} a^i_{y_j}(y)\,\xi_i\xi_j \geqslant \sum_{i=1}^{n} m_i^2\,(y_j)\,\xi_i^2, \tag{1.11}$$

where the functions $m_i(t)$ are such that $M_i(t) = \int_0^t m_i(\tau)\,d\tau$ increase strictly monotonically.

In the present paper condition (1.11) is replaced by the more general condition (1.2) or (1.1), and the numbers $p_1, \ldots, p_n$ can be different: they must satisfy only the condition $2p_i \geqslant \beta = \max\{2, p_1, \ldots, p_n\}$, $i = 1, \ldots, n$.

§2. Existence of a weak solution of the Dirichlet problem (0.1), (1.10)

To prove the existence of a solution from $L^1_{\mathrm{p}}(G)$ to the Dirichlet problem (0.1), (1.10) it is necessary to require the fulfillment of one more condition: there exist numbers $\nu > 0$ and $\mu_1 \geqslant 0$ such that for all $y \in E^n$

$$(a\,(y),\ y) \geqslant \nu \sum_{i=1}^{n} |\,y_i\,|^{p_i} - \mu_1. \tag{2.1}$$

For brevity we introduce the notation

$$D_{\mathrm{p}}\,(u) = \sum_{i=1}^{n} \|u_{x_i}\|^{p_i}_{p_i}, \qquad \|\,F\,\|^{\mathrm{q}}_{\mathrm{q}} = \sum_{j=1}^{n} \|\,f^j\,\|^{q_j}_{q_j}.$$

THEOREM 1. *Let conditions* (1.1), (1.3), (1.8) *and* (2.1) *be fulfilled. Then a solution of the Dirichlet problem* (0.1), (1.10) *exists for any function* $g(x) \in L^1_{\mathrm{p}}(G)$, *and the inequality*

$$D_{\mathrm{p}}(u) \leqslant \frac{2}{\nu} \mu_1 \operatorname{meas} G + C\{\|g\|_{\mathscr{L}^1_{\mathrm{p}}(G)} + D_{\mathrm{p}}(g) + \|F\|^{\mathrm{q}}_{\mathrm{q}}\}, \tag{2.2}$$

where the constant C depends only on n, μ, ν, p, and meas G, *is fulfilled for it.*

Proof. Let $\{\varphi^j(x)\}$ be a basis in $\overset{\circ}{L}{}^1_{\mathrm{p}}(G)$, where $\varphi^j(x) \in \overset{\circ}{C}{}^\infty(G)$. An approximate solution $u^N(x)$ of problem (0.1), (1.10) will be sought in the form of a sum $v^N(x) + g(x)$, where

$$v^N(x) = \sum_{j=1}^{N} \xi_j \varphi^j(x), \tag{2.3}$$

and the coefficients ξ_j are found from the following nonlinear system:

$$(a(u_x^N), \varphi_x^j)_G = (F, \varphi_x^j)_G, \qquad j = 1, \ldots, N. \tag{2.4}$$

Let us show that the approximate solution $u^N(x)$ exists, i.e. a solution $\xi = (\xi_1, \ldots, \xi_N)$ of system (2.4) exists. To do this we examine the mapping $\xi \longrightarrow P(\xi)$, $P = (P_1, \ldots, P_N)$, from E^N into E^N, defined by the formulas

$$P_j(\xi) = (a(u_x^N) - F, \varphi_x^j)_G, \; j = 1, \ldots, N.$$

This mapping maps E^N into E^N continuously. For it we have

$$(P(\xi), \xi) \equiv \sum_{j=1}^{N} P_j(\xi)\xi_j = (a(u_x^N) - F, v_x^N)_G.$$

Lemma. *Let $\xi \longrightarrow P(\xi)$ be a continuous mapping from E^N into E^N. If a number $R > 0$ exists such that*

$$(P(\xi), \xi) \geqslant 0 \tag{2.5}$$

for any ξ for which $|\xi| = R$, then a solution of the equation $P(\xi) = 0$ exists in the ball $K = \{\xi: |\xi| \leqslant R\}$.

The proof of this lemma occurs in [4] and, for example, in [2].

Let us show that condition (2.5) is fulfilled in our case, i.e. that there exists an $R > 0$ such that for any ξ, $|\xi| = R$,

$$(a(u_x^N) - F, v_x^N)_G \geqslant 0. \tag{2.6}$$

First of all, using condition (2.1), we obtain

$$(a(u_x^N), v_x^N)_G \geqslant \nu D_{\mathrm{p}}(u^N) - \mu_1 \operatorname{meas} G - |(a(u_x^N), g_x)_G|. \tag{2.7}$$

We estimate the last integral in (2.7) by the Hölder inequality, and we take advantage of (1.3):

$$|(a(u_x^N), g_x)_G| \leqslant \sum_{j=1}^{n} \|a^j(u_x^N)\|_{q_j} \|g_{x_j}\|_{p_j} \leqslant \mu \sum_{j=1}^{n} (\operatorname{meas} G + D_{\mathrm{p}}(u^N))^{1/q_j} \|g_{x_j}\|_{p_j}. \tag{2.8}$$

Inequality (2.8) is only sharpened if its right-hand side is replaced by the sum

$$\mu\varkappa\|g\|_{\mathscr{L}^1_{\mathrm{p}}(G)}+\mu\sum_{j=1}^{n}(D_{\mathrm{p}}(u^N))^{1/q_j}\|g_{x_j}\|_{p_j},\tag{2.9}$$

where $\kappa = \max_j(\text{meas } G)^{1/q_j}$. The last term in this sum does not exceed, by virtue of Young's inequality [5], the sum

$$\varepsilon\mu n D_{\mathrm{p}}(u^N)+\mu\varepsilon^{-p_0}D_{\mathrm{p}}(g),\tag{2.10}$$

where ϵ is an arbitrary number satisfying the inequality $0 < \epsilon < 1$.

From (2.7)–(2.10) we obtain

$$(a(u_x^N),v_x^N)_G\geqslant(\nu-\varepsilon\mu n)D_{\mathrm{p}}(u^N)-\mu_1\,\mathrm{meas}\,G-\mu\varkappa\|g\|_{\mathscr{L}^1_{\mathrm{p}}(G)}-\mu\varepsilon^{-p_0}D_{\mathrm{p}}(g)\tag{2.11}$$

Further, by applying the Hölder-Young inequality, we obtain the following chain of inequalities:

$$\begin{aligned}(F,v_x^N)_G|&\leqslant\sum_{j=1}^{n}\|f^j\|_{q_j}\|v_{x_j}^N\|_{p_j}\leqslant\sum_{j=1}^{n}\|f^j\|_{q_j}\{\|u_{x_j}^N\|_{p_j}+\|g_{x_j}\|_{p_j}\}\\&\leqslant\varepsilon D_{\mathrm{p}}(u^N)+\varepsilon^{-\alpha}\|F\|_{\mathrm{q}}^{\mathrm{q}}+D_{\mathrm{p}}(g)+\|F\|_{\mathrm{q}}^{\mathrm{q}},\end{aligned}\tag{2.12}$$

where $\alpha = \max\{q_1, \ldots, q_n\}$, $0 < \epsilon < 1$.

Thus from (2.11) and (2.12) follows the inequality

$$\begin{aligned}(a(u_x^N)-F,v_x^N)_G&\geqslant(\nu-\varepsilon\mu n-\varepsilon)D_{\mathrm{p}}(u^N)-\mu_1\,\mathrm{meas}\,G\\&-\mu\varkappa\|g\|_{\mathscr{L}^1_{\mathrm{p}}(G)}-(\mu\varepsilon^{-p_0}+1)D_{\mathrm{p}}(g)-(1+\varepsilon^{-\alpha})\|F\|_{\mathrm{q}}^{\mathrm{q}},\end{aligned}\tag{2.13}$$

from which it is easy to obtain, for $\epsilon = \nu/2(\mu n + 1)$, the inequality

$$(a(u_x^N)-F,v_x^N)_G\geqslant\frac{\nu}{2}D_{\mathrm{p}}(v^N)-C_1,\tag{2.14}$$

where C_1 denotes all terms not depending upon $v^N(x)$.

It is easily seen that the function $\Phi(\xi) = D_{\mathrm{p}}(v^N)$ is continuous in E^N and does not vanish on the unit sphere $|\xi| = 1$. We denote its minimum on this sphere by κ_0. Then, if $|\xi| > 1$,

$$\Phi(\xi)\geqslant\varkappa_0|\xi|.\tag{2.15}$$

From (2.14) and (2.15) we see that the mapping $\xi \longrightarrow P(\xi)$ satisfies the lemma's hypotheses, and consequently the equation $P(\xi) = 0$ has a solution, i.e. system (2.4) has a solution. From (2.4) and (2.13) it follows that the approximate solution $u^N(x)$ for any $N = 1, 2, \ldots$ satisfies the inequality

$$D_{\mathrm{p}}(u^N)\leqslant\frac{2}{\nu}\mu_1\,\mathrm{meas}\,G+C\{\|g\|_{\mathscr{L}^1_{\mathrm{p}}(G)}+D_p(g)+\|F\|_{\mathrm{q}}^{\mathrm{q}}\}.\tag{2.16}$$

From (2.16) we see that the sequence $\{u^N(x)\}$ is bounded in $L^1_{\mathrm{p}}(G)$; therefore from it we can select a subsequence which will converge weakly in $L^1_{\mathrm{p}}(G)$ to a function $u(x)$ from $L^1_{\mathrm{p}}(G)$, for which (2.16) is fulfilled. From (1.3) and (2.16) it follows that each sequence $\{a^i(u_x^N)\}$ is bounded in $L_{q_i}(G)$, $i = 1, \ldots, n$; therefore from them too we can pick out subsequences which converge weakly in $L_{q_i}(G)$ to some functions $b^i(x)$ from $L_{q_i}(G)$. By performing these operations in sequence it is easy to show that we can pick out a subsequence of the sequence $\{u^N(x)\}$ which will possess all the above-mentioned properties. For simplicity we shall reckon that $\{u^N(x)\}$ and $\{a^i(u_x^N)\}$ already possess these properties.

Passing to the limit in (2.4) as $N \longrightarrow \infty$, we obtain

$$(b(x), \varphi_x^j)_G = (F, \varphi_x^j)_G, \quad b = (b^1, \ldots, b^n),$$

which is valid for any $j = 1, 2, \ldots$; and consequently the equality

$$(b(x), \eta_x)_G = (F, \eta_x)_G \tag{2.17}$$

is valid for any function $\eta(x) \in \overset{\circ}{L}{}_p^1(G)$. On the other hand, for any function $\varphi(x)$ of the form

$$\varphi(x) = \sum_{i=1}^{N} \alpha_i \varphi^i(x) + g(x)$$

we have

$$(a(u_x^N), u_x^N - \varphi_x)_G = (F, u_x^N - \varphi_x)_G.$$

Hence by virtue of the monotonicity condition

$$(a(u_x^N) - a(\varphi_x), u_x^N - \varphi_x)_G \geqslant 0$$

it follows that

$$(F, u_x^N - \varphi_x)_G - (a(\varphi_x), u_x^N - \varphi_x)_G \geqslant 0.$$

Passing to the limit as $N \longrightarrow \infty$, we obtain

$$(F, u_x - \varphi_x)_G - (a(\varphi_x), u_x - \varphi_x)_G \geqslant 0, \tag{2.18}$$

for any function $\varphi(x)$ such that

$$\varphi(x) - q(x) \in \overset{\circ}{\mathcal{L}}{}_p^1(G).$$

In (2.18) we set $\varphi = u - \epsilon\eta$, $\epsilon > 0$, $\eta \in \overset{\circ}{L}{}_p^1(G)$; then

$$(F, \eta_x)_G - (a(u_x - \varepsilon\eta_x), \eta_x)_G \geqslant 0. \tag{2.19}$$

From (2.17) and (2.19) it now follows that

$$(b(x), \eta_x)_G - (a(u_x - \varepsilon\eta_x), \eta_x)_G \geqslant 0.$$

Passing to the limit as $\epsilon \longrightarrow 0$, we obtain

$$(b(x), \eta_x)_G = (a(u_x), \eta_x)_G. \tag{2.20}$$

From (2.17) and (2.20) it follows that $u(x)$ is a weak solution of (0.1), while by construction it also is a solution of the Dirichlet problem (0.1), (1.10).

The theorem is proved.

§3. Differential properties of the weak solutions of equation (0.1)

First of all we make several remarks. If equation (0.1) satisfies the ellipticity condition (1.2) and the quadratic form $A(y; \xi)$ is symmetric for all $y \in E^n$, then the inequalities

$$(a_y^i(y), \xi) \leqslant \sqrt{a_{y_i}^i(y)}\sqrt{A(y;\xi)}, \qquad i = 1, \ldots, n, \tag{3.1}$$

are fulfilled for all y and ξ from E^n. Usually (for example, see [1] or [5]), in case $p = p_1 = \ldots = p_n$ we require, along with (1.3), the fulfillment of the following condition: a number $\mu > 0$ exists such that for any $y \in E^n$

$$| a_{y_j}^i(y) | \leqslant \mu (1 + |y|)^{p-2}, \qquad i, j = 1, \ldots, n. \tag{3.2}$$

In what follows we shall require the fulfillment of the following conditions instead of (3.1) and (3.2): a number $\mu_2 > 0$ exists such that the inequalities

$$| a^i(y) - a^i(z) | \leqslant \mu_2 \sqrt{\Phi(y, z)} \left(1 + \sum_{j=1}^n |y_j|^{p_j} + \sum_{j=1}^n |z_j|^{p_j}\right)^{1/2 - 1/\beta}, \quad i = 1, \ldots, n, \tag{3.3}$$

where

$$\beta = \max\{p_1, p_2, \ldots, p_n, 2\}. \tag{3.4}$$

are fulfilled for any y and z from E^n. Conditions (3.3) are called the *subordination conditions.*

Let e_k be the unit vector in the direction of the x_k-axis, $k = 1, \ldots, n$, and let $\Delta_k(h; u)$ be the difference, of first order in x_k, with step h of the function $u(x)$, i.e.

$$\Delta_k(h; u) = u(x + he_k) - u(x).$$

$\Delta_k(h; a^i(u_x))$ is defined analogously:

$$\Delta_k(h; a^i(u_x)) = a^i(u_x(x + he_k)) - a^i(u_x(x)).$$

By G_δ, $\delta > 0$, we denote the subregion of G consisting of those points not closer than δ to ∂G. It is obvious that if the function $u(x)$ has been defined in G, then the differences $\Delta_k(h; u)$ and $\Delta_k(h; a^i(u_x))$ are defined in G_δ for any h, $|h| < \delta$.

Let $\eta = \eta(x)$ be a nonnegative continuously differentiable function not exceeding unity, equalling unity in G_δ and zero outside $G_{\delta/2}$. We set

$$M(\delta) = \max_{x \in G} | \eta_x(x) |. \tag{3.5}$$

Further, let

$$\varphi(t) = (1 + t^2)^{p_k/\beta - 1}. \tag{3.6}$$

In what follows, if some function $f \in L_{\mathbf{p}}^1(\Omega)$, we shall denote its generalized Dirichlet integral by $D_{\mathbf{p}}(f; \Omega)$, i.e.

$$D_{\mathbf{p}}(f; \Omega) = \int_\Omega \sum_{j=1}^n | f_{x_j}(x) |^{p_j} dx.$$

LEMMA. *Let the vector-valued function $a(y)$ in equation* (0.1) *satisfy the monotonicity condition* (1.1), *the growth condition* (1.3), *and the subordination condition* (3.3), *and let the right-hand side $f(x)$ in* (0.1) *satisfy* (1.7); *in addition, for any $\delta > 0$ let*

$$\begin{gathered} f(x) \in \mathscr{L}_{\mathbf{r}}^1(G_\delta), \quad \mathbf{r} = (r_1, \ldots, r_n), \\ 1/r_i = 1 - 2/\beta + 1/p_i, \quad j = 1, \ldots, n. \end{gathered} \tag{3.7}$$

If $u(x) \in L_{\mathbf{p}}^1(G)$ is a weak solution of equation (0.1) *and $2p_i > \beta$, $i = 1, \ldots, n$, then the inequalities*

$$\left(\Delta_k(h; a(u_x)), \varphi\left(\frac{1}{h}\Delta_k(h; u)\right)\Delta_k(h; u_x)\right)_{G_\delta} \leqslant h^2 C\{1 + D_p(u; G) + D_r(f; G_{\delta/4})\}, \quad k = 1, \ldots, n, \tag{3.8}$$

where the constant C depends only on $n, p_k, \beta, \mu_2, \delta$, *and* G, *are fulfilled for any* $\delta > 0$ *and for any* h, $|h| < \delta/4$.

PROOF. Let the function $v(x) \in \overset{\circ}{L}{}^1_p(G)$ have a support in $G_{\delta/2}$, and let $|h| < \delta/4$. It is obvious that $v(x - he_k) \in \overset{\circ}{L}{}^1_p(G)$. If $u(x)$ is a weak solution of (0.1), then according to definition (1.5)

$$(a(u_x(x)); v_x(x - he_k))_G = (f(x), v(x - he_k))_G. \tag{3.9}$$

Let us prolong the functions $a^i(u_x(x))$, $i = 1, \ldots, n$, and $f(x)$ outside G by zero and make a change of variables; then (3.9) takes the form

$$(a(u_x(x + he_k)), v_x(x))_G = f(x + he_k), v(x)_G \tag{3.10}$$

Now from (3.9) and (3.10) it follows that

$$(\Delta_k(h; a(u_x)), v_x)_G = (\Delta_k(h; f), v)_G \tag{3.11}$$

for any function $v(x) \in L^1_p(G)$, concentrated in $G_{\delta/2}$, and for any h, $|h| < \delta/4$.

For brevity in what follows we shall everywhere write $\Delta a, \Delta f$ and Δu, respectively, instead of $\Delta_k(h; a(u_x))$, $\Delta_k(h; f)$, and $\Delta_k(h; u)$. In (3.11) we set

$$v = \eta^2 \Delta u \varphi(\Delta u/h), \tag{3.12}$$

where $\eta = \eta(x)$ and $\varphi(t)$ are the functions defined above (see (3.5) and (3.6)). Since

$$v_{x_i} = 2\eta\eta_{x_i}\Delta u\varphi(\Delta u/h) + \eta^2 \Delta u_{x_i}\psi(\Delta u/h),$$

where $\psi(t) = \varphi(t) + t\varphi'(t)$, with due regard to our stipulations we obtain from (3.11) the relation

$$(\eta^2 \Delta a, \psi \Delta u_x)_G = (\Delta f, v)_G - 2(\eta \Delta a, \varphi \eta_x \Delta u)_G, \tag{3.13}$$

where for brevity we have dropped the argument $t = \Delta u/h$ in the functions ψ and φ.

It is easy to see that

$$\psi(t) \geqslant (2p_k/\beta - 1)\varphi(t);$$

therefore, by virtue of the monotonicity condition (1.1), from (3.13) we obtain

$$(2p_k/\beta - 1)(\eta^2 \Delta a, \varphi \Delta u_x)_G \leqslant |(\Delta f, v)_G| + 2|(\eta \Delta a, \varphi \eta_x \Delta u)_G|. \tag{3.14}$$

From the definition (3.12) of $v(x)$ it follows that

$$|(\Delta f, v)_G| \leqslant |h|^{2(1-p_k/\beta)}(|\Delta f|, \eta^2 |\Delta u|^{2p_k/\beta - 1})_G,$$

while, since $0 \leqslant \eta \leqslant 1$ and $(2/\beta) - (1/p_k) = 1 - (1/r_k)$,

$$|(\Delta f, v)_G| \leqslant |h|^{2(1-p_k/\beta)}(|\Delta f|, \eta^{2p_k/\beta} |\Delta u|^{p_k(1-1/r_k)})_G.$$

We estimate the last integral by the Hölder inequality; then

$$|(\Delta f, v)_G| \leqslant |h|^{2(1-p_k/\beta)} \|\eta \Delta f\|_{r_k} \|\eta \Delta u\|_{p_k}^{p_k(1-1/r_k)}.$$

By estimating the integrals of the differences in terms of the norms of the corresponding derivatives, we finally obtain

$$|(\Delta, fv)_G| \leqslant h^2 \|f_{x_k}\|_{L_{r_k}(G\delta/4)} \|u_{x_k}\|_{p_k}^{2p_k/\beta-1}. \tag{3.15}$$

Let us now estimate the second integral in the right-hand side of the inequality. From the subordination condition (3.3) it follows that

$$|\Delta a^i| \leqslant \mu_2 \sqrt{(\Delta a, \Delta u_x)}\, J(u)^{1/2-1/\beta}, \tag{3.16}$$

where

$$J(u) = 1 + \sum_{j=1}^{n} |u_{x_j}(x)|^{p_j} + \sum_{j=1}^{n} |u_{x_j}(x+he_k)|^{p_j};$$

therefore

$$2|(\eta \Delta a, \varphi \eta_x \Delta u)| \leqslant 2\mu_2 \eta \sqrt{(\Delta a, \Delta u_x)} \cdot \varphi |\Delta u| \sum_{i=1}^{n} J(u)^{1/2-1/\beta} |\eta_{x_i}|. \tag{3.17}$$

It is evident that the last product does not exceed the sums

$$\varepsilon^2 \eta^2 (\Delta a, \Delta u_x) \varphi + \varepsilon^{-2} \mu_2^2 \varphi |\Delta u|^2 |\eta_x|^2\, n J(u)^{1-2/\beta}, \tag{3.18}$$

where ϵ is an arbitrary positive number, while since (see (3.5))

$$|\Delta u|^2 \varphi \leqslant |h|^{2(1-p_k/\beta)} |\Delta u|^{2p_k/\beta} \tag{3.19}$$

and $J(u) \geqslant 1$, from (3.17)–(3.19) we obtain, after an integration over G,

$$\begin{aligned} 2|(\eta \Delta a, \varphi \eta_x \Delta u)_G| &\leqslant \varepsilon^2 (\eta^2 \Delta a, \varphi \Delta u_x)_G \\ &+ n\varepsilon^{-2} \mu_2^2 |h|^{2(1-p_k/\beta)} \int_G |\Delta u|^{2p_k/\beta} |\eta_x|^2 J(u)^{1-2/\beta}\, dx, \end{aligned} \tag{3.20}$$

where $\beta = \max\{p_1, \ldots, p_n, 2\}$.

Let us estimate the last integral in (3.20):

$$\begin{aligned} \int_G |\Delta u|^{2p_k/\beta} |\eta_x|^2 J(u)^{1-2/\beta}\, dx &\leqslant \left(\int_G |\eta_x|^2 |\Delta u|^{p_k} dx\right)^{2/\beta} \left(\int_G |\eta_x|^2 J(u)\, dx\right)^{1-2/\beta} \\ &\leqslant |h|^{2p_k/\beta} \|u_{x_k}\|_{p_k}^{2p_k/\beta} (M^2 \operatorname{meas} G + 2M^2 D_{\mathrm{p}}(u; G))^{1-2/\beta}. \end{aligned} \tag{3.21}$$

We obtain the inequality we need from (3.14), (3.15), (3.20) and (3.21):

$$\begin{aligned} \left(\frac{2p_k}{\beta} - 1 - \varepsilon^2\right) (\eta^2 \Delta a, \varphi \Delta u_x)_G &\leqslant h^2 \|f_{x_k}\|_{L_{r_k}(G_{\delta/4})} \|u_{x_k}\|_{p_k}^{2p_k/\beta-1} \\ &+ h^2 \varepsilon^{-2} n \mu_2^2 \|u_{x_k}\|_{p_k}^{2p_k/\beta} (M^2 \operatorname{meas} G + 2M^2 D_{\mathrm{p}}(u; G))^{1-2/\beta}. \end{aligned}$$

To obtain (3.8) it suffices to take an ϵ such that $0 < \epsilon^2 < (2p_k/\beta) - 1$. The lemma is proved.

THEOREM 2. *Let the vector-valued function $a(y)$ in equation* (0.1) *satisfy the monotonicity condition* (1.1), *the growth condition* (1.3), *and the subordination condition* (3.3), *and let the right-hand side $f(x)$ in* (0.1) *satisfy conditions* (1.7) *and* (3.7). *In addition, let* *the inequality*

$$2p_i > \beta = \max\{p_1, \ldots, p_n, 2\}$$

be fulfilled for any $i = 1, \ldots, n$. *If* $u(x) \in \mathring{L}^1_{\mathbf{p}}(G)$ *is a weak solution of* (0.1), *then the functions* $a^i(u_x)$, $i = 1, \ldots, n$, *have in* G *generalized derivatives in Sobolev's sense, and the inequalities*

$$\left\| \frac{\partial a^i(u_x)}{\partial x_k} \right\|_{L_{r_k}(G_\delta)} \leqslant C\{1 + D_{\mathbf{p}}(u; G) + D_{\mathbf{r}}(f; G_{\delta/4})\}, \quad i, k = 1, 2, \ldots, n, \tag{3.22}$$

where

$$1/r_k = 1 + 1/p_k - 2/\beta \tag{3.23}$$

are fulfilled for them.

PROOF. From (3.23) it follows that if $\beta = 2$, then $r_k = p_k$, while if $p_k = \beta$, then $r_k = q_k$.

From (3.23) we see that $1 < r_k \leqslant 2$ always. It is easily seen that $r_k = 2$ if and only if $p_k = \beta$.

We shall use the notation introduced in the lemma's proof. In addition, the indices i and k will be fixed during the proof of the theorem, and we shall write q instead of r_k. Then from the subordination condition (3.3) it follows (see (3.16)) that

$$\| \Delta a^i \|_{L_q(G_\delta)} \leqslant \mu_2 \| \sqrt{(\Delta a, \Delta u_x)}\, J(u)^{1/2 - 1/\beta} \|_{L_q(G_\delta)}. \tag{3.24}$$

If $q = 2$, from (3.24) we obtain

$$\| \Delta a^i \|_{L_q(G_\delta)} \leqslant \mu_2 \sqrt{(\Delta a, \Delta u_x)_{G_\delta}}.$$

Now let $q < 2$. Let us apply the Hölder inequality with indices $2/q$ and $2/(2-q)$ to the right-hand side of (3.24); then

$$\| \Delta a^i \|_{L_q(G_\delta)} \leqslant \mu_2 \sqrt{(\varphi \Delta a, \Delta u_x)_{G_\delta}}\, B(u)^{1 - q/2}, \tag{3.25}$$

where

$$B(u) = \int_{G_\delta} \left(1 + \left(\frac{\Delta u}{h}\right)^2\right)^{\left(1 - \frac{p_k}{\beta}\right)\frac{q}{2-q}} J(u)^{\frac{2q}{2-q}\left(\frac{1}{2} - \frac{1}{\beta}\right)} dx.$$

To estimate $B(u)$ we also apply the Hölder inequality with appropriate indices:

$$B(u) \leqslant \left(\int_{G_\delta} J(u)\, dx\right)^{\frac{2q}{2-q}\left(\frac{1}{2} - \frac{1}{\beta}\right)} \left(\int_{G_\delta} \left(1 + \left(\frac{\Delta u}{h}\right)^2\right)^{p_k/2} dx\right)^{\frac{2q}{2-q}\left(\frac{1}{q} - 1 + \frac{1}{\beta}\right)}.$$

It is easy to see (see (3.16)) that

$$\int_{G_\delta} J(u)\, dx \leqslant \operatorname{meas} G + 2D_{\mathbf{p}}(u; G).$$

Further,

$$\int_{G_\delta}\left(1+\left(\frac{\Delta u}{h}\right)^2\right)^{p_k/2}dx\leqslant 2^{p_k/2}\{\operatorname{meas}G+\|u_{x_k}\|_{p_k}^{p_k}\}.$$

Consequently

$$B(u)\leqslant 2^{\beta/2}\{\operatorname{meas}G+2D_{\mathbf{p}}(u;G)\}. \tag{3.26}$$

From (3.25) and (3.26) we finally obtain the inequality

$$\|\Delta a^i\|_{L_q(G_\delta)}\leqslant 2^{\frac{\beta}{2}\left(1-\frac{q}{2}\right)}\sqrt{(\varphi\Delta a,\ \Delta u_x)_{G_\delta}}\{\operatorname{meas}G+2D_{\mathbf{p}}(u;G)\}^{1-q/2}, \tag{3.27}$$

which is valid for any q, $1<q\leqslant 2$.

From (3.27) and (3.8) it follows that

$$\|\Delta a^i\|_{L_q(G_\delta)}\leqslant C|h|\{1+D_{\mathbf{p}}(u;G)+D_{\mathbf{r}}(f;G_{\delta/4})\}. \tag{3.28}$$

By virtue of the well-known properties of generalized derivatives from L_q, $q>1$ (for example, see [6]), it follows from (3.28) that the function $a^i(u_x(x))$ has a first-order generalized derivative with respect to x_k, belonging to $L_q(G_\delta)$, $\delta>0$, and, in particular, inequality (3.22) is fulfilled for it.

The theorem is proved.

We remark that under appropriate assumptions the results of the present paper generalize to an equation of the form

$$-\operatorname{div} a(u_x,\ u,\ x)=f(x),$$

and also to systems of such equations.

BIBLIOGRAPHY

1. G. N. Jakovlev, *On the solutions of a class of second order quasi-linear elliptic equations*, Dokl. Akad. Nauk SSSR **202** (1972), 1020–1023 = Soviet Math. Dokl. **13** (1972), 255–258. MR **45** #715.

2. ———, *The first boundary value problem for quasilinear elliptic equations of second order*, Trudy Mat. Inst. Steklov. **117** (1972), 321–340 = Proc. Steklov Inst. Math. **117** (1972), 381–403.

3. S. L. Sobolev, *Applications of functional analysis in mathematical physics*, Izdat. Leningrad. Gos. Univ., Leningrad, 1950; English transl., Transl. Math. Monographs, vol. 7, Amer. Math. Soc., Providence, R. I., 1963. MR **14**, 565; **29** #2624.

4. J.-L. Lions, *Quelques méthodes de résolution des problèmes aux limites non linéaires*, Dunod; Gauthier-Villars, Paris, 1969. MR **41** #4326.

5. O. A. Ladyženskaja and N. N. Ural′ceva, *Linear and quasilinear equations of elliptic type*, "Nauka", Moscow, 1964; English transl., Academic Press, New York, 1968. MR **35** #1955; **39** #5941.

6. S. M. Nikol′skiĭ, *Approximation of functions of several variables and embedding theorems*, "Nauka", Moscow, 1969; English transl., Die Grundlehren der math. Wissenschaften, Band 205, Springer-Verlag, Berlin and New York, 1974. MR **46** #9714.

Translated by N. H. CHOKSY